Geography, Resources, and Environment

Geography, Resources, and Environment

Volume II

Themes from the Work of Gilbert F. White

Edited by

Robert W. Kates
and Ian Burton

The University of Chicago Press
Chicago and London

Robert W. Kates is professor of geography and Research Professor at Clark University and holds a MacArthur Prize fellowship. Ian Burton is professor of geography and former director of the Institute for Environmental Studies, University of Toronto, and vice-chairman of the International Federation of Institutes for Advanced Study. Kates and Burton, who have published widely, are coauthors, with Gilbert F. White, of *The Environment as Hazard* (1978) and coeditors of *Readings in Resource Management and Conservation* (1965).

The University of Chicago Press, Chicago 60637
The University of Chicago Press, Ltd., London

Printed in the United States of America
95 94 93 92 91 90 89 88 87 86   5 4 3 2 1

Library of Congress Cataloging-in-Publication Data

White, Gilbert Fowler, 1911-
Themes from the work of Gilbert F. White.

(Geography, resources, and environment ; v. 2)
Bibliography: p.
Includes index.
1. Water resources development—Addresses, essays, lectures. 2. Anthropo-geography—Addresses, essays, lectures. I. Kates, Robert William. II. Burton, Ian. III. Title. IV. Series.
GF75.G46 vol. 2   333.7 s   85-20829
[HD1691]   [333.91'15]
ISBN 0-226-42576-2
ISBN 0-226-42577-0 (pbk.)

Contents

Figures

Tables

Introduction

For the third time in a decade, families along a Texas river have been driven from their homes by flood waters. Twice before, they have returned to their waterlogged and muddy possessions with no alternative but to repair and rebuild and hope that the next flood will be long delayed. This time a federal agency, in cooperation with the state government, will enable them to relocate off the floodplain. Their neighbors on a higher site, subject to less frequent flooding, are able to purchase federally sponsored flood insurance through the private insurance industry, with active state and local support that also restricts new development in high-hazard areas.

As the sun rises on an East African village, the people stir into their early morning activities. There has recently been an important change. Instead of watching the women and children set out on an hour-long walk to the stream (downhill with empty pots and cans, uphill with full ones), we now can see them crowd around communal standpipes and washhouses. The water is clean and plentiful, and the sense of quiet satisfaction in the air reflects the villagers' knowledge that their preferences were followed by the water supply engineers from the district office, and that they contributed labor and cash. They also now help with maintenance and vigilantly safeguard the system from misuse. A young man from the village has received a short course of training in the operation of the system and can carry out some needed repairs without help.

A major international river system in Indochina is being managed and developed for multiple-purpose use by the nations that share in the basin territory. Despite strong differences, open conflicts, and periodic fighting and social upheaval, some cooperation continues. It is understood that where the river is concerned, its development in a coordinated fashion will bring greater benefits to all the nations than if each were to act in its own short-term self-interest.

In government and university laboratories in many countries, some of the most eminent environmental scientists are now modeling and synthesizing large data sets to gain more precise understanding of the biogeochemical planetary environment in a state of dynamic equilibrium. There is a lack of knowledge of how and to what degree these cycles—carbon, nitrogen, and others—interact with each other and with the sources and sinks of the system. There are concerns that irreversible changes may be generated by human action, but without models of much greater precision it is often impossible to distinguish between natural variations and human perturbations. Now a worldwide scientific effort is underway through SCOPE (Scientific Committee on Problems of the Environment, established by the International Council of Scientific Unions), with the objectives of providing better knowledge and a more secure basis for management.

These scenes of people working to expand their understanding of environment and their relation to it are all examples of efforts that have been stimulated in part by the research, teaching, and leadership of Gilbert F. White. In these areas and in others described in the papers that follow, White has worked to improve the ways in which society uses natural resources, essentially by changing our world view and by inspiring dedication to a public service that makes effective use of the results of scientific research.

The thirteen papers in this volume were written in appreciation of the example that Gilbert White has provided, and also to help celebrate his years of teaching, research, and leadership. The contributors were chosen from among his students, colleagues and coworkers. It quickly became apparent in the early gestation period of this enterprise that no ordinary festschrift would suffice. The editors were excited by the idea of linking a set of essays on the themes of Gilbert White's work with a selection of his own papers. Hence this two-volume set.

We have tried to make volume II more than a set of distinct papers. The topics were chosen to provide a status report on what has been achieved and what challenges lie ahead across the professional concerns of White's career; the authors were asked to focus on the sorts of questions he persistently raised. A diligent reading of the papers is rewarded with a sharpened sense of how far we have come and where we now stand in the integration of social science with natural science in the development of management strategies and policy for resources and environment. Other recurrent themes that emerge are the debunking of myths and misconceptions through precise, parsimonious use of language; the need for interdisciplinary understanding and holistic approaches that nevertheless respect the essential contributions of specialized empirical observation and experimentation in field and laboratory; the central importance of human values; and the need to achieve management practices that are respectful of nature and people and are not content with technological ingenuity or economic efficiency for their own sakes.

These and other recurrent themes have been continually pondered and pursued by Gilbert White. The essays mark a stage in the chase that shows how much has been gained and how much remains to be accomplished.

We suggested that each essay might consider the period between White's first publication on the subject to the present. The longest time span is covered by the first essay, on water supply, by M. Gordon Wolman and Abel Wolman, his father. White's first professional paper dealt with the shortage of public water supplies in the United States during 1934. It foreshadowed two major interests that were to prove continuing preoccupations for the next several decades—the management of water resources, especially their supply, and the hazard of insufficient water occasioned by drought. Myths and misconceptions abound in the field of water supply, and it is this theme that is taken up by Wolman and Wolman. Their paper follows the broadening of White's interest in water to industrial water use, the role of water in arid lands, and then to the major problems of water supply for domestic use in developing countries. These interests are reflected in selections 7, 8, 16, 20, and 24 in volume I.

Notwithstanding his contributions to the problems of water in developing countries, Gilbert White is probably best known for his work on floods in western industrial countries as well as the USSR and China. His doctoral dissertation at Chicago, *Human Adjustment to Floods,* is the source of several selections in volume I and is the second topic in this volume. Rutherford Platt (chap. 2) provides a historical perspective, with some biographical detail, of the development of the flood research agenda from the period of *Human Adjustment to Floods* to the present. The progress is substantial and, although major problems remain, the evidence is suggestive of how much worse they easily might have been without White's persistent attack. Practical action still lags behind knowledge, but there can be little doubt that the general direction of a beneficial flood policy is being followed. Other related papers in volume I include selections 6, 9, 14, 22, and 27.

Gilbert White's contributions to the study of natural resources have combined the work of a specialist on water supply and floods with the development of broad integrative concepts. The continuing need to find and maintain a balance between specialization and integration in natural resource development is the theme taken up by Marion Clawson in the third paper. As Clawson notes, there are powerful forces pushing toward specialization in all professions. Although specialized knowledge is admittedly needed, a stronger thrust toward integration has been a concern of Gilbert White and a very small set of other scholars. Clawson reviews some of the major attempts at integration and neatly links these with selections 3 and 11 in volume I.

The arid and semiarid regions of the world provide a severe testing ground for methods of resource use and development that spring from modern experience in more humid regions. Beginning in the 1950s, Gilbert White became involved in the UNESCO Arid Lands Programme, a continuing interest represented in volume I by selections 8 and 18. Here, this interest is reflected in the fourth essay. Douglas Johnson provides a broad-ranging survey of the present status and potential solutions of resource use in dry lands; once again the need for holistic approaches emerges as a continuing issue.

The editors invited each contributor to this volume to consult widely with colleagues of his or her choosing and to involve coauthors where appropriate. J. C. Day pursued this option vigorously and, with the help of four others, put together a survey of river basin development in the United States, Canada, Europe, the Soviet Union, Latin America, Africa, and Asia (chap. 5). The survey takes as its points of departure Gilbert White's 1957 paper, "A Perspective of River Basin Development" (vol. I, selection 4) and his preface to the UN Report on Integrated River Basin Development (selection 17).

For a scholar seeking to apply research to public policy there is no escape from the complexities of decision making. Rather than contribute to the field by theorizing about how decisions ought to be made, White has sought to provide rational descriptions of how decisions are made in practice and thus to identify opportunities for incremental improvement. This practical approach is taken up by Howard Kunreuther and Paul Slovic in the sixth paper, which nevertheless manages to be the most theoretical in the volume. In treating both individual and public decision making, the authors address the questions raised by White in his 1961 paper, "The Choice of Use in Resource Management" (vol. I, selection 9), and his 1969 monograph *Strategies of American Water Management* (selection 16).

Elliot J. Feldman (chap. 7) focuses on a dilemma of all socially responsible scholars (a dilemma articulated especially by Max Weber)—Where is the line to be drawn between scholarship and citizenship? Feldman provides an account of the way some institutions in the United States have grappled with the dilemma by providing for education in public affairs. In an essay more biographical than most, he describes Gilbert White's contributions as president of

Haverford College, in his development of the High School Geography Project, and in his effort at peacemaking in the Mekong Basin. Selections 5, 10, 12, 13, 21, 23, 25, 26, and 28, in volume I, demonstrate how White at various times sought to resolve, for himself at least, Max Weber's question. It is the nature of social dilemmas that resolutions are temporary and new structures are continually developed to deal with them. The fact that the most recent twists and turns in the development of education for public efforts go unrecorded is not of paramount importance, however. The underlying problem and Gilbert White's approach to it stand out clearly in timeless fashion in Feldman's paper.

The word *climate* does not appear in the title of any of Gilbert White's writings included in volume I. Not surprisingly, then, "Awareness of Climate," (chap. 8), by F. Kenneth Hare and W. R. Derrick Sewell, is the only one, apart from the editors' own, that does not specifically refer to any of White's work. The contributions of White's ideas to climate are nevertheless real, appearing indirectly in his papers on related topics such as arid lands, natural hazards, weather modification, and also through the work of his students. As Hare and Sewell relate, climatology has been a cinderella science, languishing in the care of geography and looked down upon by the practitioners of meteorology. The recent transformation of climatology into a socially important and scientifically more reputable field has increased the need for contributions in the White tradition. In large measure this need is being met by White's students trained in water resources, arid lands, and natural hazards research.

The development of the study of public perceptions has been one of the major contributions to resource and environmental management. Gilbert White has stimulated that study, especially in relation to natural hazards; his 1966 paper "Formation and Role of Public Attitudes" (vol. I, selection 15), was an early landmark. Anne V. T. Whyte has been an active participant and, at times, a severe critic of perception research in geography. In her paper (chap. 9) for this volume, the criticism is muted, and her evaluation of research in hazard perception prompts her to suggest that the present period of conceptual introspection will lead back into the fold of human ecology. It is an intriguing suggestion because, as many students and colleagues of Gilbert White know, an important antecedent for many of the ideas developed since 1935 is to be found in "Geography as Human Ecology," Harlan Barrows's well-remembered paper of 1923 (cf. chap. 2 below).

As flood and drought research blossomed into a general approach to all natural hazards, so in turn natural hazard research and the environmental crisis (see vol. I, selections 11 and 12) have led to a new set of geographical contributions to risk assessment and technological hazards. This recent development is chronicled in Timothy O'Riordan's paper, "Coping with Environmental Hazards" (chap. 10), which he sees as belonging "firmly in the behavioral field of human ecology." The tradition of research which began with *Human Adjustment to Floods* (vol. I, selection 2) has now broadened into a multiplicity of interconnected problems. As happens so often in research, each potential solution gives rise to more unanswered questions. Small wonder that on this research frontier the cautious optimism engendered by Gilbert White gives way to a less confident, almost pessimistic note from Timothy O'Riordan.

Water supply was the topic of White's first paper, and natural hazards probably has been the topic receiving the most sustained attention. Certainly the topic of least attention has been energy resources and supply, which is covered in the eleventh paper, by long-time associate Kenneth Boulding. As the author notes, "neither Gilbert White nor I could claim that energy or even energy policy is our major field of competence or interest." Nevertheless, both were drawn by the growth of public interest into the national policy debate about energy in the 1970s—reason enough for the inclusion of a typically thoughtful and provocative essay.

In recent years, Gilbert White's position as a senior statesman of resource and environmental research has led him to major roles in international organizations such as SCOPE and to become an adviser and confidant to senior political figures and international civil servants. Thus global environmental prospects have become part of White's repertoire of interests as shown in selections 25, 26, 28, and 29 of volume I. The theme is reflected in R. E. Munn's essay (chap. 12) in this volume.

The same theme encouraged the editors to contribute an essay on the theme of the great climacteric—the dangerous epoch in human history extending approximately from 1798 to 2048. Our excuse is that such sweeping visions, although by no means typical of White's own style, are nonetheless stimulated by his probing. In three selections in volume I (26, 28, and 29), White has turned in a reflective mood to such global issues.

Gilbert White has shown us, his students and colleagues, the importance of identifying our intellectual antecedents and of assessing critically what has been achieved. He has demonstrated the need to identify tasks clearly and to make explicit the often hidden linkages among science, nature, and society. These are lessons that we and our fellow authors have sought to apply, however waveringly, throughout the volume. The essays are highly individualistic, but in them certain common trends emerge strongly.

In Gilbert White's work neither the concepts of mastery over nature nor a vunerable dependency are accepted; neither human beings nor nature are fixed or finite. White shows how nature and human society interact to produce both resources and hazards. Constraints and obstacles to choice are always encountered; limits, never.

Through half a century of scientific research and teaching, White has organized detailed studies that exemplify an understated philosophy of cautious "empirical stewardship." He has eschewed grand theorizing and polemics for painstaking empirical research, guided by a deep concern for the well-being of people and nature. Much of his effort has gone into the distillation and synthesis of findings in a way that can be appreciated and used by those who make policy as well as those who live and work with hazards and resources, demonstrating that such research beginning at the local level can change the world.

Two other important qualities illuminate this record. First, White has always worked closely with others, inspiring and teaching while engaged in the pursuit of original insight and knowledge. He has demonstrated in diverse organizational settings that collective scientific enterprises, embracing different disciplines and nations, can do good scientific research and be more widely heard and respected in the chambers of the power brokers at national and international levels. Second, despite the ease with which these achievements appear to flow in public, they ultimately depend upon a firm moral backbone. Enthusiastic espousal of new ideas and the casting off of old ways that protect position or privilege do not come easily in any society. When attempts have been made to stifle or ignore the results of research or to misapply them for less worthy ends, Gilbert White has not failed to take a stand and to put himself and his position into public scrutiny.

The first ten essays in this volume all consider how professional inquiry is linked to public policy, what has been achieved and what remains to be done. As White would have it, the focus is on the problem area—not his contribution alone but the collective effort.

The essays as a whole are, in keeping with that spirit, also a collective effort. Although there are only thirteen papers and twenty-one authors, wide consultation with friends, associates, students, and Gilbert's wife Anne has taken place. The acknowledgments listed at the conclusion of the papers express appreciation to a further forty people. Altogether at least sixty-one people have shared in the preparation of these papers.

While benefiting from a common orientation, the papers are not a dry or factual survey of the "inventory and prospects" type. The authors were not constrained to be all-inclusive and have been selective in their proposals for research. Each paper expresses some of the individuality and imagination of its creator or creators. Yet, taken together, they make it possible to understand and appreciate how Texas floodplain dwellers, East African villagers, Indochinese governments, and international scientists have all been influenced to the good by the persistent demonstrations of one highly unusual "scholarly juggler." They also permit a new realization that such expanding waves of efficacious effort do indeed extend to all corners of the globe.

How has this been achieved? One answer is to be found in the concluding sentence of the paper by Wolman and Wolman. Gilbert White has recognized the need for "much experimentation, evaluation, and evolutionary adaptation," and he has recognized what little warrant exists for "the absolute, the grand plan, or instant satisfaction." To adopt this uncomfortable stance leaves scholars like Gilbert White in a state of tension and uncertainty. In all these essays there is a questioning about what is intellectually valuable, what is needed for public policy, and what is valued in the moral economy.

The conflict among intellectual validity, the efficacious, and the just is not susceptible to a final or stable resolution. One meaning of Gilbert White's diverse intellectual and public service concerns that we have tried to reflect in these essays is that it is necessary to seek to increase truth, efficacy, and justice all and at the same time. An increment of progress is more valuable if shared among these three than if expended upon one alone.

IAN BURTON
ROBERT W. KATES

I Water Supply: Persistent Myths and Recurring Issues

M. Gordon Wolman and Abel Wolman

Perhaps because every human being requires water each day to sustain life, the words "need," "require," "demand," and "use" have been more confused in dealing with water than with other resources. The same confusion prevails in many discussions of minerals or energy resources, but the rhetoric and confusion surrounding the provision of water appear particularly striking. Gilbert White, along with many others, has dealt with these words and the ideas, facts, and fancies that lie behind them. We will attempt a brief exploration of some facets of water supply: recurring issues of national, local, and industrial supplies and persistent myths of shortages and absolute values. While the view is historic, history is tied to the present and to some prospects for the future. Favorite themes of Gilbert White relating to the provision of water supply, including problems of value and processes of planning and decision making, are an essential part of such a review and analysis.

A need is the minimum quantity necessary to sustain human life. For drinking, this may be less than several liters per day, and for washing, 20–100 per day, depending upon accessibility. Need can also be viewed as the expression of a desire to use water for a particular purpose that someone deems a need, as in "I need more water." The word "requirement" conveys much the same meaning as need, with an added professional note and a sense of urgency. In contrast, "demand" has an economic meaning, a quantity of water that would be used at a given price. "Use" is simply a general expression meaning the utilization of water for a particular purpose, whereas "consumptive use" is the water actually used up or not

M. Gordon Wolman is Professor of Geography and Chairman of the Department of Geography and Environmental Engineering, Johns Hopkins University. His principal interests are in the field of natural river processes and hydrology. He is coauthor of *Fluvial Processes in Geomorphology*. Abel Wolman is Emeritus Professor, School of Engineering and School of Hygiene and Public Health, Johns Hopkins University. He has performed advisory and investigative services for more than fifty foreign governments. Among his numerous awards and honors are the U.S. National Medal of Science, the Tyler Ecology Award, and the Lasker Foundation for Public Health Award. He is the author of several books, including *Water, Health, and Society*.

returned to the source from which it was withdrawn. Thus consumptive use is the difference between the amount of water withdrawn for use and the return flow from that use.

There is nothing new in these definitions, but confusion continues to reign in their use and in the meanings given to them by the public, the press, and even professionals. If Senator Hayden had been right when he said, "If the Central Arizona Project is not built, Arizona will dry up and blow away," one might have said that water was a necessity to stabilize the landscape of Arizona. What he had in mind, of course, was that cheap water was needed in parts of Arizona to maintain economic growth. This may or may not have been true, but a more appropriate measure in that case would have been some measure of the value of water applied to certain alternative uses in Arizona.

More recently the adequacy of water supplies for energy development in the semiarid and arid regions of the western United States has been questioned, particularly the adequacy of water for energy conversion. Once again comparisons are made between the quantities of water in free-flowing streams, not yet claimed under the doctrine of "first in time—first in right," and estimates of the projected use of water for energy conversion. A simple balance is then struck on the assumption that the amount of water that could be made available represents only the remaining difference between some definition of a hydrologic quantity and existing rights. Again no mention is made of the possibility that water may move from one use to another if the price is right, even though this may take some time and negotiation, or of complex storage and technological options.

Partly as a result of the confusion of terms, and partly from the existence of a lingering environmental determinism, a variety of myths has grown up about water use and water supply. We will note a number of these in examining the history of certain common issues of water supply. We use a rough spatial scale, beginning with the nation and considering rural and household water supply, industrial water use, and water supply and planning in urban and metropolitan regions. Concepts of the value of water that transcend discussion of water use at every level are dealt with separately.

The Recurring Presumed National Water Shortage

Writing in the early 1950s, the Paley Commission noted in *Resources for Freedom* (1952) that the projected rapidly rising curves of water use, associated with rising levels of economic activity, presaged serious potential shortages. The commission's projections showed not only a progressive increase in total consumption but also an increase in per capita consumption. In some ways it merely repeated warnings of the 1930s.

All projections require assumptions about the economy, technology, and social preferences for some substantial periods. As White (1969) pointed out, a striking change occurred in the approach toward a national shortage of water as a result of the studies by Nathaniel Wollman for the Senate Select Committee on National Water Resources (U.S. Congress 1961). These indicated that quality rather than quantity was likely to pose more serious issues in a number of regions. Projections of the quantities of oxygen-demanding wastes against desirable standards of water quality, in terms of minimum levels of dissolved oxygen, highlighted the need for added low flow for dilution as well as much higher levels of waste treatment by the year 2000.

A careful study by Piper (1965), using Wollman's projections of expected needs for water (again not scaled to supply) and more conservative assumptions about available supply, indicated that, by the year 2000, maximum projected commitments would exceed potential assured supply in sixteen of nineteen major drainage basin regions. The reverse was true in New England, the South Atlantic–Eastern Gulf region, and the Ohio basin. "In 7 of the 16 regions of seeming water deficiency" (1965, p. 20), total streamflow would equal or exceed projected total commitment, and hence commitments could be met by full regulation. For the remaining nine basins, reductions in use, including conservation, conversions among uses, and augmented supplies, would appear to be needed. In most of the West, consumptive use in irrigated agriculture dominated the projections, as it continues to do. In three regions of projected shortage, fish and wildlife instream habitats to be created by man were major projected uses.

By the time of the second national water assessment of the U.S. Water Resources Council (1978), projections for selected water withdrawals begin to show a marked reduction (p. 29). As illustrated in table 1.1, projections by the Senate Select Committee show large expected increases in total withdrawals and in consumption to the year 2000, significantly larger than those projected by the Water Resources Council in 1968 and in 1978. The differences are particularly striking for consumptive use in manufacturing (although some of the discrepancy may be due to the apparent inclusion of some manufac-

Table 1.1 Actual and projected withdrawals and consumption of fresh water in three successive studies (billions of gallons per day)

	Actual			Projected					
	1954	1965	1975	1980		1985	2000		
	Senate[a]	WRCI[b]	WRCII[c]	Senate	WRCI	WRCII	Senate	WRCI	WRCII
Manufacturing									
Withdrawals	32	46	51	102	75		229	127	24
Consumption	3	4	6	9	6	9	21	10	15
Irrigation agriculture									
Withdrawals	176	111	159	167	136	166	184	150	154
Consumption	104	65	86	104	82	93	126	90	93
Total: All uses									
Withdrawals	300	270	338	559	443	421	888	805	306
Consumption	110	78	107	119	104	121	156	128	135

[a] Senate Select Committee 1961.
[b] Water Resources Council 1968.
[c] Water Resources Council 1978.

turing in the municipal category in the Senate committee's study). Projections of growth in population and in the GNP have been lower in recent years, and these determinants of water use, along with greater conservation, may be contributing to the reduction in projected uses. The Senate committee's projections of withdrawals for 1980 exceed by roughly 30 percent the "actual" values given by the Water Resources Council (1978), and estimates of total consumptive use are about 10 percent higher. Emphasis here should not be placed on accuracy of projection; clearly projections should improve as the moving present approaches (Ascher 1978). Rather, comparison of projections reveals two important elements: the changing interests of society over time, and the multiplicity of social, economic, and technological adjustments that are made to accommodate new and changing demands. The projections, then, rather than being self-fulfilling, reflect the stimuli that lead to new ways of coping. None of this either implies an inexhaustible supply of resources or demonstrates that curves depicting projected quantities of uses must necessarily cross fixed supplies of water, thus automatically signaling "shortages."

Successive projections clearly reflect the dominance of new perspectives or interests in society. Before 1960, water quantity was the primary focus. In the later 1960s, water requirements for low-flow dilution and water quality received special attention. And by 1978, in its assessment, the Water Resources Council called special attention to instream-flow uses. Basing instream-flow needs primarily on the dominant instream use, the council noted: "In all subregions, the fish and wildlife use is the dominant instream-flow use" (1978, p. 42).

To the extent that all these projections require important assumptions about both the use side and the supply side of the equation, it is not surprising that a growing economy, projected against a static or nearly static source of supply, has customarily resulted in repetitive projections of shortages. However, projected consumptive uses and depletions for the nation as a whole by 2000 are likely to be about 10–15 percent of available stream flow. Thus in the summary of the 1978 report, the Water Resources Council (1978, p. 2) states clearly: "Nationally, the United States has an ample supply of water from both surface and underground sources." The key word, of course, is "national."

The myth of a national water shortage, however, is unlikely to disappear from the public press or from the speeches of politicians. Projections are not self-fulfilling predictions. First, the nation is not homogeneous in terms of either human activity or hydrology. Areas of surplus water remain, as activities are concentrated either in regions of high rainfall (as in New England) or where uses are low. Even in such regions, water is short of expectations in some dry years. Second, in areas of modest precipitation, demands for water continue to be highly localized. In most instances, for the "oases" by which Walter Prescott Webb (1957) characterized much of the Plains region and parts of the Southwest, water is available but at some distance. Third, economic transformations of water use are taking place as a result of changing economies in the Sun Belt of the East and, more importantly, even in the Southwest, where rights to water exceed available supplies. Much of this transformation is apparently related to changing ideas about the relative value of water for different purposes (discussed below). Fourth, despite significant opposition in regions such as the Northwest, the possibilities still exist for continuing interbasin transfer of water, including water from Canada. The technical obstacles to such transfer are minimal. The matter is, rather, one of institutional and regional ideas about the value of keeping water at its place of origin, or the importance of reserving water for potential uses in areas where development is contemplated. At the international level, interbasin transfers become more complex international issues.

For a variety of reasons one can expect that the myth of a national water shortage will continue to be fueled because of the simplicity of its statement and because recurring shortages such as those experienced in California in 1977 tend to be generalized to the country as a whole. The lead sentence in the February 1979 special report of the *Wichita Eagle* and the *Wichita Beacon* reads, "Water. The lack of it is reaching a crisis stage in parts of Kansas," and on 22 August 1976, the Sunday *New York Times* headlined, "Fight for Water in West Grows." Again, in April 1985, *ASCE News* announced, "National Alliance Seeks to Avert Water Crisis" (*ASCE News*, p. 6) These issues are not illusory. Specific water problems, including the need to provide supplies of adequate quantity and quality, will continue to pose problems at the local and regional scale. The Water Resources Council (1978, p. 3) estimates serious problems of inadequate surface-water supplies by 2000 in roughly 10 percent of the subregions of the country. The national water shortage will probably continue to remain a myth for some time; there is no national shortage of water. As suggested below, many opportunities exist for adjustments to meet competing demands.

To the extent that the problem of water supply is national, variety rather than uniformity should characterize the solution. Regional, metropolitan, and drainage basin cooperative management efforts will clearly increase. The perennial search for a national water policy must focus on the processes of decision making and on the variety of solutions to specific problems rather than upon a grand design.

Water for Individuals and Households

The United States

Rural water supply has traditionally been a private and individual decision. Individual farmers, using their own knowledge, that of their neighbors, and, more recently, information from state and federal agencies, have located a well to supply the needs of their homesteads. White (1969a) notes that the amount of information made available to such rural private decision makers by governmental agencies has often been quite small. Some states and the U.S. Department of Agriculture provide technical bulletins on the availability of groundwater and methods of drilling wells and information on methods of using both surface and groundwater. For a smaller number of areas, information is available at a level of both detail and simplicity appropriate to individual users. In addition, some state surveys and the U.S. Geological Survey provide various levels of consultation for rural water development. On public lands, the U.S. Geological Survey, in cooperation with the Bureau of Land Management, has often been responsible for the location of wells for stock tanks and small reservoirs, primarily for agricultural use.

Despite the resources available and the presumed existence of planning, in many suburban regions where half-acre lots or larger have been the preferred mode of development there is less information available to prospective home owners, even from developers, than is available to farmers. In many metropolitan areas, suburban development of water supply continues to depend upon the drilling of wells by prospective home owners. In some communities it is still considered a remarkable advance to require that a well of proven yield under specified pumping conditions be demonstrated on a given property before a new house can be built or a lot for a house sold. As in some rural areas, sophisticated upper-class home owners continue to seek the advice of water dowsers in helping to locate wells on their property. White (1969a) noted that this is one of the few areas in which significant private decisions, often by the educated as well as the less informed citizenry, are still made by methods considered witchcraft by the scientific community. The mysterious character of groundwater occurrence appears to be related not simply to a lack of knowledge about the hydrologic cycle. Individuals apparently turn to water dowsers in arid areas where water is scarce and the best scientific techniques are uncertain, and in regions where groundwater is nearly ubiquitous and dowsers are both canny and likely to succeed (Vogt 1953). Public information available to rural and suburban dwellers may reduce the mystery somewhat, although "trying anything" in desperate circumstances is likely to remain an option.

What has changed is the increasing attention of public authorities to the quality of water in private systems. Drinking water standards established by state and county authorities must be met by wells in rural areas. Because high nitrate concentrations in some wells have been the source of methehemoglobinemia in infants, the suggestion that many wells in agricultural areas would be contaminated with nitrates as a result of the infiltration of nitrates from excessive fertilization, particularly in Illinois, produced a wave of concern. This fear has not been supported by subsequent study (Walker 1973). Although proper concern continues to ensure that both rural residents and those living in small communities are assured a safe, clean water supply that meets exacting standards, outbreaks of waterborne diseases from untreated wellwater continue. It is not unlikely that infiltration of chemical wastes from old disposal sites will also pose problems as houses spread across the land. In general, the largest number of reported outbreaks of waterborne disease continues to be associated with small systems and individual households (Craun and McCabe 1973; Craun, McCabe, and Hughes 1976). This problem is discussed below in relation to concepts of the value of water.

The Developing World

In contrast to conditions in the United States and most of the industrialized nations, in less developed countries nearly two-thirds of the population do not have access to safe and ample supplies of water. A survey by the World Health Organization indicates that 70 percent of the urban population and 12 percent of the rural population, accounting for one-quarter of the total population, were served by house connections or standpipes in 1970. To account for the total world population in both the developed and developing countries, White (1974, pp. 39–40) distinguishes between urban and rural populations and estimates values from personal observation for the People's Republic of China (which was not covered by the WHO study). He suggests that the "People's Republic of China has provided standpipe service for 90% of its city dwellers . . . and about 70% of the rural population." Recognizing the very uncertain quality of the data, he estimates that for the world as a whole, supplies are relatively adequate for city dwellers in the developed nations, inadequate for the squatter settlements surrounding tropical urban centers, and "thoroughly adequate for only about half of the rural dwellers." Rural dwellers include both small settlements and dispersed rural populations.

Distinctions between city, peripheral, and clustered and dispersed rural populations, between types of service such as house connections and standpipes, and between incomes or wealth in various parts of the world are important in evaluating the prospects for development of adequate water supplies. Categorizing nations as developed or less developed, of course, masks major disparities within countries. Even in the United States much confusion has been generated by failure to distinguish between a finding that many communities are served by public supplies that fail to meet approved standards of quality and the fact that 90 percent of the population of the United States is provided with water which meets the standards (see Craun et al. 1976). Many supplies of small size tend to be less than satisfactory according to standards considered appropriate in the United States, although some question the significance of the standards themselves. The problem of standards is considered briefly below in relation to the concept of the value of water.

Although progress has not kept pace with the need for water in the vast rural areas of the world, it is important to note that in Latin America the population served with water rose from 66 million in 1960 to 196 million in 1976. Surprising to some, 70 percent of the 6 billion dollars (U.S.) invested in water and sewerage came from national and local funds and 30 percent from outside sources, including the InterAmerican Development Fund and the World Bank. Estimates of requirements for the coming decade are at once astronomical and staggering, with annual costs to supply water estimated to be at least $8 per capita in urban areas and $3.50 for rural water throughout much of the developing world. Additional amounts are required for excreta disposal, perhaps 2.5 times as much for complete sewerage and sewage treatment works, and 25 to 60 per cent of the water supply figure for privy systems in rural or other regions (F. W. Montanari, Agency for International Development, pers. comm., April 1980).

The ten-year projections of need follow upon the objectives enunciated at the United Nations Mar del Plata Conference on Water in March 1977. These in turn derive from the findings of the United Nations Conference on Human Settlements held in Vancouver (Canada) in 1976: that nearly two-thirds of the population in the less developed countries do not have reasonable access to safe and ample water supply, and an even greater proportion lack the means for hygenic disposal of human waste. While the rhetoric giving water supply top priority among the needs of the developing countries is well intended, and the need heartfelt, there is a tendency for the resolve to melt in the face of the large sums of money that appear to be needed—and more importantly, upon recognition of the administrative or institutional demands of such large efforts.

History suggests that the overwhelming character of the task of supplying water to mankind can be made less formidable, if not less demanding, by recalling past experience. First, virtually all who have dealt with the problem point to the need for commitment at the local level, if safe and clean water is to be provided to small villages and dispersed communities and individuals. Such commitment recognizes that only with indigenous leadership and community support, including the management and political skills required, will adequate facilities be built and maintained. Second, community involvement demands that leadership be exercised, informed, and assisted by dedicated professionals available to guide and to train the previously unskilled. Indeed, remarkable people are needed, but every village contains them. Third, costs have been neither overwhelming nor unsupportable where communities have committed themselves to supplying or paying for the clean water they need and want, though governmental assistance is needed to initiate progress in the poorest areas. Fourth, the technology and the standards for quality and quantity must be appropriate to the resources and stage of development of the communities involved, with provision for later change and improvement. Fifth, while all these requisites have a ring of newness about them, virtually all of them have characterized the development of water supply in the developed and developing regions of the world.

While the principles are easy to state, progress in rural water development, like all development efforts, continues to be frustratingly slow in many areas (Saunders and Warford 1976). Although there are similarities with the history of development in the developing countries, there are also marked differences in financial resources, in the availability or potential availability of trained individuals, and in expectations. Water supplies can be constructed and maintained in the developing world (Wolman, Hollis, and Pineo 1972; Pan American Health Organization, 1979); but despite the existence of adequate technology, failure rates remain distressingly high, perhaps 50 percent in many rural water supply projects in the last decade (Henry 1978, p. 372). Thus recent research stresses operation, maintenance, and community participation. Studies demonstrate that household users, government officials, and communities have different perceptions of what constitutes the proper location, type, and quality of water supply (Whyte and Burton, 1977). Further, in many communities women may be seen as drawers of water but are less likely to be accepted as carpenters, pipefitters, or plumbers—that is, as maintainers of water supplies.

Although perhaps the most important and difficult task is the encouragement of local leadership and, with it, community support and commitment, repetitive identification of these difficulties may inhibit action. There are no handbooks available on the day-by-day tasks of motivating, cajoling, convincing, and organizing individuals to provide for themselves clean, safe water. Nevertheless, as some agricultural development efforts have shown, precepts of community organization comparable to the kind emerging in some urban neighborhoods in the United States can be taught and successfully adapted to rural development.

A valid emphasis on community efforts, however, should not obscure the need for governmental action. In a developing world dominated by authoritarian governments, and in the face of extreme poverty and the absence of rudimentary institutions, skilled technicians are needed to assist in installing and maintaining water supplies and in training others to assume these tasks. Money may also be needed for materials and equipment in the poorest areas. Unfortunately, despite the practitioners' interest in operation and maintenance, politicians in the developing countries, like their counterparts elsewhere, are more interested in new projects and capital investments. Nonetheless, the requisites for providing water supplies in developing nations are not primarily technical but managerial. Innovation is required in the application of technology and in the development of institutions and their managers. The process is likely to be slow and to trace out again the steps that societies everywhere have taken in providing themselves with what they have considered to be a better quality of living.

Water for Industry

While industrialists concerned with locating industries and officials responsible for providing water have learned to discount water supply as a unique determinant, geographers may have been slower to do so because of lingering environmental determinism. White (1969b) helped to lay this myth to rest. Some very large industries, such as power plants that use but do not consume massive quantities of water for cooling purposes, continue to locate adjacent to water bodies, if once-through cooling is their choice of technology. But "Rocky Mountain Spring Water" and "Land of Sky Blue Waters" are, of course, advertising slogans, not geographic determinants.

Two kinds of observations have gradually eroded the view that water is a primary determinant of industrial location. First, careful documentation reveals that the quantity of water used per unit of product by producers of the same product varies enormously from place to place. The prime example was the Fontana Division of Kaiser Steel Company in California, which used 1,400 gallons of water per ton of steel produced, in contrast, for example, to a number of large eastern steel mills that used about 50,000 gallons per ton.

Water consumed per ton of product in the making of papers, chemicals, canning, and other primary and secondary industries is exceedingly variable, which suggests that the quantity of water consumed is a function of a number of variables and that "efficient" or reduced use of water was determined by different production functions, including the availability of water in different locations.

Second, surveys of industry's ranking of factors deemed important to site location suggest that water availability often ranks behind such factors as supply of raw material, transportation, availability of labor, and markets. Neither variability in water use nor ranking with respect to location factors reduces the importance of a satisfactory water supply, but there are a variety of opportunities for adapting industrial processes to produce a given product with varying inputs of water. The range of technology available to supply suitable water at different costs for a variety of industries provides additional economic flexibility.

Detailed studies by Bower (1966), Löf and Kneese (1969), Abbey (1979), and others have shown the wide variety of options available for altering water use, consumption, and effluents. These studies point to the same production-function characteristics illustrated by the geographic data. That is, economic incentives (sometimes determined by regulatory requirements) result in adoption by industry of more efficient water use commensurate with the costs of many inputs to the production process. Such savings, barring those mandated by law, are economically determined in most instances and not directed toward saving water per se, because water costs are often less than 1 to 3 percent of all production costs.

In Saudi Arabia, where fresh water is scarce, very large expenditures in new construction may be made to save water. Thus the unit cost of water may be exceedingly high by American standards—more than one dollar per gallon. Nevertheless, the percentage of the total cost of constructing hotels, hospitals, and industries is likely to remain modest, and though availability of water makes development more complex and costly, its scarcity has not deterred development in the face of other overwhelming social and economic forces.

A relatively new pressure leading to the adoption of water conservation in industry comes from mandatory controls on effluents designed to improve water quality, as well as from the Conservation Act of 1977. Information developed from studies for the National Commission on Water Quality indicates that since passage of the Clean Water Act in 1972 (PL 92–500), a number of industries have significantly reduced both withdrawals and consumptive use of water (reflected in table 1.1). These reductions accompany efforts to redesign industrial processes, within both old and new plants, to minimize the total cost of water used, and to improve water effluent. Thus Minnesota Mining and Manufacturing Company completely altered a process used to make a glued tape to eliminate costly treatment of an effluent including complex organic chemicals. By redesigning both the product and its manufacture, less material was wasted and the effluent made more tractable to treatment. (By itself, a smaller and more concentrated effluent need not be less costly to treat—indeed, it may be more costly). Löf and Kneese (1969) noted that the percentage of recirculated versus intake water in beet sugar processing plants ranged from 0 to as high as 90.

The concept of reuse includes use of waste water after disposal from one source by another, as well as within a single industrial plant. It is estimated that on a major river such as the Ohio, water may be reused three to five times. Such indirect reuse is, of course, the rule rather than the exception. In semiarid or arid regions, return flows from major urban areas may constitute the sole base flow of a river. Direct reuse, such as the use of wastewater from a sewage treatment plant of Baltimore by the Bethlehem Steel Company, indicates that, while the cost of water in the production of steel may be relatively small compared to other inputs, a cheap source of large volumes of water may produce significant savings. The pipelines and treatment facility constructed by Bethlehem Steel begun in 1941 cost $7.5 million initially and have a present value of about $50 million. Bethlehem obtains some 100 mgd of treated sewage effluent from the Baltimore treatment plant at a price of 0.13¢ per thousand gallons. No other source of comparable volume is available today. Annual savings amount to several million dollars, an economic opportunity beneficial to both supplier and consumer. The city profits by avoiding millions of dollars of investment and operation costs otherwise necessary to provide both additional treatment and the extension of its outfall to a more appropriate location for disposal. While the number of such opportunities is growing, the Bethlehem Steel operation begun in 1941 is still the largest such exchange in the United States, and perhaps in the world.

To the extent that industrial concentration, population growth, and demands for clean water continue, one may expect increasing efforts to reduce water demands. Products such as beverages, where considerable water is included in the product, must reach a fixed minimum level of consumptive use. However, even in the canning and beverage industries, the amount of water in the product is perhaps 10 percent of the total required for all uses including processing, washing, and plant maintenance.

Changing production functions will undoubtedly determine future trends. There is no reason to believe that the cost of water in most industrial operations will determine industrial location. New technology will permit different products to be made with smaller inputs of water. At the opposite end of the spectrum, however, in one form or another, dissipation of waste heat from power plants will demand a minimum quantity of water determined by thermodynamic considerations. The economic assumption sometimes made that a silk purse can be made from a sow's ear by substitution clearly has limits.

Water Supply and Planning in Metropolitan Regions

As one moves from the household or industry to the regional scale, it is common and easy to note the desirability and the necessity of planning for the provision of public works, including water supply. This call need not suppose a grand design at the national level—in some developing countries the search for such a design has indeed inhibited the initiative of regional and local authorities. The presumed necessity to provide water supply and wastewater facilities simultaneously and of equal sophistication may also create barriers to action.

In the United States it is clear that urban planning and development are tied to the provision of public works. The question whether the provision of water supply and sewerage "determines" the pattern of urban growth or responds to other social forces remains a subject of lively debate—the country has not yet made up its mind about how much, or how strongly, it wishes governments to control decisions about land use. The courts have permitted limitations on the provision of water supply or waste disposal as growth control measures when these are part of an overall zoning or planning design. Other moratoria on construction of water systems have been temporary, or wells and small treatment plants have permitted urban expansion (as in the Baltimore region). The evidence indicates that social pressures determine where the pipes are laid.

The high cost of energy, the burden of commuting, and a new set of values characterizing the good life may set a new course for urban living, and the providers of water may respond. Political or institutional factors will determine not only the mode or structure for the provision of water but also the rate at which society will respond to claimants with different interests and values. The current conjunction of research in optimization, multiobjective planning, and social choice suggests that new avenues are needed to assist society in arriving at timely decisions reconciling economic efficiency with political equity. We must document the success and failure of institutional and administrative structures in the provision of water supply as Blake (1956) has done for Baltimore, Boston, Philadelphia, and New York. The rarity of such evaluations suggests that practitioners are not particularly eager to have the record reviewed and to see new seekers declare past experience irrelevant.

The Value of Water to Society

The demand for water has been shown to be a function of many variables: the availability of alternative sources, competing uses, economic well-being, historical precedent, and cultural, political, and economic perceptions. To many, the value of water to individuals and to society is self-evident. Since antiquity it has been a centerpiece in the evolution of civilization, in agricultural development, and in the aesthetic quality of many of the world's great cities (Frontinus, 1899; Robins 1946). In the United States, provision of public water supplies for many major eastern cities was motivated by a combination of objectives including dust control on dirt streets, fire protection, and the provision of water for commerce and for individual homes. In this country and in Europe it was not until the turn of the century that arguments for public water supplies began to rely heavily on their vital importance for health.

Because water has often been looked upon as a free good, the careful evaluation of costs and benefits derived from varying quantities of water has been primarily the concern of individual industries seeking to lower costs. Responsible private or public municipal suppliers were required to calculate costs and revenues, including charges for water, in order to operate efficiently to assure that stockholders received a profit or that ratepayers were not overcharged. Because publicly owned water systems were regulated and a part of the political process, charges required justification on the basis of the costs of providing water and expanding the system. After the initial success in selling the concept of publicly available water, debate centered on the proper mode of fixing charges.

In developed countries, over the last several decades, debate has centered on the rate at which expansion should proceed, the way in which it should be financed, the relative financial responsibility of old and new users, and the design criteria to be used. None of these is a new issue. Since early this century, charges have customarily been based on some combination of an assessment or flat fee to cover installation of the service to an area and a schedule of charges for the quantity of water used. In many areas, sewer charges were later made proportional to water use.

An assault has been made on these procedures on two principal grounds: first, that many municipal systems were overbuilt, that is, large capital expenditures for major dams, reservoirs, and distribution systems were made prematurely (Hirshleifer, Milliman, and De Haven 1969); and, second, that charges were too low, leading to a waste of water, and were not designed to reflect the significantly higher costs associated with providing high peak demands of residential customers. It has been difficult to sustain the first argument in its original form, because demands for water have risen rapidly since the Second World War and inflation has made some premature investments look farsighted, at least to public officials and citizens. Discussion has moved from financial issues to a broader consideration of financial, economic, and social questions. This shift has been stimulated by what White might call changing public perceptions and values.

Until recently the use of reservoir sites for public water supply was viewed as inevitable and logical. The availability of water in a river was based upon a calculation of an estimated safe yield determined by the lowest flows on record or extrapolated from records on comparable rivers or streams. Little attention was given to the frequency of such low-flow periods and to the consequences of assuring supplies for droughts or low-flow periods of different frequencies and severity. The costs of droughts have only recently begun to be carefully studied (Russell, Arey, and Kates 1970). Part of the impetus for examining alternatives has come from social pressures to maintain the natural flow of rivers, to preserve dam and reservoir sites for alternative environmental values, and to eliminate or greatly reduce the drawdown of reservoir levels in order to maintain recreation, wildlife, and aesthetic values.

In metropolitan areas such as Washington, D.C., the potential difficulties of locating adequate reservoir storage have led to more careful analyses of the probable consequences of planning for periods of tight water supplies, from small limitations on peak uses for lawn watering and car washing to sanctions on commercial and industrial use when availability is more severely limited. Such a range of options has been formulated for the Washington metropolitan region (Ecological Analysts 1977) in a form that allows decision makers to choose from a mix of drought management options and capital expenditures for storage over a period of years. Sufficient quantities and pressures must, of course, be maintained for fire fighting. With growing demand, additional storage will often be required in any case, but the options for drought management provide a basis for explicit consideration of the economic costs of bearing various levels of water shortage.

Dangers are involved in the assumption that management decisions can be sufficiently fine-tuned to assure that periods of low supplies can be handled with calculated and agreed-upon levels of disruption. The precise characteristics of the probability distributions cannot be known, although margins of safety can be allowed for, and deferring of reservoir construction can be extended to the point that the rate of demand expansion exceeds the rate at which provision for expanded supplies can be provided. The Thames Water Authority, in the wake of the summer drought of 1975 in England, noted, "It is at times such as these that water suppliers are glad to have plenty of stocks in reserve. London has appreciated as never before the foresight of the former Metropolitan Water Board who built the vast storage reservoirs in the Thames and Lee Valleys" (Thames Water Authority 1976). Whether or not more careful management of available supplies will lead to unacceptable risks remains to be seen. Much depends not simply on the weather but on the way in which the public accepts various levels of conservation, cutbacks in use, and control strategies. The frequency with which the public is asked to make adjustments, the magnitude of the adjustments, and the way in which little and big belt-tightening and emergencies are perceived are likely to influence the response. It is by no means obvious that the system will be so skillfully managed as to meet the above requirements promptly and appropriately.

Studies of the impact of price on the demand for water in various uses, of the effect of metering on water use, and of technological alternatives available for water conservation have all been part of the attempt to define more precisely how and why water is used in various urban activities in different regions. Surprisingly few detailed studies have been made of the amounts and determinants of urban water use. Earlier assumptions that household demands were inelastic have given way to the recognition that price can influence the behavior of household consumers. Howe and Linaweaver (1967) found that while domestic demand is relatively inelastic (−0.23), dropping from 250 gpd/dwelling unit at 20¢/1000 gal. to 225 gpd/du at 40¢/1000 gal., summer sprinkling demand in the humid eastern United States is elastic (−1.6), ranging from 460 gpd/du at 20¢/1000 gal. (roughly one-half of the actual charge) to 155 gpd/du at 40¢/1000 gal. Summarizing a variety of studies, Hanke (1978) and Bauman et al. (1979) show that estimated elasticities for urban residential water vary from about −1.0 to −0.1. The studies cover a variety of climatic and economic regions and are temporal and cross-sectional in design. Elasticities vary greatly among different uses although smaller values appear to characterize average residential use.

In the United States, at least, price may have limits as an effective tool for limiting or regulating consumption. Debate continues on the necessity or desirability of installing meters in individual dwelling units in apartment complexes or other facilities where the cost of metering may exceed the returns. Time-of-day and seasonal pricing, while presumably economic approaches to paying for the added costs of meeting peak demands, may not be universally applicable or economically sound where reductions in water use lead to reduced flows in sewerage systems, impairing efficient operation. Thus the benefits derived from increase in price may not exceed the costs of metering and administration or the costs required to offset the negative effects of low-volume sewage-water flows. It should be remembered that strength or concentration of sewage material requiring treatment is not affected by reduced water use.

There has been a growing recognition that consumers' concepts of the value of water in different uses are not fixed. In their migration from the humid eastern to the dry western United States, suburbanites carried with them their desire for green lawns. As the price of maintaining such lawns increases in the future, some assume or hope that residents will find new delight in desert flora for their yards. Indeed, a *New York Times* report of 8 June 1980, suggests that they have done so—in the past four years average daily water use in Tucson, Arizona, has dropped 30 percent.

Attempts to define the value of water in more precise economic terms have been stimulated in the western United States by interest in new activities involving outdoor recreation and urban expansion. Thus Wollman (1962) demonstrated that the economic value added by an acre-foot of water in recreation was five to six times ($250/af) and in industry about eighty times ($3500/af) that generated by added increments of water in irrigation ($45/af). More recently, new pressures have been created by the prospects for energy development in the West, where the marginal value of water for energy conversion is about $870–$5,360 per acre-foot as opposed to $0.53–$22.20 in irrigation (Spofford, Parker, and Kneese 1979). This emphasis on valuation derives from several sources. Changes in social and economic priorities have increased interest in these new uses for water in the West. This, in turn, has been accompanied by a concern that water has been undervalued and underpriced through public subsidies and sustained by appropriation law which slows the rate of transformation of water use as the interests of society change. The evidence suggests that, while the process may be slow, transfer of water to new uses is taking place as the economy of many arid and semiarid regions expands (see Abbey 1979).

An entirely different way of valuing urban water actually stimulated the growth of municipal water systems. When beneficial health effects were found to be associated with provision of municipal supplies a correlation was noted between the increase in community water supplies and the decline of the incidence of typhoid fever. Typhoid deaths declined from 36 per 100,000 in 1900 to less than 0.05 in 1967 (Gorman and Wolman 1939; Metropolitan Life Insurance Co. 1973). The correlation, once demonstrated, became a powerful weapon in the effort to get cities, towns, and villages to build water systems. The decline in typhoid fever often preceded the adoption of formal water-quality standards, and the introduction of community water supplies was among the many changing social and economic factors that contributed to the typhoid decline (Chamberlin et al. 1979, p. 44).

In the developing world, improvement in public health has accompanied the provision of private and public water supplies, but the incidence of diarrhea and other intestinal diseases has sometimes risen following installation of water supplies in villages and poor urban areas. These failures have been used as arguments against the premise that health accompanies provision of water supply. The evidence, dating from Snow's Pump, does not refute the premise but shows simply that "a poorly designed and poorly constructed water system capable of being contaminated by human waste can become an efficient mechanism of disease transmission" (Kawata 1978a, p. 2116). Moreover, inadequate design, poor and crowded living conditions, minimal or inoperable facilities, and failure to use the system overwhelm the effects of the presumed "availability" of water to the expected users (Kawata 1978b; World Bank 1979). In all cases there has been a fundamental failure to break the link between anus and mouth.

On balance, experience indicates real improvement in health to many populations, although, as White, Bradley, and White (1972) note, both the costs associated with poor health and hence the benefits of improved health are nearly impossible to quantify in economic terms. These authors conclude (p. 204), "Assuming that industrial supplies will be provided as they are needed, the role of domestic water would seem to be in assuring as far as practicable the health of the people and the satisfaction of their needs for amenities." In this light, Feachem's (1978, p. 360) pithy observation is particularly apt: "We should not conduct research into the health impact of water supplies in order to know whether to build more water supplies. Water supplies will continue to be built, irrespective of evidence for health benefits, because they fulfill the legitimate political objectives of many governments. We should conduct research in order that water supplies may be built better and may have a greater impact on health."

While the effort to rationalize expenditures of public funds for various purposes in the interest of efficiency and the accompanying refinement of benefit-cost analysis must generally be acknowledged as desirable, attention is often focused on the less developed nations, where some minimum needs are obvious, and not on the United States, where a search for purity is sometimes seen as unconstrained by scarce resources.

In the United States, vast expenditures are now being required under the Safe Drinking Water Act to assure "safe" water supplies to millions of people already enjoying the benefits of the safest and best public supplies found anywhere in the world (Russell 1978). Poor operation and maintenance of some systems result in regular outbreaks of viral hepatitis and gastro-intestinal illnesses (table 1.2). In public systems these usually result from faulty treatment and cross-connections or contamination in distribution systems. In industrial societies, with the use of increasingly complex materials, a variety of new and sometimes exotic substances will find their way into raw water sources. Control of wastes, as well as more sophisticated treatment of potential water supplies, will be needed, and the costs of supplying water of adequate quality will rise. Thus far, it appears that new technology and careful operation of water treatment plants can remove small amounts of substances potentially deleterious to human health at reasonable costs, particularly in view

Table 1.2 Outbreaks of waterborne diseases in the United States

Years	Number	Cases per outbreak
1938–40	45	583
1941–45	39	201
1946–50	23	121
1951–55	8	139
1956–60	7	121
1961–70	14	114
1971	19	272
1972	29	57
1973	26	68
1974	25	334

Sources: 1938–60: Weibel et al. 1964; 1961–70: Craun and McCabe 1973; 1971–74: Craun, McCabe, and Hughes 1976.

of the very low costs at which water is currently provided to society in the industrial world. It remains important, however, that the perceived benefits to health be thoughtfully evaluated, particularly in light of the difficulty of lessening already low health risks and the even more important task of using scarce resources in the most effective way to reduce such risks.

Industry, similarly, is expected to reduce materially its waste water discharges to ambient waters via process changes and in situ residuals treatment (Wolman 1983).

Some Needs and Cautions

In research into the provision of water supply there is a full spectrum ranging from the purest of scientific research to the behavior of human beings and social institutions. The necessity for inquiry is fueled both by the persistence of some myths about water supply (table 1.3) and the continual emergence of new issues. The distinction between the needs of the developing and developed countries provides an additional way of looking at the quest for information.

Myths of need cross the boundary between developed and developing regions. Thus in the technologically advanced countries there is a pressing need to develop projections of water demands at the national level from the bottom up, that is, from the disaggregated or regional level to the nation as a whole. Further, such projections must begin to reflect more strongly the complex technological, economic, and social elements likely to enter into and alter water use,

Table 1.3 Water supply: persistent myths

Myths of need	Clarification, potential corrective, or needed research
Need:	
Water needs are absolute requirements	Clear definition may help
There is an impending national water shortage in the United States	Uncorrectable: a political necessity
Water is a free good and must be continuously supplied, as shortages are intolerable	Determination of response to incentives such as rising cost (and price) for water combined with evaluation of politically understandable tradeoffs and limits of tolerance to shortages
Clean water is worth any price	Elucidation of the bases for water-quality standards and the benefits and costs of approaches to purity
An adequate village water supply is the same as any village water supply	Assurance of well-maintained along with well-planned rural and village water supplies; studies of tradeoff between water constituents and increased disease
Water as determinant of location and growth:	
Urban expansion follows water supply	Only where it is required to do so, as the suburban spread of septic tanks attests
Water determines the location of industry	Ask industry: examine the quantity of water used per unit of product and the fractional cost of water among the factors of production
The past as irrelevant:	
The problems of water supply in developing countries are new and unprecedented	Recognition that all the developed countries developed; rising expectations make time and poverty, not magnitude, the issues

thus providing a better base for evaluating projections of water use. This is not a search for perfection but, rather, a more explicit recognition of the variable character of demand. Similarly, there is a pressing need to understand the way in which small quantities of potentially deleterious substances may accumulate in the body and the physiological mechanisms that may cause chronic exposure to be dangerous or benign.

In the developing world, careful analysis of the design and operation of village water systems is needed to assure that adequate, rather than any, community water supply becomes the objective. Yet problems of operation and maintenance of small water systems are not unique to the developing world, and training and supervision of operating personnel pose administrative problems everywhere. Analogies between the problems of the developing and the developed nations may warrant more careful inquiry.

Organizing to provide villages throughout most of the world with a continuous supply of water, adequate in quantity and quality, poses challenges no less demanding than those of providing adequate management of water systems for the complex demands of large industrial and urban populations. The developing countries are clearly aware that expansion of urban areas results from a myriad of social forces. While vast areas of squatter communities on the periphery of cities need water, its absence did not deter settlement. For different, but equally complex reasons, the availability of water supply through expansion of metropolitan systems has not determined the pattern of growth in metropolitan areas in the developed world. Prospective inquiry is needed into the way in which growth management, using public works as a tool, serves as a lever for land use planning. It can only do so if the society is committed to land use control. The myth that water is the determinant of the location and magnitude of economic growth perhaps requires testing in the developing world. Water is essential to economic development, but decisions about where and what to develop perhaps should precede and determine the way in which water will be provided to enhance such development.

Continuing study is needed to identify those elements that will enable communities throughout the world to improve their health and well-being by the provision of water supplies adequate in quantity and quality. Such research cannot be isolated from action; it requires participation with the doers as well as cooperation among natural and social scientists and engineers often unaccustomed to such symbiosis. Yet caution may be warranted. Study, research, and analysis of the desirable can impede rather than promote achievement of the necessary. While the expenditures of vast sums on ill-conceived plans for community water supplies can hardly be commended, it would be unfortunate if developing countries' search for knowledge of a kind currently essential to evaluation and decision making in developed countries diverted and delayed needed action. History of the provision of water supply for human needs suggests the need for much experimentation, evaluation, and evolutionary adaptation—and little warrant for the absolute, the grand plan, or instant satisfaction.

Acknowledgments

The authors acknowledge the perceptive and valuable reviews provided by David Bradley, F. W. Montanari, and Theodore Schad.

References

Abbey, D. 1979. "Energy Production and Water Resources in the Colorado River Basin." *Natural Resources Journal* 19: 275–314.

ASCE News, April 1985. "National Alliance Seeks to Avert Water Alliance." Vol. 10, no. 4, p. 6.

Ascher, W. 1978. *Forecasting: An Appraisal for Policy-Makers and Planners*. Baltimore.

Baumann, D. D., J. J. Boland, J. H. Sims, B. Kranzer, and P. H. Carver. 1979. "The Role of Conservation in Water Supply Planning." For the Institute for Water Resources, U.S. Army Corps of Engineers, Fort Belvoir, VA (draft).

Blake, N. M. 1956. *Water for the Cities*. Syracuse, N.Y.

Bower, B. T. 1966. "The Economics of Industrial Water Utilization." In *Water Research*, ed. A. V. Kneese and S. C. Smith, pp. 143–73. Baltimore.

Chamberlin, C. E., J. Boland, A. Malik, and H. Shipman. 1979. "Wholesome and Palatable Drinking Water: A Background Paper on Water Quality Aspects of Water Supply." Washington, D.C.: U.S. Agency for International Development.

Craun, G. F. and L. J. McCabe. 1973. "Review of the Causes of Waterborne Disease Outbreaks." *Journal of the American Water Works Association* 45: 73–84.

Craun, G. F., L. J. McCabe, and J. M. Hughes. 1976. "Waterborne Disease Outbreaks in the United States, 1971–1974." *Journal of the American Water Works Association* 48: 420–24.

Ecological Analysts, Inc. 1977. *Water Supply Study for Montgomery and Prince George's Counties, Maryland*. For the Washington Suburban Sanitary Commission, Hyattsville, MD.

Feachem, R. 1978. "Domestic Water Supplies, Health and Poverty." In *Water Supply and Management* 2:357–62. London.

Feachem, R., E. Burns, S. Cairneross, A. Cronin, P. Cross, D. Curtis, M. K. Khan, D. Lamb, and H. Southall. 1978. *Water, Health and Development*. London.

Feachem, R., M. McGarry, and D. Mara, 1977. *Water, Wastes, and Health in Hot Climates*. New York.

Frontinus, S. J. [A.D. 97] 1899. *The Water Supply of the City of Rome*, ed. Clemens Herschel (1899). Boston.

Gorman, A. E., and Abel Wolman. 1939. "Water-Borne Outbreaks in the United States and Canada, and Their Significance." *Journal of the American Water Works Association* 31: 225–373.

Hanke, S.H. 1978. "A Method for Integrating Engineering and Economic Planning." *Journal of the American Water Works-Association* 70: 487–91.

Henry, D. 1978. "Designing for Development: What is Appropriate Technology for Rural Water and Sanitation?" In *Water and Society, Conflicts in Development: The Social and Ecological Effects of Water Development in Developing Countries*, ed., C. Widstrand, pp. 365–72. New York.

Hirshleifer, J., J. Milliman, and J. De Haven. 1969. *Water Supply: Economics, Technology, and Policy*. Chicago.

Howe, C., and F. P. Linaweaver, Jr. 1967. "The Impact of Price on Residential Water Demand and Its Relation to System Design and Price Structure." *Water Resources Research* 3: 13–32.

Kawata, K., 1978a. "Water and Other Environmental Interventions—The Minimum Investment Concept." *American Journal of Clinical Nutrition* 31: 2114–23.

———. 1978b. "Of Typhoid Fever and Telephone Poles: Deceptive Data on the Effect of Water Supply and Privies on Health in Tropical Countries." *Progress in Water Technology* 11: 37–43.

Löf, G. O. G., and A. V. Kneese. 1969. *The Economics of Water Utilization in the Beet Sugar Industry*. Baltimore.

Metropolitan Life Insurance Co. 1973. "The Impact of Waterborne Diseases on Mortality." *Statistical Bulletin*, July 1973, pp. 3–6.

New York Times, Sunday, 8 June 1980. "The People of Tucson are National Leaders in Conserving Water." P. 52.

Pan American Health Organization, Directing Council. 1979. *Strategies for Extending and Improving Potable Water Supply and Excreta Disposal Services during the Decade of the 1980s*. Pan American Health Organization Science Publication no. 390.

Piper, A. M., 1965. *Has the United States Enough Water*? U.S. Geological Survey Water Supply Paper 1797.

Resources for Freedom. 1952, Report of the President's Materials Policy Commission, 5. Washington, D.C. (Paley Commission).

Robins, F. W. 1946. *The Story of Water Supply*. London.

Russell, C. S., ed. 1978. *Safe Drinking Water: Current and Future Problems*. Resources for the Future. Washington, D.C.

Russell, C. S., D. G. Arey, and R. W. Kates. 1970. *Drought and Water Supply*. Baltimore.

Saunders, R. J., and J. J. Warford. 1976. *Village Water Supply: Economics and Policy in the Developing World*. Baltimore.

Spofford, W. O., Jr, A. L. Parker, and A. V. Kneese, eds. 1979. "Potential Impacts of Energy Development on Stream Flows in the Upper Colorado River Basin." In *The Impact of Energy Development on the Waters, Fish, and Wildlife in the Upper Colorado River Basin*, chapter 6. Resources for the Future. Washington, D.C.

Thames Water Authority. 1976. *Thames Water Annual Report and Accounts for the Year Ended 31st March 1976*. London.

U.S. Congress. 1961. *Report of the Select Committee on National Water Resources*. 87th Cong. 1st Sess., Report no. 29.

U.S. Water Resources Council. 1968. *The Nation's Water Resources: The First National Assessment*. Washington, D.C.

———. 1978. *The Nation's Water Resources 1975–2000: Second National Water Assessment*, 1, Summary. Washington, D.C.

Vogt, E. Z. 1953. "Water Witching: An Interpretation of a Ritual Pattern in an Rural American Community." *Science Monthly* 75: 175–86.

Walker, W. W. 1973. "Ground-water Nitrate Pollution in Rural Areas." *Ground Water* 11, no. 5. Illinois State Water Survey.

Webb, W. P. 1957. "The American West as Perpetual Mirage." *Harpers* 214: 25–31.

Weibel, S. R., F. R. Dixon, R. B. Weidner, and L. J. McCabe. 1964. "Waterborne Disease Outbreaks." *Journal of the American Water Works Association* 36: 947–57.

White, G. F. 1969a. *Strategies of Water Management*.Ann Arbor, Mich.

———. 1969b. "Industrial Water Use: A Review." *Geographical Review* 50: 412–30.

———. 1974. "Domestic Water Supply: Right or Good?" In *Human Rights on Health*, Ciba Foundation Symposium, 23: 41–59.

White, G. F., D. J. Bradley, and A. U. White. 1972. *Drawers of Water: Domestic Water Use in East Africa*. Chicago.

Whyte, A., and I. Burton. 1977. "Water Supply and Community Choice." In *Water, Wastes, and Health in Hot Climates*, ed. R. Feachem, M. McGarry, and D. Mora. New York.

Wollman, N. 1962. *The Value of Water in Alternative Uses*. Alburquerque.

Wolman, A. 1983. "Reflections, Perceptions, and Projections." *Journal of the Water Pollution Control Federation* 55:1412-16.

Wolman, A., M. D. Hollis, and C. S. Pineo. 1972. "A Generation of Progress in Sanitary Engineering Facilities and Services for Latin American and Caribbean Countries." *Bol. de la Oficina Sanitaria Panamericana* 6: 9–25.

World Bank. 1979, *Eight Case Studies of Rural and Urban Fringe Areas in Latin America: Appropriate Technology*. For the Water Supply and Waste Disposal in Developing Countries Series, P.U. Rept. No. Res. 23.

2 Floods and Man: A Geographer's Agenda

Rutherford H. Platt

Introduction—Two Floods

In April 1927, the lower Mississippi River reclaimed its floodplain. Rainfalls of twelve to twenty-four inches generated floodflows which breached the levees in some 200 locations. Eighteen million acres of bottomland were inundated in six states. Property damage was estimated at $284 million. At least 313 lives were lost. In their classic work on floods, Hoyt and Langbein (1955, p. 261) characterized this event as a turning point in national flood policy: "Few natural events have had a more lasting impact on our engineering concepts, economic thought, and political policy in the field of floods. Prior to 1927 control of floods in the United States was considered largely a local responsibility. Soon after 1927 the control of floods became a national problem and a federal responsibility."

Congress, through the Lower Mississippi Flood Control Act of 1928, established a dominant federal role in the reconstruction of levees and the provision of diversion floodways. The role of nonfederal interests was reduced largely to the furnishing of land for rights-of-way. This was a prelude to a nationwide program of federal flood control structures under the Flood Control Act of 1936.

On 5 February 1978, a howling northeaster struck the coast of New England. Inlanders will forever recall the event as the "Blizzard of '78," which dumped up to forty-eight inches of snow on Providence and Boston. Far more destructive, however, was a combination of high tides and waves driven by hurricane-force winds. The storm surge was estimated to reach a seventy-five-year level in Boston harbor. Two consecutive high tides battered the outer beaches of Cape Cod. Coastal towns on Massachusetts Bay north and south of Boston, especially Hull, Revere, and Scituate, lost 339 dwellings,

Rutherford H. Platt is Professor of Geography and Director of the Land and Water Policy Center, University of Massachusetts at Amherst. A lawyer as well as a geographer, he has conducted extensive research on policy issues concerning floodplains, wetlands,and coastal areas.

with another 1,012 suffering major damage. Another 5,503 homes sustained minor damage. Seawalls proved ineffective; several toppled into the sea. At Scituate, a seawall stood firm but the homes behind it were demolished by overtopping waves.

The Massachusetts coastal flood of 1978 may well rank with the 1927 Mississippi event as a second turning point in American flood policy. As recently as 1962, another New England coastal storm had fostered predictions that "one billion dollars could be spent on coastal protection by the federal government alone in the next fifteen to twenty years" (Burton and Kates 1964, p. 368).

No such specter arose in 1978. For possibly the first time, a major flood disaster prompted few demands for federal structural protection. Instead, state and local officials sought federal assistance for nonstructural responses. Actions undertaken or proposed have included: flood insurance, acquisition of shore property and relocation of its occupants, new kinds of disaster assistance including a grant for a special mental health program for flood victims, and enforcement of state and local building codes and land use regulations. The lieutenant governor of Massachusetts created a special unit to promote sensible policies in reconstruction of devastated areas. A bill was introduced (but not adopted) in the legislature seeking $10 million to acquire shoreline properties. The Federal Insurance Administration (FIA) in its first major coastal disaster processed 1,663 claims totaling $20 million. Both FIA and the Federal Disaster Assistance Administration attempted to enforce new federal requirements regarding mitigation of future losses. Even the Chief of the New England Division of the Army Corps of Engineers promised the Corps would carefully review its policy on seawalls in view of their proven unreliability (New England River Basins Commission 1978).

The floods of 1927 and 1978 thus delimit the rise and fall of the era of massive federal efforts to control floods through engineering structures. During this era at least $14 billion[1] have been spent by federal agencies to construct about 900 flood control projects in every major watershed in the continental United States. These include more than 260 reservoirs containing 100 million acre-feet of storage capacity, over 6,000 miles of levees and floodwalls, and 8,000 miles of channelization. Some 1,100 "small watershed projects" with a cost of $1.37 billion have been approved (U.S. Water Resources Council 1968, 5-2-2). Table 2.1 lists major floods and government responses from 1927 to 1983.

This national commitment to the taming of rivers and coastal waters ranks among the foremost undertakings of mankind, equiv-

Table 2.1 Chronology of major floods and public response

Year	Major flood disasters	Significant events in national response
1925	Lower Mississippi—1927 New England—1927	Lower Mississippi Flood Control Act of 1928
1930		Tennessee Valley Authority Act of 1933 Report of Water Resources Committee of National Resources Board, 1934
1935	Kansas River—1935 Upper Susquehanna—1935 Eastern United States—1936 Ohio/middle Mississippi—1937 New England—1938	Flood Control Act of 1936 Flood Control Act of 1938
1940		Flood Control Act of 1944
1945		
1950	Kansas and Missouri rivers—1951 New England—1954	President's Commission on Water Resources Policy—1950 Watershed Protection and Flood Prevention Act of 1954
1955	New England—1955	Flood Insurance Act—1956
1960	Gulf Coast—1960 Southwest and Midwest—1961 Atlantic coast—1962 Louisiana—1964	Floodplain Information Program, Corps of Engineers—1961
1965	Mississippi-Louisiana—1965 Upper Mississippi—1965 Upper Mississippi—1969 Mississippi-Louisiana—1969	Southeastern Hurricane Disaster Relief Act of 1965 Water Resources Planning Act of 1965 HUD Study on Flood Insurance—1966 Report of Task Force on Federal Flood Control Policy—1966 Executive Order 11296—1966 National Flood Insurance Act of 1968
1970	Rapid City, South Dakota—1972 Hurricane Agnes—1972 Upper Mississippi—1973	Flood Disaster Protection Act of 1973 Water Resources Development Act of 1974 Federal Disaster Assistance Act of 1974
1975	Mid-Atlantic—1975 Massachusetts coasts—1978 Southern California—1978 Pearl River, Mississippi—1979 Red River—1979 Texas Gulf—1979	Executive Orders 11988, 11990 (1977) Creation of Federal Emergency Management Agency—1979
1980	Hurricane Frederic	OMB Directive on Post-Flood Mitigation Assessments—1980
1983	Hurricane Alicia	Coastal Barrier Resources Act of 1982

alent to the pyramids of Egypt, the Great Wall of China, and the moon program. It is now in the process of joining them as past history. It bequeaths upon future generations a mixed legacy. Continuing flood prevention benefits from structures now in existence have been estimated to average $1 billion annually (U.S. Water Resources Council 1968, 5-2-7). Most large-scale projects confer additional benefits, including water supply, navigation, hydroelectric power, and recreation. On the adverse side, flood control reservoirs and channelization have obliterated or irreversibly altered many hundred reaches of scenic and fertile stream valleys. Human alteration of natural stream regimen promotes erosion and siltation and disrupts fish and wildlife habitats.

The federal flood control program has been superseded not because it is worthless but because it has failed. Despite heroic efforts, average annual flood losses continue to rise. Hoyt and Langbein placed yearly flood losses at $500 million in 1955. Property losses due to floods in 1975 reached $3.4 billion; by 2000, current trends indicate average annual loss (in 1975 dollars) to reach $4.3 billion (U.S. Water Resources Council 1978, p. 68). Loss of lives has decreased in major river valleys, where forecasting and evacuation procedures have improved (and where flood control structures may be most useful). But flash floods in tributary creeks and streams are causing an average annual loss of 180 lives and have been called the nation's "most dangerous weather phenomenon" (Cressman 1978, p. 1). Coastlines exposed to hurricanes also possess great potential for flood loss (White and Haas 1975). In short, the nation continues to suffer grievously from floods.

The very nature of floods—spasmodic, localized, and devastating—long obstructed the evolution of a "scientific" approach to their management. In place of rational consideration of alternative objectives and means, each crisis was treated on an ad hoc basis, and the technological fix prevailed. It was noted as early as 1937 by a "young officer of the National Resources Planning Board" that this treatment threatened to attract further encroachment upon floodplains unless accompanied by state and local land use restrictions (White 1969, p. 132). A flood exceeding the design limits of the protective structure would then cause greater damage than would have occurred in its absence. To this observation "the public reaction was hostile."

But the "young officer" persisted. From a modest beginning in the Executive Office of the President (where he allegedly occupied the largest office of his career), Gilbert F. White survived service as an ambulance driver and relief worker in Europe, returned after the war to become president of Haverford College, moved on to the University of Chicago Department of Geography, and went thence to the University of Colorado Institute of Behavioral Science. These major epochs have been interspersed by countless special assignments, responsiblities, and honors throughout the world. But through it all, the call for a rational approach to floodplain management expressed in 1937 has been sounded repeatedly with clarity, vigor, and eventually profound effect.

Origins of National Flood Policy, 1934–40

The "Rosenwald White House"

In 1933, University of Chicago geographer Harlan H. Barrows was invited by Public Works Administrator Harold L. Ickes to serve on a newly created resource planning body, the Mississippi Valley Committee (Colby and White 1961). The following year, at Barrows's suggestion, committee chairman Abel Wolman invited Gilbert White, then a graduate student at Rosenwald Hall (the home of the geography department at the University of Chicago until 1970), to come to Washington and serve as the committee's staff executive. The offer was accepted, and, according to Abel Wolman (pers. comm., 13 November 1978), White made himself immediately and immensely valuable. The Mississippi Valley Committee in 1934 became the Water Planning Committee of the National Planning Board. Barrows and White served with this committee under a succession of national planning units culminating with the National Resources Planning Board (NRPB) in 1939. Another Chicago geographer, Charles C. Colby, served on the parallel Land Committee (which White also served as staff).

The selection of Barrows and, by implication, White was an inspired choice. Barrows in the 1930s was the chief architect of what William Pattison terms the "man-land commitment" in the field of geography. A member of the first generation of American geographers to be trained at Chicago, Barrows and a fellow student, Carl Sauer, proceeded to define very different agendas for the fledgling discipline. Both agreed that purely physical concerns should be left to geologists. But where Sauer propounded a cultural basis for geographic inquiry, Barrows advocated the study of economic interaction between man and his natural habitat as an organizing framework (Pattison, n.d.). He stated in his 1923 presidential address to the Association of American Geographers: "Geography will aim to make clear the relationships existing between natural environments, and the distribution and activities of man. . . . Through a comparative study of human adjustment to specific natural environments, certain reliable generalizations or principles have been worked out, while many others have been suggested tentatively. These are the requisites of any science: a distinctive field, and a controlling point of view by means of which its data may be organized with reference to the discovery of general truths or principles" (Barrows 1923, pp. 3, 7).

Barrows's intellectual influence upon White is symbolized in the latter's thesis title: "Human Adjustment to Floods." But beyond the man-land legacy, Barrows epitomized the ethos of geography in public service, which would become White's special province. Between 1934 and 1941, Barrows served as consultant to and member of various resource management entities while also serving as Chairman of the University of Chicago's Department of Geography. During this time he contributed substantially to the development of national policy on river basin planning. According to Abel Wolman, then chairman of the Water Planning Committee, Barrows's preface to the committee's 1934 report was "as good a philosophical statement on water resources management as could be written today." Both Barrows and White were to contribute some of their finest prose to the cause of anonymous committee and task force reports.

A further Barrows legacy was his writing style. According to Colby and White (1961, p. 398), while Barrows was a member of the national water planning committees, he "insisted upon exacting standard of writing for the public reports which poured out. His style was direct, accurate, unequivocal. He slashed the Federalese, the sodden engineering description, and the politically ambiguous prose that came within his reach."

The NRPB and its predecessors afforded a superb forum for the talents of Barrows, White, and their colleagues. The Board's primary function, according to charter member Charles E. Merriam (1944, p. 1076), was "that of advice to the President on problems of long-range planning." The scope of such planning ranged broadly across the entire government to include economic and relief policies, natural resources, social and health programs, and public works. In essence, NRPB combined many of the functions today performed by the Council on Environmental Quality, the Water Resources Council, the Council of Economic Advisers, the Office of Technology Assessment, and the Library of Congress. But unlike these agencies, NRPB was not spread all over Washington with functions divided among jealously competing staffs. It was housed both administratively and physically in the Executive Office of the President, with offices across the street from the White House. It was thoroughly an expression of Franklin Roosevelt's concept of a strong executive branch (which would eventually lead to the board's demise at the hands of Congress in 1943).

The principals of the national planning units were few in number and seasoned authorities in their respective fields (Clawson 1981). Charles E. Merriam, the University of Chicago political scientist, and Frederic A. Delano, the President's uncle and a member of the New York Regional Plan Association, joined the National Planning Board in 1933 and remained until the 1940s. Charles W. Eliot II, the Boston landscape architect, served as director of the planning staff throughout the period. Planning tasks were assigned to a series of standing and special committees assisted by a small research staff and outside consultants.

White's role was central to the purpose of the Water Planning Committee. On the one hand, he was charged with the preparation of daily summaries on pending resource legislation and other matters for the president.[2] On the other hand, according to Eliot, the staff was responsible for involving relevant federal agencies as well as state and local governments in the work of the board. White has told Marion Clawson that this involved some delicate diplomacy in dealing with such personalities as Henry Wallace and Harold Ickes. Altogether, service with the NRPB surpassed the usual learning experience of a graduate student in geography.

The modus operandi of the national planning units was especially congenial to White. The following description of the NRPB aptly depicts his own later style, as reflected for instance in the Workshops on Natural Hazards Research Applications held annually in Boulder, Colorado, beginning in 1977:

> A continuing activity of the Board was clearance of information, projects, and ideas between (1) federal agencies, (2) federal, state, and local regional agencies and (3) public and private agencies. . . . Conferences and consultations were held for this purpose both in Washington and in the field. Special efforts were made by the Board in the direction of consultations with business, labor, agriculture, and with the special federal agencies concerned with aspects of plan-

ning. . . . Many of these repeated conferences were of the very greatest value in the determination of lines of national policy. (Merriam 1944, p. 1077)

Deluge and Democracy

Floods during the 1930s were to command more attention from the Executive Branch and Congress than at any other time in the nation's history. In ironic juxtaposition with the worst drought in American experience, a succession of disastrous floods significantly worsened the nation's economic and social malaise. It started in 1935 with major flooding in Colorado, Nebraska, Texas, and New York. The following spring, an extraordinary downpour upon melting snowpack caused what is still the flood of record for many streams in New England and Appalachia. The Connecticut River at Hartford was 8.6 feet higher than had been experienced since settlement of the area, three centuries before (Hoyt and Langbein 1955, p. 383). January 1937 produced the nation's most damaging flood in the Ohio and Central Mississippi valleys. More than 500,000 persons were displaced from homes, and the central business districts of Cincinnati, Louisville, and many smaller cities were devastated. The year 1938 witnessed the great hurricane which struck Long Island and New England with incredible fury.

Altogether, during the period 1935–38, flood caused at least 1,573 deaths and over $1 billion in property losses throughout the United States. This was a severe blow to an economy with a gross national product that stood at only $72 billion in 1935. In the face of these events, Congress and the President clashed on the issue of flood control.

Federal interest in flood control did not of course begin with Franklin D. Roosevelt's administration. Its roots reach back to the era of navigational improvements in the mid-nineteenth century. Navigation gradually began to subsume flood control as a federal concern with the founding of the Mississippi River Commission in 1879 and the Flood Control Act of 1917.

By 1933, three elements of a national policy on flood control and water resource management existed, at least in preliminary form. The first element was recognition that flooding is a national problem calling for significant federal initiative and investment in the initiation, planning, design, construction, and operation of large-scale projects where needed. Precedent for this proposition was afforded by the

Lower Mississippi Flood Control Act of 1928, although the nature of the federal role in other areas of the country remained unclear, as did the exact relationship between federal and nonfederal interests in the execution of such activities. The second element was the use of the river basin as a logical and appropriate geographic scale for water management planning. Basinwide planning for flood control had begun with the creation of the Miami Valley Conservancy District under Ohio State legislation following disastrous floods in that region in 1913. It was further refined in the basinwide reports prepared by the Army Crops of Engineers under authority of House Document 308 of the 69th Congress. One of these "308" reports underlay the Tennessee Valley Authority Act of 1933. The third element was the idea of accomplishing multiple objectives, such as flood control, water supply, navigation, irrigation, and hydroelectric power, through a single structure or system of structures. This was tentatively established in the proposals for Hoover Dam, the Tennessee Valley Authority, and in the "308" reports generally (White 1969, chapter 3).

These signs of rationality in water resource planning, however, were but straws in the wind. There was no binding national commitment or policy toward the planning of flood control and related projects, except in the Lower Mississippi. The massive public works program proposed by Roosevelt upon his election set off a pork-barrel stampede in Congress, with each member competing to get a share of the funds. As Roosevelt related, there was little prior planning to guide such decisions: "When the Federal Government shortly after March 4, 1933, decided to undertake a vast program of public works in order to provide and stimulate employment, there was practically nothing in existence in the way of planning and research, upon which an adequate comprehensive program of public works could be based. In the emergency the work had to be done quickly; and the projects had to be developed and carried through on a more or less individual basis" (Rosenman 1941, 6:32).

It was precisely this void that prompted Roosevelt to create the National Planning Board on 20 July 1933 as a division of the Public Works Administration. But national planning was a new and alarming specter in 1933, particularly to members of Congress, who viewed it as a threat to their prerogatives. As described by Arthur Maass (1951, p. 62), public works projects were traditionally allocated on the basis of annual "Rivers and Harbors" appropriations determined through a process of interaction between the Corps of Engineers and congressional committees. The presidency had exercised little or no influence on this process.

Roosevelt therefore attempted to introduce to Congress and the nation the idea of national planning by stressing the two aspects that were relatively unassailable: sound fiscal managment and quality of life. His public statements created the impression that national planning would simply expedite and amplify the natural tendency of Congress to spend money "in the interest of the general welfare." Little was said about the implications of comprehensive studies by the Water Planning Committee that were reviewing the entire flood problem and related river basin issues. The selection of one project over another however suggests the application of planning skills and detailed research in preference to the congressional seniority system of pork-barrel allocations. A fundamental issue between the President and Congress was thus barely disguised.

The Key Issues Emerge

Beginning in January 1935 there ensued a period of high drama and confrontation between the President and Congress in the arena of water resources managment that lasted until the outbreak of war in Europe in 1939. As noted, these years were punctuated by flood disasters unparalleled in frequency and magnitude in the nation's history. These events lent urgency and stridency to the public debate of the issues involved. But even before the floods began, the conflict was well in progress. The precipitating event was a request by Roosevelt addressed to Congress on 26 January 1935 that $4 billion be authorized for public works programs to be undertaken during the following eighteen months. As a percentage of gross national product, this was equivalent to a request for $48 billion in 1970 or 60 percent of the entire defense budget of that year. Congress generally concurred in the need for massive public spending to revive the national economy in 1935, but how and where these funds would be spent remained very much in question.

Armed with the newly published reports of the National Resources Board and the Mississippi Valley Committee which he transmitted to Congress with his 26 January message, Roosevelt made his own position quite clear:

> For the first time in our national history we have made an inventory of our national assets and the problems relating to them. For the first time we have drawn together the foresight of the various planning agencies of the Federal Gov-

> ernment and suggested a method and a policy for the future. . . . For the coming eighteen months I have asked the Congress for four billion dollars for public projects. A substantial portion of this will be used for objectives suggested in this report. As years pass the Government should plan to spend each year a reasonable and continuing sum in the development of this program. It is my hope, for example, that after the immediate crisis of unemployment begins to mend, we can afford to appropriate approximately five hundred million dollars each year for this purpose. Eventually this appropriation should replace all appropriations given in the past without planning. (Rosenman 1941, 6:61).

In other words, the price of national public works was acceptance of national planning within the Executive Branch.

Congress was not appeased. As documented by Maass (1951), the lines were drawn between Congress and the Corps of Engineers on the one hand and the President and his National Resources Board on the other.[3] The major legislative landmarks of the period were the Flood Control Acts of 1936 and 1938. Out of the debate that centered on these acts and related legislation, there emerged certain key issues of water resources management of which some remain controversial today.

Federal Cost Sharing

An initial issue involved the allocation of the costs of flood control projects between federal and nonfederal authorities. Before the Lower Mississippi Flood Control Act of 1928, most flood control works were undertaken by states, local governments, special districts, and private interests. Federal investment was directed largely to navigational improvements except where a federal role in flood protection was declared by Congress, as with the Sacramento and the Lower Mississippi. Where federal participation did occur, it was usual for nonfederal interests to provide land for the project and absorb certain other costs, including maintenance.

The concept of nonfederal participation in the cost of water resource and other public investments is a well-established principle of fiscal administration: other things being equal, those who benefit by a public investment should bears its costs in proportion to benefits received. It is assumed that if states and local entities must bear part of the cost of a project, they will be more scrupulous in determining that it will in fact benefit them. Dissociation of costs from benefits, it is commonly believed, promotes the construction of unjustified federal projects and thereby wastes national resources.

The mood of Congress in the mid-1930s however was to waive substantial requirements for nonfederal contribution so as to move ahead rapidly with the national public works program. Where employment was needed most, nonfederal governments were in the worst fiscal condition and therefore could contribute little or nothing to the proposed projects. Section 3 of the bill (HR 8455) that eventually became the Flood Control Act of 1936 required nonfederal interests only to: "(a) provide without cost to the United States all lands, easements, and rights of way necessary for the construction of the project . . . ;(b) hold and save the United States free from damages due to the construction works; and (c) maintain and operate all the works after completion."

The position of the Water Planning Committee had been stated as early as July 1935:

> The policy of allocation of costs proposed by the bill is inequitable and in the long run will tend to foster abuses of sound conservation principles. There is no logical reason for making the local and State contributions equal to and not exceed the costs of rights of way, easements, and damages. In some projects these costs borne by local interest would amount to one-half the total cost of the project, and in other projects the local contributions would be two or three percent only. These percentages bear no relation to the actual benefits received by local interests. Moreover, such a system of cost allocation would promote consideration of small-scale levee projects in which land damage costs are relativey low, rather than of larger reservoir and other projects having higher land costs. *The costs should be shared in part by local interests, but in proportion to the benefits received.* (Quoted in Maass 1951, pp. 153–54)

Congress in 1938 took a position still more contrary to the committee's view of cost allocation. The Flood Control Act of that year waived the requirement for nonfederal contribution of land, easements, and rights-of-way in the case of flood control dam and reservoir projects. Nonfederal interests thenceforth preferred massive flood control dams and reservoirs built entirely at federal cost (and in someone else's taxing jurisdiction) to local protection works for which they must still provide the site. Notwithstanding the opposition of the Water Planning Committee, the flood control program was thus substantially federalized. By 1950, the nonfederal share of the combined costs of local works and reservoirs constructed by that time amounted to only 7 percent of the total flood control budget (Leopold and Maddock 1954, p. 102).

Economic Justification

The necessity for economic justification of the proposed flood control projects posed even more complex difficulties. Customarily, the decision to build a particular project was made by the Rivers and Harbors bloc in Congress in collaboration with the Corps of Engineers and other construction agencies. The "308" reports prepared by the Corps afforded a semblance of basinwide planning as a basis for such projects, but little or no review was performed by the Executive Branch or any other outside authority. As already noted, one of Roosevelt's key objectives in establishing the Mississippi Valley Committee and the National Resources Board was to develop a capacity within the Executive Branch for reviewing and ultimately planning and scheduling river basin projects in accordance with scientific rather than political priorities.

The essence of rational appraisal of water resource projects, like any other public or private investment, lies in the evaluation and comparison of costs and benefits that the project involves. This is basically a three-stage process involving identification of relevant costs and benefits, quantification of such elements including the reduction of future costs and benefits to present values, and comparison of total benefits with total costs. A preliminary test of economic justification requires that the benefits expected to accrue from a project will exceed its costs. Where alternative projects or measures are under consideration, as is usual in the river basin context, justification is indicated by comparison of their benefit-cost ratios.

The application of this process to the task of allocating several billion dollars for water resource projects is formidable indeed. In the 1930s, there was little experience in its use. In his first journal article on floodplain management, Gilbert White (1936, p. 133) wrote: "Notwithstanding the huge expenditures for flood protection during the past 20 years, only a slight amount of attention has been paid to the economic justification for the works constructed. Detailed study of flood protection has been restricted for the most part to determination of engineering practicability."

There was little agreement on an accepted method or set of principles for the performance of benefit-cost studies. White cited the questions of "social criteria by which benefits and costs may be evaluated, and of discounting future benefits." In the absence of a "uniform method for estimating the limit of economic justification for flood protection" there is no assurance that a given set of data regarding a proposed project in the hands of an "honest and competent technician" will clearly indicate whether or not the project is justified.

Remarkably, despite the novelty and imprecision of benefit-cost analysis, Congress embraced it in the Flood Control Act of 1936 as a threshold test for federal funding: "It is the sense of Congress . . . that the Federal Government should improve or participate in the improvement of navigable waters or their tributaries, including watersheds thereof, for flood-control purposes if the benefits to whomsoever they may accrue are in excess of the estimated costs, and if the lives and social security of people are otherwise adversely affected" (33 USCA sec 701a). This provision was perhaps the nation's first legislative mandate requiring objective assessment of the impacts of a project while it was still in the proposal stage. It foreshadowed the environmental protection measures of the 1960s such as the National Environmental Policy Act. But as with the environmental impact assessment process required by the latter, it was one thing to legislate benefit-cost analysis and quite another to apply it as a sensitive tool in public decision-making. White in his 1936 paper and later in his doctoral thesis greatly helped to elucidate the technical issues involved. Among those that have plagued the use of benefit-cost analysis may be cited three.

Benefits and costs. The potential benefits and costs of a resource project are of many kinds; existing and potential, general and special, direct and indirect, tangible and intangible. A special problem arises in the consideration of alternative future uses of the downstream floodplain that is theoretically protected by a project. Should future development of such areas be counted as a benefit, with the possible result that the project is "bootstrapped" into a favorable benefit-cost ratio? Further difficulty is presented by intangible effects that are not readily converted into dollars such as destruction of scenery and wildlife habitats or the provision of recreation.

Selection of discount rate and time horizon. A perennial issue in benefit-cost analysis is the selection of an appropriate discount rate for reducing future benefits and costs to present values. The discount rate represents the rate of interest that a given sum of money invested today would earn so as to yield an amount equal to the estimated benefit or cost at a given time in the future. Thus, the lower the selected discount rate, the greater is the present value of future costs and benefits. Since the benefits of water resource projects are largely in the distant future while the costs are more immediate, a low discount rate tends to overweigh benefits as compared with costs and contributes to a favorable benefit-cost ratio. This problem has occupied economists over the years, most notably Otto Eckstein and John V. Krutilla.

Closely related is the question of a suitable time horizon to establish the duration of benefits and costs related to a particular project; "There is no generally accepted method of adjusting the time period for discounting benefits to the time period for capitalizing costs" (White 1936, p. 146).

National perspective. The Flood Control Act of 1936 requires that benefits must exceed costs "to whomsoever they may accrue." This inclusive test obscures questions involving the treatment of costs and benefits that arise at various geographic scales or within particular units of political and legal authority. Thus a floodwall may confer benefits upon the community that it protects but may cast corresponding costs upon the community across the stream that receives increased flood flows due to the wall. Clearly both must be considered, but even if the net result is favorable, is it fair to disregard the fact that one community is being disadvantaged for the sake of another? Besides such equity considerations, there is the problem of transfer effects. A shopping center developer announces that he will locate in the floodplain of community A if a flood control project is undertaken, but will build on an upland site in community B if the project is not undertaken. Community A's loss is B's gain. Theoretically, the benefits and costs "to whomsoever they may accrue" should cancel out. Benefit-cost analysis thus involves a national perspective. Regional and local effects are counted only to the extent that they contribute to or detract from national economic development.[4]

Multiple Adjustments

President Roosevelt delivered a succinct summary of the idea of multiple adjustments to a press conference at the height of the Ohio Valley flood on 26 January 1937:

> We had an editorial this morning that pointed out that whenever we have a flood we have three or four different groups who rush to the Government to get money for this, that, or the other thing. There are the people downstream, who want more and better levees; and then the next group that want

> dams in the rivers; and another group that want to go up into the headwaters and plant trees; and another group that says it is entirely a question of soil erosion. So you get all these different groups that say their own particular pet theory will stop the flood. I have come to the conclusion that we have to pursue all of these things simultaneously. They all tie in to a general picture; and, for the first time, we have in the last three or four years been developing a synchronized program to tie in the entire field of flood prevention and soil erosion. (Rosenman 1941, 6:23–4).

There are two sides to the coin of multiple adjustments. One is disclosed in the above statement, namely the variety of means available to cope with a given problem such as the management of flood losses. The other involves the achievement of multiple objectives, e.g. flood control, irrigation, water supply, and recreation, through a given project. White (1969, p. 34) notes that the latter concept was well established by the 1930s but the former was to incur continuing resistance:

> Single purpose construction strategies evolved in two important directions beginning in the late 1920's. Most dramatic and widespread was the shift from single-purpose to multiple-purpose public construction, as exemplified by the great systems of dams serving joint aims in the Columbia, Missouri, Ohio and Tennessee basins. Less visible and accepted more slowly was the effort to apply different means to gaining a single goal as revealed in the management of flood losses. Enthusiasm for the multiple purpose strategy caught on in the mid-1930's, probably reached a zenith thirty years later, and then became the subject of somewhat jaundiced reappraisal. Application of multiple means did not receive a full national test until action of flood loss management in 1966, and its fate still hangs in the balance.

The slow acceptance of multiple means was not due to lack of advice on the subject. The list of available means for the management of flood losses was well developed in various reports of the Water Planning Committee extending back to 1934. A number were considered in White's seminal work *Human Adjustment to Flood* (1945). (1) Land elevation means raising structures above the level of expected floods through "borrow and fill." (2) Flood abatement consists of measures taken in the upstream watershed to reduce overland flow to stream channels, including soil conservation, tree planting, forest fire control, and land treatment. (3) Flood protection includes levees and floodwalls, channel improvements, channel diversions, and flood control dams and reservoirs. (4) Emergency measures may include removal of persons and property from the path of an approaching flood, flood fighting by means of sandbagging and other temporary measures, and rescheduling of public utilities and other services. (5) Structural adjustments are the modification of public and private structures to provide increased resistance to flood damage, e.g. by the use of water-resistant building materials, the plugging of apertures below the level of expected floods, the elevation of utilities within the structure, and anchoring to prevent the structure from floating away. These are collectively referred to as "floodproofing." (6) Land use is essential: the impact of floods may be reduced by limiting the usage of floodplains to activities that will not be substantially harmed by occasional inundation. Floodplain zoning upon which great reliance is now placed was listed tentatively as a novel approach to the avoidance of flood losses. Acquisition and relocation were recommended to clear areas of repeated flood losses. (7) Public relief, the transfer of certain costs of floods from the victims to some wider public, can take place through charitable or public assistance.(8) Insurance can play a major role. Private insurance companies by the 1930 had largely ceased offering coverage against flood losses due to the high probability of loss in known floodplains and the need to maintain very large cash reserves against catastrophes such as the 1927 flood. To meet these problems, a federal program of flood insurance was proposed: "To be successful, a national system of flood insurance would require national coverage, so that all major areas of flood loss would be represented and so that the various zones of hazard in each flood plain would be involved also would require some governmental guarantee to bear the indemnities that would have to be paid if a great flood were to occur in a densely-settled flood plain during the early operation of the system" (White 1945, p. 203).

Several of these proposals, especially flood insurance and floodplain zoning, required decades before they became accepted. Their subsequent experience will be considered below. Only two of the above measures were seriously considered in the 1930s: flood protection and flood abatement.

The debate between the proponents of land management and flood control, sometimes termed "upstream" and "downstream" measures, has been fully discussed by Leopold and Maddock (1954).

The advocates of upstream land treatment, soil conservation, and reforestation bore a somewhat moralistic resemblance to the champions of floodplain zoning and other nonstructural measures at the present time. Their position was greatly strengthened by the publication in January 1936 of a National Resources Committee report, *Little Rivers,* and by a national Conference on Upstream Engineering held later that year. (White (1969, p. 38) summarizes the debate as follows:

> The basic difference between the Department of Agriculture and the Corps of Engineers in early flood control investigations was the difference between engineering and ecology. The engineers saw themselves as being the valiant and competent technicians who set out to curb a stream on rampage, to keep the father of waters from invading the homes of the sons of man along his shores, in short, to harness a recalcitrant nature. The Department of Agriculture people saw themselves essentially as trying to harmonize man's actions with what they regarded as ecological principles so long as the measures were those which the Department customarily proposed to land owners.

The Flood Control Act of 1936 resolved the issue, for the time at least, by assigning distinct responsibilities to each agency for different portions of watersheds. This outcome suggests a reason for the nonacceptance of other forms of public response to floods. Quite simply, they lacked political clout. In the absence of a public constituency on the one hand and a bureaucratic sponsor on the other, they were to remain abstract and obscure for three decades. The eventual awakening of the American public and Congress to these alternatives—floodproofing, floodplain zoning, acquisition, flood warning, and flood insurance—was a direct result of White's patient refinement and elaboration of these concepts through his own efforts and those of his students and colleagues.

Executive versus Legislature

Eclipsing the more technical issues of cost allocation, benefit-cost analysis, and multiple adjustments was the overall issue of national water policy—where should the decision-making power lie, in the Executive Branch or in Congress? In the late 1970s, this issue emerged in President Carter's veto of construction of certain western water resource projects, an action furiously opposed by many members of Congress. In the 1930s, the issue centered on Franklin D. Roosevelt's concept of a permanent National Resources Planning Board and hierarchy of lower-order planning authorities.

Following the creation of the National Planning Board in 1933, state planning agencies were established in forty-six states by the end of 1935. To bridge the gap between the National Planning Board and its state counterparts, the president on 3 June 1937 proposed legislation to create seven regional planning authorities corresponding to the nation's major drainage systems. These together with the Mississippi River Commission were to provide an intermediate level of planning and review of proposed water and land resource projects within their respective regions. The question naturally arose as to whether the president envisioned that these institutions would serve as Tennessee Valley–style authorities with plenary operational responsibilities for the development of natural and social resources. In his initial statement to Congress, such a possibility is distinctly implied; by 6 October of that year the President was taking a more moderate position, denying that the regional authorities would ever be regarded as "little TVAs."

Despite the President's soothing words, Congress viewed the entire national-regional-state planning structure as a threat to its prerogatives in the field of water resource development. By the summer of 1937, a congressional counterattack was underway.

The vehicle for congressional response was a proposed joint resolution[5] to assign to the War Department and the Department of Agriculture (by then united in common cause) responsibility for preparing and maintaining a comprehensive national plan for the control of floods on all major rivers and their tributaries. According to Maass (1951, p. 88): "Debate on the resolution in the House [revealed] that it was drafted and supported with the intention of requiring that the Engineering Department report directly to Congress, thus protecting the Department against any interference by the planning agency, and of insuring that the water development functions of the Army Corps be protected against any legislation authorizing additional valley authorities."

The proposed resolution was vetoed by the president on 13 August 1937 on recommendation of the Water Planning Committee expressed in a staff memorandum prepared by Gilbert White. The memo, which was forwarded by Interior Secretary Harold L. Ickes to Roosevelt on 4 August stated in part:

> In his message of June 3, 1937, the President outlined a program for planning of national resources on a regional basis. The essential features were a group of regional planning bodies which were to be integrated through a national planning board and the Bureau of the Budget. If the joint resolution were to be approved in its present form, the authority for such planning would in effect be delegated to the Secretary of War. He would report directly to Congress. The planning would not be organized on a regional basis. It would not "start at the bottom" with the state and local groups as suggested by the President. The results would not necessarily be coordinated with the work of more than a few other agencies. There would be no provision for revision and checking the program in the light of national budgetary considerations and of national planning policies. Once authorized, this new planning venture would be used as an argument against the regional organization proposed by the President. (Nixon 1957, 2:97)

Congress lost the battle but eventually won the war when the National Resources Planning Board was abolished in 1943. With the nation's energies turned to other directions than water resources management (and with few major floods until the 1950s), the federal flood control policy established in the Acts of 1936 and 1938 prevailed for three decades. The goal of a national planning structure for water resource projects was indefinitely deferred. Only with the establishment of the U.S. Water Resources Council under the Water Resources Planning Act of 1965 was the concept of an evaluative capacity independent of Congress and the major federal bureaus finally resurrected, albeit in pale comparison with the hopes of New Deal planners. (This capacity was, however, dismantled by the Reagan administration in 1981 with the dissolution of the Water Resources Council and the six federal-state basin commissions established under the 1965 act.)

Interlude: A Harvest of Research

If measured by congressional innovation, the two peaks of national interest in flood policy have occurred during the periods 1917–38 and 1968 to the present time. During the intervening thirty years, the only flood legislation of any novelty was the Watershed Protection and Flood Prevention Act, better known as PL 566. This act authorized the Department of Agriculture to undertake flood control projects in upstream watersheds of less than 250,000 acres. Otherwise, Congress merely passed its annual Rivers and Harbors acts, approving a seemingly endless series of structural flood control projects. By 1960, the Army Corps of Engineers had spent $3.8 billion of a total authorization of $8.8 billion for protective works.[6] Conceding that flood losses were continuing to rise, the corps recommended that even this magnitude of spending should be increased by another $6 billion for riverine flood control and $500 million for coastal hurricane protection (U.S. Congress 1961, p. 98).

Unlike 1933, however, the 1960s presented an effective challenge to the proponents of a purely structural approach to the flood problem. This challenge did not arise from any political constituency or public outcry for nonstructural measures. It arose from a harvest of books, articles, and other scholarly and professional publications that lent growing weight to the view that the wise use of floodplains must involve multiple adjustments rather than any single approach.

The seedbed of this literature was White's *Human Adjustment to Floods*. The full impact of this book was perhaps delayed by the war and the circumstances of its publication. Originally submitted in 1942 as a doctoral dissertation to the Department of Geography of the University of Chicago, it was printed and circulated on a limited basis in 1945 and finally appeared in the department's new research series in 1953.

Human Adjustment begins with a comprehensive summary of the state of the art of managing flood losses, circa 1940. Three streams of public and private action are identified: engineering, forecasting, and public relief. All three, in White's view, contribute to further encroachment upon flood hazard areas:

> Taking into account all phases of public action and inaction, the policy in essence is one of protecting the occupants of flood plains against floods, of aiding them when they suffer flood losses, and of encouraging more intensive use of flood plains. By providing . . . protective works, the Federal government . . . reduces this flood hazard for the present occupants and stimulates new occupants to venture into some flood plains that otherwise might have remained unsettled or sparsely settled. . . . The Federal forecasting system tends to encourage continued use of flood plains by reducing the expectancy of loss and discomfort from flood disasters. Public relief is now so widespread that the threat of flood, while

> not pleasant, has lost many of its ominous qualities. (White 1945, pp. 32–33)

Piecemeal efforts to make the flood hazard area safer thus contain the seeds of their own failure as greater exposure of investment to catastrophic loss results. This hypothesis (later confirmed empirically in the Chicago studies) leads to the proposition that a more rational approach to flood losses involves consideration of the uses of floodplain itself by man:

> From the three converging streams of public action with respect to the flood problem, and from corollary fields of action, such as land-use planning, we may draw an approach to this problem more comprehensive than any one of them. It is a view which considers all possible alternatives for reducing or preventing flood losses; one which assesses the suitability of flood-protective works along with measures to abate floods, to evacuate people and property before them, to minimize their damaging effects, to repair the losses caused by them, and to build up financial reserves against their coming. It is a view which takes account of all relevant benefits and costs. . . . It seeks to find a use of the flood plain which yields maximum returns to society with minimum social costs, and it promotes that use. (Ibid, p. 34)

The balance of *Human Adjustments* elaborates upon this theme. Section 3 contains probably the most comprehensive typology of flood losses ever compiled. Section 4 considers the various adjustments (listed earlier in this paper) that are available for the mitigation of flood losses.

The early 1950s saw several further contributions to the incipient literature on floods. Arthur Maass's *Muddy Waters* (1951) gives a critique of the role of Army Corps of Engineers in its rivers and harbors activities. A preface by Harold L. Ickes expresses longstanding frustration with the corps dating back to the National Resources Board. Maass carefully documents the struggle for control of the nation's water resource planning process and builds upon White's discussion of benefit-cost analysis. This proved to be the first in an important series of studies at Harvard dealing with the economics of water resource planning.

A different perspective was taken in a book sponsored by the Conservation Foundation in 1954, *The Flood Control Controversy: Big Dams, Little Dams, and Land Management,* by Luna B. Leopold and Thomas Maddock, Jr., both hydrologists with the U.S. Geological Survey. Their specific concern was with the debate between proponents of upstream and downstream flood control measures, respectively the Department of Agriculture and the Corps of Engineers. The conceptual basis and empirical experience of each approach is reviewed in detail. The authors conclude that both are flawed. A major weakness in both programs is that they purport to protect lives and property while in fact serving to promote land development in hazardous areas.

Another pair of Geological Survey hydrologists wrote a classic work in the field, *Floods,* published in 1955. With felicity of style and compelling illustrations, William G. Hoyt and Walter B. Langbein humanized the subject while supplying exhaustive factual information on the nature of floods and human response to them and summaries of U.S. flood experience organized chronologically and by river basis.

In the field of law, the seminal study of floodplain regulation, published in 1959, was written by Allison Dunham, a University of Chicago colleague and Hyde Park neighbor of the Whites. At Gilbert White's invitation, Dunham prepared his paper initially for a conference in December 1958 at the university involving state officials, planners, engineers, and geographers. A resolution adopted by the conference urged states to adopt floodplain regulations. Yet practical experience with such techniques was rare, and few judicial decisions had explicitly addressed floodplain zoning. There was even precedent that a private party cannot be protected from the results of his own folly through the police power. After review of available literature and constitutional principles, Dunham offered three grounds upon which floodplain zoning might be upheld: protection of unwary investors, protection of other owners from possible increase in flooding due to encroachment, and protection of the public from the expense of rescue and rehabilitation of the unwise floodplain occupant. The Dunham analysis has proven to be crucial in establishing the constitutionality of floodplain and shoreland zoning measures. It has been directly cited in a landmark decision by the highest court of Massachusetts.[7]

The most important scholarly legacy of *Human Adjustment to Floods* in terms of scope, quantity, number of contributors, and subsequent influence has been the series of research projects undertaken, written up, edited by White, and published at the Department of Geography of the University of Chicago. According to John R. Sheaffer, *Human Adjustment* served as a blueprint for a fifteen-year program of doctoral research at Chicago, whereby specific "adjustments" were subjected to intensive investigation and the results published. Sheaffer (1960) dealt with floodproofing, Murphy (1958) with floodplain regulation, Burton (1962) with agricultural adjustment, and Kates (1962) with hazard perception and adoption of mitigation measures. Other volumes in the series edited by White collectively addressed key issues in national flood policy: *Changes in Urban Occupance of Flood Plains* (1958) and *Choice of Adjustment to Floods* (1964).[8]

If one sentence could summarize the collective impact of this research upon a reluctant public and Congress it would be the following from *Changes in Urban Occupance:* "Persistent human indecision appears to be the dominant characteristic of recent urban occupance of flood plains of the United States" (p. 203).

The Turning Point: House Document 465 and Beyond

The attention of Congress was redirected to flood issues by a devastating series of hurricanes along the Atlantic and Gulf coasts: Carol and Hazel in 1954, Connie and Diane in 1955, Audrey in 1957, Donna in 1960, the "Northeaster" of March 1962, Hilda in 1964, and Betsy in 1965. Taking more than 1,100 lives and causing several billion dollars in property damage, these represented a new kind of flood threat against which the riverine flood control program was largely ineffective. Federal disaster assistance spiraled from $52 million in 1952 to $374 million in 1966. In the Southeast Hurricane Disaster Relief Act of 1965, Congress directed the Department of Housing and Urban Development (HUD) to undertake a study of the feasibility of a national flood insurance program as an alternative to ever-increasing outlays for disaster relief. Nine months were allowed for the study. Marion Clawson of Resources for the Future was retained by HUD as study director,

The HUD study was too limited in focus and duration to accomplish a sweeping review of national flood policy. For this purpose the Bureau of the Budget requested Gilbert White to undertake a parallel study to make broad recommendations to Congress concerning future options for national policy. White agreed to chair a small task force of specialists.[9] James Goddard, originator of floodplain management at TVA, offers the following recollections on the work of the task force:

> We had a broad objective and from the beginning agreed that policies and constraints of members' agencies should not limit nor guide our considerations. Experiences and needs were reviewed and innovations and/or revised approaches were evaluated. Members acted as professional, experienced individuals rather than as agency mouthpieces, although each was able to and did present the thinking and experience of his respective agency. Each also acted as the contract for obtaining agency review and comment on the report. Basic data, cost estimates, and special information on selected items were obtained from appropriate Federal agencies. Representatives of other agencies and selected specialists were invited to meetings for discussion of their specialties and/or special programs.
>
> . . . Each member acted as liaison in presenting the final draft to his respective agency (having kept his agency abreast and aware of the direction the study was taking) and obtaining official comments. Gilbert, as chairman, presented the draft to the Director of the Bureau of the Budget, discussed it with him, and arranged a dinner meeting with BoB to discuss the report and various implications and effects. Gilbert then polished the report and the secretary prepared it for onward transmission, by BoB, to the President and then to Congress.[10]

The task force thus relied heavily upon federal officials, clearing its proposals with relevant agencies throughout the process. Finally published as House Document 465 (U.S. Congress 1966a), each agency felt that it had written the section relating to its own field and was thus inclined to support it. There is little doubt, however, as to the principal authorship of House Document 465. Its opening echoes the White approach: "The nation needs a broader and more unified national program for managing flood losses. Flood protection has been immensely helpful in many parts of the country—and must be continued. Beyond this additional tools and integrated policies are required to promote sound and economic development of the flood plains. Despite substantial efforts, flood losses are mounting and uneconomic uses of the Nation's flood plains are inadvertently encouraged. The country is faced with a continuing sequence of losses, protection and more losses."

The report identified three shortcomings of existing flood policy: inadequate recognition of the nature of the flood threat and limitations of engineering works, virtual exclusion of means other than river control, and inadequacy of cost sharing by nonfederal beneficiaries. It called for an "integrated flood loss management program" involving federal, state, local, and private initiative. The report offered recommendations under five headings: (1) to improve basic knowledge about flood hazard; (2) to coordinate and plan new developments on the flood plain; (3) to provide technical services to managers of flood plain property; (4) to move toward a practical national program for flood insurance; and (5) to adjust federal flood control policy to sound criteria and changing needs.

The task force did not completely discredit structural in favor of nonstructural measures for floodplain management. The former would be retained in their proper place as part of a unified national effort. Further, the report proposed no sweeping alteration of the prevailing distribution of authority regarding the control of land use in floodplains. No preemption of local prerogative was proposed. Instead, the report focused on the need for federal and state technical assistance to local governments to enable them to improve their management of flood hazard areas.

The task force devoted little attention to flood insurance, leaving that task to Marion Clawson, with whom they were in close contact. What is said in House Document 465 on the subject of flood insurance is cautious:

> A flood insurance program is a tool that should be used expertly or not at all. Correctly applied, it could promote wise use of flood plains. Incorrectly applied, it could exacerbate the whole problem of flood losses. For the Federal Government to subsidize low premium disaster insurance or provide insurance in which premiums are not proportionate to risk would be to invite economic waste of great magnitude. Further, insurance coverage is necessarily restricted to tangible property; no matter how great a subsidy might be made, it could never be sufficient to offset the tragic personal consequences which would follow enticement of the population into hazard areas. (U.S. Congress 1966a, p. 17)

The task force proposed that a national flood insurance program be initiated on a limited trial basis in selected areas to determine its impact upon floodprone areas and the allocation of losses.

A stronger endorsement of the concept of national flood insurance was expressed in the Clawson report (U.S. Congress 1966b, p. 6) dated 12 August 1966, two days later than the task force report. The Clawson report recommended that such a program would supply an appropriate middle ground between the extremes of private risk-bearing and total assumption of losses by taxpayers. Other recommendations dealt with the importance of federal delineation of flood hazard areas, the estimation of actuarial rates according to differential zones of risk within the floodplain, and the exercise of land use powers by state and local governments to limit further encroachment upon hazard areas.

In its deliberations on the proposed legislation to establish a national flood insurance program, Congress was closely guided by the Clawson report and by supportive testimony by HUD undersecretary Robert C. Wood (U.S. Congress 1967). The overall structure of the program as finally adopted 1 August 1968 (PL 90-448, Title XIII) indicated that Congress had learned much on the subject of comprehensive management of floodplains since 1936.

The National Flood Insurance Program

The National Flood Insurance Program (NFIP) is now a primary vehicle of federal policy with respect to flood losses. It is intended to serve two interrelated objectives. First, it seeks to reallocate a portion of the burdens of flood losses to all occupants of flood hazard areas through the mechanism of insurance premiums. Second, it seeks to mitigate rising flood losses by discouraging additional development and investment in floodplains. This second objective is sought to be achieved through appropriate land use controls to be exercised by state and local government, higher insurance premiums for new structures than for existing structures in floodplains, and selective acquisition of floodprone property and relocation of victims after a flood.

NFIP was established as a joint venture between the private insurance industry and the federal government.[11] The federal government subsidizes insurance premiums for existing structures and provides "reinsurance coverage" as a reserve against catastrophic losses. It also administers the floodplain mapping studies, establishes minimum standards for floodplain management, and sets "chargeable rates" for new structures in floodplains. The marketing of insurance has been conducted through private insurance companies and independent agents. The NFIP was originally assigned to the Federal Insurance Administration of HUD. In 1979, it was moved to the Federal Emergency Management Agency (FEMA) created by President Carter to deal with all natural and wartime disaster needs.

Congress was mindful of the warnings of the task force and HUD studies. When it established NFIP in 1968, Congress imposed two limitations on the sale of flood insurance. First, it could be sold only in communities that satisfied the floodplain management standards to be set by FIA. Second, while premiums for existing structures would be lowered by federal subsidy, new structures were required to pay actuarial rates based on the real risk associated with their location and elevation. These limitations were high-minded but unworkable. During its first year, of twenty thousand communities with flood problems,[12] exactly four became eligible for the sale of insurance and twenty policies were sold.

Subsequent amendments have vastly magnified the level of NFIP activity. It was soon realized that most communities lacked detailed maps upon which to base floodplain regulations. Furthermore, no one had yet devised a means for calculating actuarial rates for new structures. Congress responded to the impasse in 1969 by suspending "regular program" requirements during a temporary "emergency phase" pending the preparation of detailed floodplain maps and rate studies for each community by the FIA. The exception quickly swallowed the rule: by mid-1973, over two thousand communities had opted to enter the NFIP, most of them in the "emergency phase." Such communities were exempted from adopting full-scale floodplain management requirements, and new structures could be insured at subsidized rates, albeit within specified limits of coverage. Were it not for the national recession of 1973–75, this might well have triggered the invasion of floodplains against which the task force had cautioned.

Another problem with the 1968 act was that NFIP was voluntary both as to community participation and as to individual purchase of coverage. This was remedied by the Flood Disaster Protection Act of 1973 (following Hurricane Agnes) which amended NFIP significantly. The new approach makes purchase of flood insurance compulsory when any federally related financing is involved in the development or acquisition of floodprone property. This applies both to direct federal loans and grants and to mortgage loans by financial institutions insured or regulated by such federal agencies as the Federal Reserve Bank and the Federal Deposit Insurance Corporation. In other words, any investment in flood hazard areas to which the federal government is in any way a party or guarantor will proceed in accordance with NFIP rules or not at all (Platt 1976).

These two modifications—emergency implementation and compulsory purchase of insurance where the federal government is involved—have elicited a tremendous response. By 30 June 1984, 17,629 communities were enrolled in NFIP of which 10,519 were in the regular phase. Within these communities, nearly two million policies were in effect, covering nearly $100 billion worth of floodprone property. This growth was helped in part by the tendency of floodplain occupants to purchase flood insurance when they experience an actual loss or a near miss (Kunreuther et al, 1978) (see table 2.2). The program, however, has not yet lived up to expectations that it would be fiscally self-sustaining. Payments on claims in 1979 were three times the revenue from premiums. In 1981, NFIP insurance rates were raised substantially to help close this gap.

NFIP is a fait accompli. Notwithstanding legislative and judicial challenges,[13] it has survived infancy and has passed from youth into adolescence. In statistical terms it has grown rapidly but it has had relatively few opportunities to be tested in actual disasters. Until it converts most communities from the emergency program to the regular program, copes with the special problem of the coastal zone, and effectively encourages and monitors local programs in limiting floodplain development, the jury will still be out on the impact of NFIP on U.S. flood losses.

Table 2.2 National Flood Insurance Program (1971–80)

Fiscal or Calendar year[a]	Premiums ($1,000)	Loss payments ($1,000)	Participating communities	Policies in effect	Coverage ($ billion)
1971	6,341	251	158	75,864	1.1
1972	7,003	2,500	637	95,123	1.5
1973	15,315	15,007	2,271	272,448	4.6
1974	25,777	36,638	4,090	385,478	8.4
1975	40,950	26,235	9,625	539,888	13.7
1976	57,524	81,359	14,502	793,779	22.7
1977	83,783	59,190	15,585	1.1 million	33.6
1978	40,235	50,887	16,000 ±	1.2 million	37.1
1978	99,456	135,568	16,000 ±	1.3 million	NA[b]
1979	117,069	482,375	16,488[c]	1.6 million	60
1980	NA[b]	NA[b]	16,957[d]	2.0 million	95

[a] Fiscal year 1971–78; the first entry for 1978 is for 1 July–December 1977; thereafter calendar year.

[b] NA = not available.

[c] As of 11 July, 1979. This figure includes 3,381 communities in the regular program and 13,107 in the emergency program.

[d] As of 15 November, 1980, including 5,571 regular and 11,386 emergency.

Source: Federal Emergency Management Agency data.

Mapping

A monumental task for NFIP has been the mapping of floodplains and preparation of flood insurance rate studies for each of the more than 20,000 floodprone communities in the nation. The mapping program called for in the 1968 act got off to a slow start until basic questions were resolved concerning the geographic scope of the regulatory floodplain and the method to be employed in delimiting it. After some initial confusion, the 100-year floodplain (that area having a probability of being flooded at least 1 percent in any given year) was selected. Initially, rough "flood hazard boundary maps" were prepared for every floodprone community to indicate which property should be covered by flood insurance, pending completion of detailed "flood insurance studies" for each community. By April 1981, FEMA had undertaken studies for 10,008 communities of which 5,818 had been completed. With the average cost per community study rising from $27,000 in 1978 to $62,000 in 1981, FEMA has been seeking cheaper and more efficient ways to continue the mapping program and to update studies already completed.

Problems of the Coastal Zone

NFIP has been least effective in coastal areas. Newspaper accounts following coastal floods commonly report rebuilding in hazardous locations using flood insurance proceeds. At Scituate, Massachusetts, one house has reputedly been destroyed and rebuilt three time with flood insurance proceeds (albeit a little higher in elevation each time). Even regular program communities such as Brookhaven, New York (on Long Island), have experienced massive construction along their oceanfronts. Resort and convention cities such as Miami Beach and Ocean City, Maryland, continue to expand their waterfront development in spite of—or because of—NFIP. H. Crane Miller (1975) has documented an actual increase in oceanfront construction in Rhode Island as a result of NFIP. This finding was later corroborated in a study of six coastal communities by the U.S. General Accounting Office (1982).

Coastal hazard management is complicated profoundly by physical, economic, legal, cultural, and psychological factors. In the physical realm, coastal hazards are of many types and sources: hurricanes along the Gulf and Atlantic coasts, northeasters in the North Atlantic, tidal waves and mudslides on the Pacific coast, and shoreline erosion around the Great Lakes. Coastal flooding is greatly affected by storm surge, a condition of unusually high water relating to wind speed and direction, configuration of the bottom and shoreline, and stage of astronomic tide. The problem is further complicated by the process of erosion which undermines structures that are technically above estimated flood levels.

Economic and legal complexity is inherent in coastal management due to the extremely high property values involved and the interaction of federal, state, local, and private jurisdiction at the water's edge. In cultural terms, disputes rage concerning the best use of shorelines—for conservation of natural ecology, for recreation, for holiday use, for year-round residents, or for industry and commerce. There is a well-known tendency for people to be attracted to water partly because of its tempestuous and unpredictable character.

NFIP is ill-equipped to deal with these multiple conflicts of the coast. Among many obstacles, three in particular may be noted; coastal flood mapping, coastal erosion, and coordination with other federal programs (Platt 1978). (In 1982, Congress adopted the Coastal Barrier Resources Act [PL 97 348] which placed certain "undeveloped" barriers off limits to flood insurance and other federal incentives to construction.)

Floodplain Management

To promote flood insurance at subsidized rates without effective land use would be an "unjustifiable giveaway program," in the words of former Federal Insurance Administrator George Bernstein (U.S. Congress 1973a, p. 10). After a shaky start, there is growing evidence that communities are accepting the burdens of NFIP standards in order to gain the benefits of insurance. FEMA has an important continuing role in monitoring community enforcement of regulations already in effect.

There is neither reason nor time for timidity in the promotion of floodplain regulations. Development in coastal hazard areas has visibly and dangerously increased since the advent of NFIP. A corresponding increase in riverine floodplain encroachment has occurred in certain places as well. The proper question then is not whether NFIP has too much power but rather whether it uses its available power too little.

Toward a Unified National Program

NFIP combines four nonstructural adjustments to floodplain occupancy: land use control, floodproofing, insurance, and land acquisition. It therefore represents a major departure from the largely structural approach to flood losses that prevailed before 1968. Yet even NFIP must be viewed not as a comprehensive solution in its own right but as a potentially important element of a "Unified National Program for Flood Plain Management" (U.S. Water Resources Council 1979). President Carter's Executive Order 11986 (of 24 May 1977) mandated a unified effort by federal agencies to mitigate flood losses, but coordination in the nonfederal sector remains elusive.

The Need for Orchestration

Ironically, the nation's response to floods has shifted from being excessively narrow as to approach and agency responsibility during the 1930s to its present state, where it is overly diffuse. The U.S. Water Resources Council (1979, 2:2) has found that the federal effort is scattered among twenty-eight agencies and nine program purposes. These cover the entire gamut of available adjustments including structural activity, planning and technical services, flood forecasting and warning, emergency relief and rescue, land acquisition, and relocation. Each element of a unified national program is thus to be found somewhere in the bureaucracy of the federal government. The problem is that there is no agency or guiding hand available to orchestrate the entire assemblage. As of early 1983, the only "contribution" of the Reagan administration has been to abolish the Water Resources Council: as in Shakespeare, the bearer of unwanted information is destroyed.

Benefit-Cost Analysis Revisited

The issues of benefit-cost analysis raised by the Water Planning Committee in the 1930s and explored further by Eckstein, Krutilla, and others in the 1950s continue to plague water resource planning efforts. Despite the new emphasis on nonstructural adjustments, flood control projects have continued to be authorized and carried out since 1960.

Despite extensive experience in the use of benefit-cost analysis to date, certain technical problems continue to impair its usefulness: identification of pertinent costs and benefits, double counting of benefits, selection of discount rate, and treatment of future land use changes to be influenced by the project. There is ample evidence of deliberate misrepresentation by sponsoring agencies in overestimation of benefits and understanding of costs.[14]

Postdisaster Mitigation

The best opportunity to reduce flood losses is immediately after a flood or other disaster. Rapid City, South Dakota, for example, following the flash flood of 9 June 1972, which claimed 238 lives, determined to prevent a repetition of the event. With the aid of a $48 million urban renewal grant from HUD, the city acquired some fourteen hundred parcels of floodprone land, relocating all residents and businesses to higher ground and converting the floodway into open space and parks.

The Rapid City experience was unique. It resulted from the confluence of several factors: strong local leadership, urban renewal plans already drawn up, national publicity, and availability of massive federal funding. Postdisaster land acquisition has been undertaken on a smaller scale in places such as Big Thompson Canyon, Colorado. But nowhere else has there been a comprehensive intergovernmental effort to eliminate the conditions that result in recurrent flood damage. Typically, public efforts have the opposite effect, namely to restore the status quo as quickly as possible (Haas, Kates, and Bowden 1977).

In 1979, FIA and the Federal Disaster Administration were combined with other disaster agencies into a new Federal Emergency Management Agency (FEMA). Postdisaster mitigation is a major priority for the new agency. In 1980, twelve federal agencies under FEMA's encouragement signed an interagency agreement to promote hazard mitigation following major flood disasters. Since 1980 hazard mitigation teams have been dispatched to each flood declared by the president to be a major disaster. Their mission is to identify ways to reduce future losses through the flood recovery process.

A related endeavor has been "quick response studies" of natural disasters by hazard researchers. Gilbert White has directly promoted this genre of research through small grants administered by his Natural Hazards Information Center. The National Research Council Committee on Disasters also has conducted some "quick response studies," for instance following the Texas floods of May 1981 and Hurricane Alicia in 1983. Collectively, these various postdisaster studies are contributing new insights to disaster management and recovery procedures.

Intergovernmental Coordination

A further consideration in the development of a unified national approach to flood losses is the need for improved coordination within and between public entities. This implies at least three dimensions of intergovernmental relations: functional coordination among federal agencies and programs relating to the overall problem of floods; vertical interaction among federal, state, regional, local, and private interests having overlapping jurisdictions; and horizontal coordination between adjoining units of authority which share a given floodplain or watershed (Platt et al, 1980).

The shift from flood control to floodplain management enhances the responsibilities of local governments. Rather than relying upon the federal government to install and maintain upstream dams and reservoirs, communities are now expected to adopt regulations to qualify for the flood insurance program. And whereas floodwalls, levees, and other local structures normally function without day-to-day attention, public regulations require detailed and continuing enforcement. The effects of local actions in floodplains may be felt in neighboring jurisdictions, upstream, downstream, or across a stream. A sensible floodplain management program in one community may be impaired or nullified by incompatible actions in neighboring areas that divert or increase flood flows. It is essential therefore that public bodies through or between which a common stream flows be able to rely upon and, if necessary, to influence each other's actions. Otherwise there is no basis for confidence in nonstructural programs for floodplain management.

Where lateral arrangements between neighboring governments are not feasible, a unified approach to floodplain management may be achieved through intervention by higher authority, often with the consent of the subordinate units. Thus, the states of Minnesota, Wisconsin, Michigan, and Maine have statewide shoreland zoning laws. Many other states have coastal and/or inland wetland laws that affect floodplains. County governments in many states may regulate land use in unincorporated areas, and regional special districts can play varied roles relating to flooding.

Conclusion

Gilbert White's quest for a rational approach to managing flood losses is thus carried forward by many agencies, programs, and individuals. But flood disasters continue with increased frequency and severity. In 1979 alone, thirty major flood disasters were declared by the President involving eighteen states, 369 counties, and two territories. One is tempted to lament, "Plus ça change, plus c'est la même chose."

There has been progress, however. Blind faith in structural protection is a thing of the past. Flood insurance is widely held, albeit generally at subsidized rates. Floodplain zoning is soundly established judicially and is growing in local acceptance. Postdisaster mitigation measures find expression at least in public rhetoric and legislation, if not yet in routine application. Thanks in part to Gilbert White's annual workshops on natural hazard research in Boulder, a new generation of public officials and academic researchers is joining in common cause rather than harboring mutual mistrust. A new agency has been created that will devote much of its attention to issues of disaster response and mitigation.

Much remains to be learned. FEMA must improve and expedite its mapping and rate studies so as to provide reasonably accurate and current data to floodprone communities. The performance of local governments in regulating new development in floodplains must be closely monitored. We should make better use of supralocal institutions such as counties, special districts, and state programs to afford mutual reassurance among neighboring jurisdictions. We must facilitate federal cost-sharing in nonstructural measures, especially land acquisition. Postdisaster mitigation efforts must be better coordinated so that diverse federal agencies do not undermine each other's efforts. Legal research is needed to underpin more stringent limitations on building in coastal high-hazard areas and in riverine floodplains outside the known flood of record. Most important, research is needed on the impact of NFIP. What are its effects upon the behavior of investors, lending institutions, floodplain occupants, and public officials? Are flood losses rising or declining as a result of NFIP? If the former, how can such perverse effects be ameliorated (National Science Foundation 1980)?

It is perhaps fitting to conclude this paper by echoing the *Boulder Daily Camera* in an editorial (23 May 1978) commemorating Gilbert White's retirement from the faculty of the University of Colorado: "White was not ahead of his time in his warnings and recommendations. Boulder has been behind the times in acting. But we are on the way—not too late, we hope." This sentiment is widely shared.

Acknowledgements

Many persons have been helpful in the preparation of this paper. I would like to thank particularly James E. Goddard, Abel Wolman, Charles W. Eliot II, Marion Clawson, Anne U. White, John R. Sheaffer, Earl J. Baker, Frank H. Thomas, Richard W. Krimm, and Jon A. Kusler for their assistance. I also thank William D. Pattison for loaning me portions of his manuscript "The Geographer's Way." Marion Clawson generously shared his thoughts and document collection concerning the National Resources Planning Board. George M. McMullen, my research assistant during 1978–79, spent many hours hunting down material in the University of Massachusetts Library.

Notes

1. Estimate by George R. Phippen, former head, Flood Plain Management Services, U.S. Army Corps of Engineers, Washington, D.C.
2. White has related to John R. Sheaffer that on one occasion he and fellow staffers appended a poem to one of their summaries to find out whether their efforts were being noted. It was promptly returned with another poem written by President Roosevelt.
3. The board included the Secretary of War as a member, albeit a frequently dissenting one.
4. The "Principles and Standards" of the U.S. Water Resources Council published in the *Federal Register,* 10 September 1973, modify this to include equal consideration of environmental quality effects, but still from a national perspective.
5. Introduced as Senate Joint Resolution 57 and House Joint Resolution 175 (75th Cong., 2st Sess.).
6. When completed, this program would result in 125 million acre-feet of flood storage in 330 reservoirs, 12,000 miles of levees and floodwalls, and 10,000 miles of channel improvements. This does not include flood control activities of other federal agencies including the Tennessee Valley Authority, the Bureau of Reclamation, and the Soil Conservation Service.
7. *Turnpike Realty Co. v. Town of Dedham* 284 N.E.2d 891 (Massachusetts, 1972).
8. Flood-related papers in the University of Chicago Department of Geography Research Series include:

1953 White, Gilbert F., *Human Adjustment to Floods: A Geographical Approach to the Flood Problem in the United States* (no. 29)
1958 Murphy, Francis C., *Regulating Flood-Plain Development* (no. 56)
1958 White, Calef, Hudson, Meyer, Sheaffer, and Volk, *Changes in Urban Occupance of Flood Plains in the United States* (no. 57)
1960 Sheaffer, John R., *Flood Proofing: An Element in a Flood Damage Reduction Program* (no. 65)
1961 White, Akin, Berry, Burton, Dougal, Goddard, Hertzler, Holmes, Kates, Renshaw, Roder, and Sheaffer, *Papers on Flood Problems* (no. 70)
1962 Burton, Ian, *Types of Agricultural Occupance of Flood Plains in the United States* (no. 75)
1962 Kates, Robert W., *Hazard and Choice Perception in Flood Plain Management* (no. 78)
1964 White, Gilbert F., *Choice of Adjustment to Floods* (no. 93)
1965 Kates, Robert W., *Industrial Flood Losses: Damage Estimation in the Lehigh Valley* (no. 98)
1968 Burton, Kates, and Snead, *The Human Ecology of Coastal Flood Hazard in Megalopolis* (no. 115)
1974 Mitchell, James K., *Community Response to Coastal Erosion: Individual and Collective Adjustments to Hazard on the Atlantic Shore* (no. 156).

9. Members of the task force included, in addition to White: James E. Goddard (TVA); Irving Hand (Commonwealth of Pennsylvania); Richard A. Hertzler (Army); John V. Krutilla (Resources for the Future); Walter B. Langbein (U.S. Geological Survey); Morton J. Schussheim (HUD); Harry A. Steele (Agriculture). John R. Hadd of the Bureau of the Budget served as staff to the task force.

10. Letter to the author dated 2 August 1978; "BoB" is the Bureau of the Budget.

11. During its first decade, it operated under a contract between FIA and a nonprofit consortium of 130 private insurance companies, the National Flood Insurers Association (NFIA). In 1978, harmony dissolved between FIA and NFIA, resulting in mutual withdrawal from their contract. FIA replaced NFIA with a data-processing firm which serves as "fiscal agent."

12. "Community" in NFIP jargon includes incorporated municipalities, counties, states, and, in a few cases, special districts.

13. In a lawsuit challenging the constitutionality of NFIP brought by a group of property owners and communities, the Federal District Court for the District of Columbia sustained the validity of the pro-

gram: *Texas Landowners' Rights Assn. v. Harris* 453 F.Supp. 1025 (1978), aff'd 598 F.2d 311 (1979), cert. den. 444 U.S. 927 (1979).

14. The *New York Times* of 26 November 1978, for instance, reported that the Tennessee-Tombigbee Waterway, which had been disallowed in the 1940s, was subsequently authorized and partially constructed as a result of misleading and erroneous estimates of benefits and costs.

References

Barrows, H. H. 1923. "Geography as Human Ecology." *Annals of the Association of American Geographers* 13 (March): 1–14.

Bosselman, F., D. Callies, and J. Banta. 1973, *The Taking Issue*. Washington, DC.

Burton, I. 1962. *Types of Agricultural Occupance of Flood Plains in the United States,* Research Paper no. 75, University of Chicago Department of Geography.

Burton, I., and R. W. Kates. 1964. "The Floodplain and the Seashore." *Geographical Review* 54:366–85.

Clawson, M. 1981. *New Deal Planning: The National Resources Board*. Baltimore.

Colby, C. C., and G. F. White. 1961. "Harland H. Barrows, 1877–1960." *Annals of the Association of American Geographers* 51 (December).

Cressman, G. P. 1978. "Flash Flood Warnings—Federal Plus Local Action." *Natural Hazards Observer,* September.

Dunham, Allison. 1959. "Flood Control via the Police Power." *University of Pennsylvania Law Review* 107:1098–1132.

Federal Insurance Administration. 1977. *Proceedings of the National Conference on Coastal Erosion*. Washington, DC.

Haas, J. E., R. W. Kates and M. J. Bowden. 1977. *Reconstruction Following Disaster.* Cambridge and London.

Hoyt, W. G., and W. B. Langbein. 1955. *Floods*. Princeton.

Kates, R. W. 1962. *Hazard and Choice Perception in Flood Plain Management*. Research Paper no. 78, University of Chicago Department of Geography.

Krutilla, J. V. and O. Eckstein. 1958. *Multiple Purpose River Development,* Baltimore.

Kunreuther, H., et al. 1978. *Disaster Insurance Protection: Public Policy Lessons*. Somerset, NJ.

Leopold, L. B. and T. Maddock, Jr. 1954. *The Flood Control Controversy: Big Dams, Little Dams, and Land Management*. New York.

Maass, A. 1951. *Muddy Waters: The Army Engineers and the Nation's Rivers*. Cambridge.

Merriam, C. E. 1944. "The National Resources Planning Board: A Chapter in American Planning Experience." *American Political Science Review* 38 (December): 1075–88.

Miller, H. C. 1975. "Coastal Flood Plain Management and the National Flood Insurance Program: A Case Study of Three Rhode Island Communities." *Environmental Comment* (November).

Murphy, F. C. 1958. *Regulating Flood Plain Development*. Research Paper no. 56, University of Chicago Department of Geography.

National Science Foundation. 1980. *A Report on Flood Hazard Mitigation,* Washington, DC.

New England River Basins Commission. 1976. *The River's Reach*. Boston.

———. 1978. *Proceedings of Workshop on New England Coastal Storm of Feb. 6–7, 1978*. Boston.

Owen, M. 1973. *The Tennessee Valley Authority*. New York.

Nixon, E. B., ed. 1957. *Franklin D. Roosevelt and Conservation*. 2 vols. Hyde Park, NY.

Pattison, W. D. Undated, "The Geographer's Way." Unpublished manuscript.

Platt, R. H. 1976. "The National Flood Insurance Program: Some Midstream Perspectives." *Journal of the American Institute of Planners* 42 (July): 303–13.

———. 1978. "Coastal Hazards and National Policy." *Journal of the American Institute of Planners* 44 (April): 170–80.

Platt, R. H., et al. 1980. *Intergovernmental Management of Floodplains*. Technology, Environment and Man Monograph no. 30, University of Colorado Institute of Behavioral Science. Boulder, CO.

Rosenman, S. I., ed. 1941. *The Public Papers and Addresses of Franklin D. Roosevelt,* 9 vols. New York.

Sheaffer, J. R. 1960. *Flood Proofing: An Element in a Flood Damage Reduction Program*. Research Paper no. 65, University of Chicago Department of Geography.

Sorenson, J. H. and J. K. Mitchell. 1975. *Coastal Erosion Hazard in the United States: A Research Assessment*. Boulder, CO.

U.S. Army Corps of Engineers. 1971. *Report on the National Shoreline Study.* Washington, DC.

U.S. Congress. 1961. *Report of the Select Committee on National Water Resources*. Report no. 29, 87th Cong., 1st sess. Washington, DC.

———. 1966a. *A Unified National Program for Managing Flood Losses*. House Document 465, 89th Cong., 2d sess. Washington, DC.

———. 1966b. *Insurance and Other Programs for Financial Assistance to Flood Victims*. Committee Print no. 43, 89th Cong., 2d sess. Washington, DC.

———. 1967. *National Flood Insurance Act of 1967*. Hearings before the Subcommittee on Housing of the House Committee on Banking and Currency. 90th Cong., 1st sess. Washington, D.C.

———. 1973a. *Expansion of the National Flood Insurance Program*. Hearings before the Subcommittee on Housing of the House Committee on Banking and Currency. 93d Con., 1st sess. Washington, DC.

———. 1973b. *Flood Disaster Protection Act of 1973*. Hearings before the Subcommittee on Housing and Urban Affairs of the Senate Committee on Banking, Housing and Urban Affairs. 93d Cong., 1st sess. Washington, DC.

———. 1975. *Oversight on Federal Flood Insurance Program*. Hearings before the Subcommittee on Housing and Urban Affairs of the Committee on Banking, Housing and Urban Affairs. 94th Cong., 1st sess. Washington, DC.

U.S. General Accounting Office. 1982. *National Flood Insurance: Marginal Impact on Flood Plain Development*. CED-82-105. Washington, DC.

U.S. Water Resources Council. 1967. *A Uniform Technique for Determining Flood Flow Frequencies*. Washington, DC.

———. 1968. *The Nation's Water Resources*. Washington, DC.

———. 1971. *Regulation of Flood Hazard Areas*. 2 vols. Washington, DC.

———. 1976a. *Guidelines for Determining Flood Flow Frequency*. Hydrology Committee Bulletin no. 17. Washington, DC.

———. 1978. *The Nation's Water Resources 1975–2000*. Second National Water Assessment. Washington, DC.

———. 1979. *A Unified National Program for Flood Plain Management*. Washington, DC.

White, G. F. 1936. "The Limit of Economic Justification for Flood Protection." *Journal of Land & Public Utility Economics*, May, 133–48.

———. 1953. *Human Adjustments to Floods*. Research Paper no. 29, University of Chicago Department of Geography.

———. 1964. *Choice ofAdjustment to Floods*. Research Paper no. 93, University of Chicago Department of Geography.

———. 1969. *Strategies of American Water Management*. Ann Arbor, MI.

———. 1975. *Flood Hazard in the United States: A Research Assessment*. Boulder, CO.

White, G. F., et al. 1958. *Changes in Urban Occupance of Flood Plains in the United States*. Research Paper no. 57, University of Chicago Department of Geography.
White, G. F. and J. E. Haas. 1975. *Assessment of Research on Natural Hazards*. Cambridge and London.
Whyte, P. L. 1949. "The Termination of the National Resources Planning Board." Master's thesis. Department of Policial Science, Columbia University.

3 Integrative Concepts in Natural Resource Development and Policy

Marion Clawson

Gilbert White's scholarship has been eclectic, inclusive, comprehensive, and integrative, reflecting his own personality. His thinking has never been constrained by narrow disciplinary boundaries, narrowly defined subject matter, or rigid conceptual considerations, and he has never been a slave to specialized methodologies. These characteristics of his professional work will be described, explicitly or implicitly, in all the other essays of this volume, but nowhere are they more evident than in the field of natural resources as a whole, the subject of this chapter. I shall argue that "natural resources" must be defined broadly and that the formation of policy in this field must integrate many diverse and specialized pieces of knowledge.

Specialization versus Integration

In natural resources, as in many other fields of human interest, in the past four decades there have been two opposing tendencies in research, writing, and scholarship generally: specialization and integration.

Within most specialized fields of knowledge, such as economics, or ecology, or silviculture, there has been a trend toward greater specialization. This has typically been associated with greater sophistication, especially in research. Problems tend to be more narrowly defined, and the theory for dealing with them is cast in narrower but more incisive terms, often becoming more mathematical in the process. Typically, models are proposed or developed, and data are

Marion Clawson is Senior Fellow Emeritus, Resources for the Future, Washington, D.C. An economist, he has conducted extensive research on natural resources, including land, forests, and water resources as used for agriculture, urban occupancy, recreation, and forest production. Among more than thirty books he has written or edited, two more recent ones are *New Deal Planning: The National Resources Planning Board* and *The Federal Lands Revisited.*

analyzed by computer. The past generation may well be described as that of the computer and the model—and the future generation is likely to continue, even to go further in these directions. This development has been highly productive in one sense, because it has made possible analyses previously unavailable; but it has also been partial, often excluding major elements of the total situation on the ground that they could not well be fitted into the model, or that data were unavailable, or that—as with political factors—they were not easily amenable to the same kind of analysis.

There have been, of course, powerful forces within every professional field, pushing most workers toward greater specialization. One has been the sheer proliferation of knowledge, the impossibility of any man or woman knowing it all. By limiting one's range of interests or at least limiting one's activities, one can more easily become familiar with the significant literature in the field, one can more easily make a new contribution to existing knowledge, and one can more readily carve out an "ecological niche" in one's profession. But in the process one may be uninformed about developments in some other field or subfield, which may render much of one's results redundant, obsolete, or unimportant.

The opposite trend and force may best be called integrative. A person or, more probably, a team, committee, or task force tackles a larger, broader problem, such as energy, water, land use, environment, or something equally large and often ill-defined around its borders. The physical, biological, economic, social, and political aspects of such a broad problem will be studied or at least considered; and typically some sort of program or recommendation is forthcoming. Indeed, it is often the need or the desire for a program or a recommendation that leads to the creation of the committee or task force and serves as it guiding force.

The strengths of the integrative approach lie in its applicability to real-life situations and in its consideration of most if not all of the factors that will be important in the real-life situation. Its weaknesses lie in its very breadth, in the difficulty if not inability of even a talented group to understand all the factors involved, to focus sharply on some factors, and to formulate relations between one kind of factor and another kind. One may intuitively feel that economic power translates into political power, and vice versa, but it seems clear that the interrelationship is not fixed and invariable—there is no formula even roughly analogous to Einstein's famous formula equating mass and energy.

In examining this balance between integration and specialization, I will begin by describing a series of landmark integrative studies and then contrast these with the increasing trend toward specialization.

Major Integrative Studies

World Resources and Integrative Research

High on any list of integrative natural resource studies must be the work of Erich W. Zimmermann, both for the imaginativeness and incisiveness of his major book (1933, rev. ed. 1951) and because he saw himself as an economic geographer. A few quotes (from the second edition) will capture some of the flavor of his thinking about natural resources: "Resources are living phenomena." "To a large extent, they are man's own creations." "The word 'resource' does not refer to a thing or a substance but to a function which a thing or a substance may perform or to an operation in which it may take part." "To understand resources one must understand the relationship that exists between MAN and nature." "It is true that nature provides the opportunity for MAN to display his skill and apply his ever-expanding knowledge. But nature offers freely only an infinitesimal fraction of her treasure; she not only withholds the rest, but seems to place innumerable and, in many cases, wellnigh unsurmountable obstacles in the way of resource-seeking and resource-creating MAN. The bulk of MAN's resources are the result of human ingenuity aided by slowly, patiently, painfully acquired knowledge and experience." "Knowledge is truly the mother of all other resources."

Zimmerman relates his views to those of other economists, such as Alfred Marshall, Arthur C. Pigou, and Wesley C. Mitchell, and to those of geographers such as Isaiah Bowman. The first part of his book develops his theory and his general approach, which in the second part is applied to agriculture and in the third part to industry (including energy and metals). To those brought up scholastically to look at resources in physical or narrowly economic terms, Zimmermann's approach at first seemed novel, even somewhat startling, but on more careful examination it became incisive, imaginative, integrative, and highly productive. (I yet remember the excitement and the stimulation of first reading Zimmermann.) Critics can doubtless find many flaws in his work, but it was truly monumental and path-breaking.

Barnett and Morse (1963), thirty years after Zimmermann's first edition, in their integrative approach to natural resource problems, both quote him (somewhat as I have done) and reach the same basic conclusions as to the importance of knowledge as the ultimate natural resource. The flavor of their findings is suggested by the following brief quotation (p. 11): "Advances in fundamental science have made it possible to take advantage of the uniformity of energy/matter—a uniformity that makes it feasible, without preassignable limit, to escape the quantitative constraints imposed by the character of the earth's crust. . . . Flexibility, not rigidity, characterizes the relationship of modern man to the physical universe in which he lives. Nature imposes particular scarcities, not an inescapable general scarcity. Man is therefore able, and free, to choose among an indefinitely large number of alternatives."

Barnett and Morse cite Zimmermann and many of the same sources he had cited. Their emphasis on the role of knowledge and/or research was as great as that of Zimmermann or any other major scholar. Critics have disputed some of their findings, particularly their general optimism, yet their approach still commands great respect.

World Population and Production

Another classic in the general resources field, approximately contemporaneous with Zimmermann's revised edition, is the Woytinskys' (1953) survey of world resources and their use. In a massive (over 1200-page) volume, they not only survey natural resources and their use, but also deal with population in a way that few others concerned primarily with natural resources have done. Their concern is the world; while unavailability of date inevitably limits their analysis of some countries and regions, the scope of their work is most impressive. While one feels that a fully adequate theory of natural resource use underlies their work—and a theory not dissimilar to that of Zimmermann—their discussion emphasizes the empirical data and relationships, not the underlying conceptual framework.

Policy Commission, Water Resources, and UN Report

The late 1940s and early 1950s were marked also by two major commissions in natural resources and water resources and an international conference on resource conservation. These were reviewed by

Gilbert White (1953) in "A New Stage in Resource History" in the *Journal of Soil and Water Conservation.* The three reports are the Paley report (President's Materials Policy Commission 1952), the President's Water Resources Policy Commission report (1950), and the report of the U.N. Scientific Conference on Conservation of Resources (United Nations 1950). White's article is more than a simple review of these three publications, although it does give the reader a reasonable idea of their contents; it also brings in other publications and, above all, shows the relations among them. The main thrust of the article, as the title suggests, is that times have changed and that this necessitates new and different ways of looking at all natural resources. The need for a comprehensive approach to water resource development is stressed. The shift in the U.S. position vis-à-vis the rest of the world in many raw materials and the prospective increases in demand for most natural resources are the focus of the review of the Paley report. The possibility of future scarcities or at least of future higher prices for natural resource commodities is a major theme. The role of both government and private business in natural resource use and development is also discussed. While the acute environmental concerns of the 1960s and 1970s are not envisioned in the report or in White's article, still there are some concerns expressed about environmental protection or conservation as it was then called.

Man's Role in Changing the Face of the Earth

A different approach, but one of great inclusiveness, is that of an international symposium organized and financed by the Wenner-Gren Foundation and the National Science Foundation (Thomas 1956). The focus here was more biological than social science, although geographers, economists, and other social scientists were involved. This was a group effort, with more than fifty authors or coauthors of papers, and with a distinguished directing panel of three, as well as editor and sponsors. Some papers were largely historical, with trulylong vistas of time; others ranged over a wide variety of resources, uses, and processes.

Social and Economic Aspects of Natural Resources

An attempt to integrate research needs was made by Gilbert White (1962) as chairman of the Social and Economic Study Group. This study is noteworthy both for its breadth and inclusiveness and for its use of numerous specialists, either as members of the committee or as advisers. The section headings in the report are suggestive of its content: Introduction; Natural Resources and Social Development; Natural Resources and the United States Economy; Special Problems of the Underdeveloped Countries with Respect to Natural Resources; Needed Research; and Organization and Public Support. There is emphasis throughout on the dynamics of changing times and the new problems and opportunities these present. The physical, biological, and technological aspects of natural resource use are certainly not neglected but, as the title suggests, the greatest emphasis is placed on social and economic values of natural resource use. The viewpoints and the analysis are far broader than would normally be forthcoming from scholars of a single professional background. A full and detailed accounting of the contents of this report is not appropriate here, but one can judge that this was a major step toward the harnessing of the nation's scientific community for the solution of national and international problems in the area of natural resources.

Resources in America's Future

In 1963, Resources for the Future also made one study of great inclusiveness and of very wide subject matter scope, though limited to the United States (Landsberg, Fischman, and Fisher 1963). This is a study by economists, not geographers, and the approach, as the title *Resources in America's Future* suggests, is national, with only modest attention to geographical or regional problems and opportunities. One of its strengths is the attempt, very largely successful, to integrate the analysis for one commodity or group of commodities with the analysis for others and to integrate supply and demand relationships for different commodities or materials. This volume too was a group effort, though the number of individuals involved was fewer than in the book edited by Thomas (1956); more important, the contributions of the various persons involved were fully synthesized by the three authors.

The Conservationists and the Doomsayers

Some attempts have been made in recent years to synthesize specialized knowledge about natural resource development and policy into some sort of coordinated or summarized overview. Some of these have been the studies of broad resource problems by committees of diverse membership, often coming from the National Academy of Science. But more have come from devout conservationists, all trying to see the whole resource situation (at least as they understand it), most concerned primarily with the future, many of them doomsayers—but some clearly optimists. Cornish (1977) lists some 120 books that he thinks have major significance in any study of the future, and many of them deal explicitly with natural resources. Included in this list are such well-known names as Jay W. Forrester (1971), Dennis Meadows (Meadows et al. 1972), and Kenneth E. F. Watt (1974). For most of these writers, integration of specialized research knowledge is selective and for a purpose, often to reach a policy conclusion that one suspects was the starting point of the analysis. But they are valiant attempts to see the natural resource world as a whole; one may accept the approach while having some doubts about the results.

To Live on Earth

A more constructive and better balanced effort, in my judgment, was Brubaker's book *To Live on Earth* (1972). Brubaker tries to look at the environment as a whole, at all uses of natural resources, and at all threats to the environment. The treatment is not highly detailed, but the book is both more readable and more easily understood because of its brevity. It is outstanding as representing the integrative capacity of a single author. Brubaker had assistance from colleagues, but the synthesis was wholly his own.

Forests for Whom and for What?

The development and use of a truly comprehensive theory or conceptual framework for analysis of natural resource problems and use are both difficult and personally demanding. I speak from experience. In my little book *Forests for Whom and for What?* (Clawson 1975) I try to develop a truly inclusive, comprehensive, and eclectic approach to natural resources, even though in that book I apply it only to forests. I mention the book because its writing forced me to think along lines that otherwise I might not have used, and because I think it is directly relevant to this matter of specialization versus integration. I argue there that the consideration of forest use and policy—and elsewhere I have argued, of any resource use and policy—must involve five different kinds of knowledge and analysis: physical and biological feasibility and consequences, economic efficiency, economic welfare or equity, social or cultural acceptability, and operational or administrativepracticality.

It is relatively easy to deal with any one of these approaches. Ecological models of energy and nutrient flows, of interspecies competition, and of many other relationships can be built, simply or with complexity, as the model builder's tastes and the availability of data permit. Or economic models can be built—of production functions, transformation functions, demand, supply macro and micro, and all the rest of it—to determine an economic efficiency result: benefit-cost, internal rate of return, etc. These relatively simple approaches are less applicable to economic equity—who pays and who gains, what is fair, and so on—but still it is possible to discuss the distribution of gains and benefits with pomposity if not with illumination. Likewise, social or cultural acceptability and operational or administrative practicality can be dealt with in their own terms by persons specializing in such matters.

The difficulty comes in reconciling these various approaches: how does one trade off some degree of economic efficiency for some degree of cultural acceptability, to use but one example? Moreover, how far is one set of considerations independent and how far influenced by another set of considerations; if something is highly profitable, may not initial cultural resistance yield in time? The defense or justification for my apparently complex approach is that both public and private decisions on natural resource use and policy must take cognizance of this range of considerations; that omission of any one of them results in an apparent solution that will not hold up; and that in the real world accommodations not only must be, but also are, made between these superficially incomparable kinds of considerations.

Man, Mind, and Land

A more ambitious attempt to develop theory was by White (1963) in his perceptive review of Walter Firey's *Man, Mind and Land: A Theory of Resource Use* (1960). Firey had written an ambitious book in which he tried to identify the roles of what he called ecological, ethnological, and economic factors in natural resource use, development, and conservation. He tried to develop theories and relationships for application of each of these concepts to a wide range of natural resource situations around the world, and above all to show how these different approaches must be reconciled or at least accommodated if a viable and operable system of resource use were to develop and operate satisfactorily. As with any ambitious attempt at a truly comprehensive theory, many critics raised many objections. White is rather more sympathetic: "The whole work commends itself as a consistent, systematic statement of a theory within which any resource use might be appraised." But he is also critical about its omissions: the book "does not directly deal with the process of decision making by which uses are selected by private or public managers . . . [and] touches only lightly upon the basic problems of risk and uncertainty as they affect choice." Nevertheless, "the reviewer is inclined to think that this is the best formulation of resource use thus far produced. . . . [It] is a landmark in the expanding terrain of thought about natural resources. . . . [It] will serve as a baseline for efforts at refined description."

Trends toward Specialization

These landmark studies are exceptional not only in their breadth but also in their relative rarity. Most research on natural resources is highly specialized. Three illustrations from my recent personal experience may serve to document the trend toward specialization.

Weather Modification

A conference in 1965 attempted to include all important human aspects of weather modification and to achieve something of an integration or synthesis of the materials (Sewell 1966). The bibliography at the end of the proceedings lists approxmately 150 books, articles, and reports that the authors of the papers thought relevant to this subject. Not more than half a dozen of them, and the conference volume itself, could reasonably be described as integrative in character; the rest are specialized.

Water Planning

Water planning has perhaps the most outstanding record of integrative analysis and planning of any natural resource. Under the leadership of Abel Wolman, Harlan H. Barrows, and Thorndike Saville, the Water Planning Committee of the Natural Resources Planning Board (and its immediate predecessor agencies) in the 1930s and early 1940s developed the idea of coordinated or integrated water planning on a watershed or river basin basis. There were two very obvious hydrological facts operating in their favor. First, geographically, water use in one part of a river basin had inescapable consequences for water use in every other part of the same basin. Second, functionally, water used for one purpose was not available for use for another purpose; reservoir capacity used or held for one purpose (such as flood protection) could not be used for another purpose (such as storage of irrigation water); and yet there are substantial economic benefits from multiple-use water development and management.

But even in this field of water planning, where integration of specialized knowledge has perhaps gone the farthest, most of the studies about water are still specialized. There has recently come to my attention, for instance, a publication from the Virginia Water Resources Center (Giles 1977), which seeks to provide an integrated framework for water planning and management. The author cites some 125 reports that he believes relevant to his analysis; nearly all would have to be considered specialized in nature and content. Howe and others (1969) provide something of a synthesis of one aspect of water use—inland transportation—yet their report lists as relevant about thirty-five reports, nearly all of which I would classify as specialized. There have been some integrative and inclusive water studies in recent years—two Senate documents (U.S. Congress 1960); a book by Nathaniel Wollman and Gilbert W. Bonem (1971); and reports of the U.S. Water Resources Council (1968; 1973).

Agricultural Policy

In early 1979, for another purpose, I reviewed the issues of the *American Journal of Agricultural Economics* and the publications of the National Academy of Science, each for the most recent five years. In them I found citations for some three hundred books, articles, and reports that seemed to have some bearing on agricultural policy. My test of relevance was subjective, not absolute; and these references were possibly only half as many as might be assembled by diligent further search in other professional sources, but they were at the least a large proportion of all writings that could reasonably have been considered agricultural policy in recent years. Nearly all of the three hundred references were specialized; only about a dozen could reasonably be described as integrative in purpose or in result: writings by senior scholars, nearly always as incidental to their position as president of the association or otherwise as an invited paper; and group or committee activities, especially from the National Academy of Sciences (1975, 1976, 1977). Integration in agricultural policy then is the province of elder statesmen or committees and appears absent from day-to-day research efforts.

Specialization and the Need for Integration Will Continue

In earlier sections, I described the conflicting forces of specialization and integration as they have operated over the past forty years or more in virtually all professional fields concerned with natural resources. But these forces are not all in the past; on the contrary, they are active today and will almost certainly continue in the future. Indeed, a question may well be asked: is there such a field, in any reasonable sense, as "natural resources?" That is, given the enormous variety of land, water, soil, geology, topography, mineral, energy, aesthetic, and other resources; given the many fields of knowledge such as climatology, geography, geology, ecology, forestry, range management, wildlife management, engineering (in all its specialized branches), economics, sociology, anthropology, political science, and others, all of which have both interest in and useful contributions to make to natural resources problems; and given the many specialized interests in natural resource in the population as a whole: is "natural resources" so vague, so inclusive, so amorphous if you will, so incapable of being understood and grasped whole by anyone or by any small team of persons, however able, that it is simply an unrealistic and unhelpful term to use? May it not be better to concentrate on some small part of this enormous territory, making a kind-of-resource choice, or a field-of-knowledge choice, or an interest-group choice, and make some useful contribution to knowledge within the self-chosen limits, if one is a researcher, or some useful contribution to private choice or public policy, if one is an actor or an adviser to actors?

If judged by actions more than words, it appears that the overwhelming majority of professional workers in all the listed fields (and in any others concerned with natural resources that I may have inadvertently omitted) have indeed chosen the small as contrasted with the large view of natural resources.

There are great advantages in specialization and powerful forces pushing professional workers toward more specialization. As I have noted, the proliferation of knowledge has made it impossible for any worker to know all about anything except a small field; sheer frustration and helplessness in the face of mounting knowledge leads many a professional worker to establish narrow fences around the pasture within which he or she will graze. In this small field, the worker may become a major figure and be able to make specialized contributions when the same application of abilities and energy would not contribute anything new in a much larger field.

There are indeed strengths in specialization. But someone, somewhere, in our society must look at natural resources as a whole, must balance up the use of one kind of resource against the use of others, must balance the demands of one user group against those of others, and must synthesize the partial truths of various specialists into a larger, broader, more inclusive truth. If one looks to the improvement of human welfare, now and in the future, one must consider all the possible resources and all the possible avenues to that improvement. This almost certainly includes physical, biological, economic, and social factors, together with political considerations and analyses and the interactions of many concerned groups within the larger body politic.

Obviously, integration is complex and difficult, and high precision—fine tuning—may be unattainable. But I argue that it is also necessary, indeed indispensable. The big question is how not whether.

Acknowledgments

In the preparation of this essay, I have been much aided by Charles J. Hitch and Emery N. Castle, of Resources for the Future; by Marion Marts, of the University of Washington; and by Andrew J. W. Scheffey, of the University of Massachusetts.

References

Barnett, Harold J., and Chandler Morse. 1963. *Scarcity and Growth: The Economics of Natural Resource Availability.* Baltimore.

Brubaker, Sterling. 1972. *To Live on Earth: Man and His Environment in Perspective.* Resources for the Future. Baltimore.

Clawson, Marion. 1975. *Forests for Whom and for What?* Resources for the Future. Baltimore.

Cornish, Edward. 1977. *The Study of the Future: An Introduction to the Art and Science of Understanding and Shaping Tomorrow's World.* Washington, DC.

Firey, Walter. 1960. *Man, Mind, and Land: A Theory of Resource Use.* Glencoe, IL.

Forrester, Jay W. 1971. *World Dynamics.* Cambridge, MA.

Giles, Robert H., Jr. 1977. *A Watershed Planning and Management System: Design and Synthesis.* Virginia Water Resources Research Center Bulletin 102. Blacksburg, VA.

Howe, Charles W., and others. 1969. *Inland Waterway Transportation: Studies in Public and Private Management and Investment Decisions.* Resources for the Future. Washington, DC.

Landsberg, Hans H., Leonard L. Fischman, and Joseph L. Fisher. 1963. *Resources in America's Future: Patterns of Requirements and Availabilities, 1960–2000.* Resources for the Future. Baltimore.

Meadows, Donella H., Dennis L. Meadows, Jorgen Randers, and William H. Behrens III. 1972. *The Limits to Growth.* New York.

National Academy of Sciences and National Research Council. 1975. *Agricultural Production Efficiency.*

———. 1976. *Climate and Food: Climatic Fluctuation and U.S. Agricultural Production.*

———. 1977. *World Food and Nutrition Study: The Potential Contributions of Research.* Washington, DC.

National Water Commission. 1973. *Water Policies for the Future.* Washington, DC.

President's Materials Policy Commission. 1952. *Resources for Freedom* (Paley Report). 5 vols. Washington, DC.

President's Water Resources Policy Commission. 1950. *A Water Policy for the American People.* 3 vols. Washington, DC.

Sewell, W. R. Derrick, ed. 1966. *Human Dimensions of Weather Modification.* University of Chicago Department of Geography Research Paper no. 105.

Thomas, William L., Jr., ed. 1956. *Man's Role in Changing the Face of the Earth.* An international symposium under the cochairmanship of Carl O. Sauer, Marston Bates, and Lewis Mumford. Chicago.

United Nations. 1950. *Proceedings of the United Nations Scientific Conference on the Conservation and Utilization of Resources*. 7 vols. New York.

U.S. Congress, Senate. 1962. *Policies, Standards, and Procedures in the Formulation, Evaluation, and Review of Plans for the Use and Development of Water and Related Land Resources*. Senate Document 97, 87th Cong., 2d sess.

U.S. Congress, Senate, Select Committee on National Water Resources. 1960. *Water Resource Activities in the United States: Future Needs for Navigation*. 86th Cong., 2d sess.

U.S. Water Resources Council. 1968. *The Nation's Water Resources*. Washington, DC.

Watt, Kenneth E. F. 1974. *The Titanic Effect: Planning for the Unthinkable*. New York.

White, Gilbert F. 1953. "A New Stage in Resources History." *Journal of Soil and Water Conservation* 8, no. 5 (September).

———. (as chairman of the Social and Economic Study). 1962. *Social and Economic Aspects of Natural Resources*. Publication 1000-G. National Academy of Sciences National Research Council, Committee on Natural Resources. Washington, DC.

———. 1963. Review of *Man, Mind and Land: A Theory of Resource Use,* by Walter Firey. *Economic Geography* 39, no. 4 (October).

Wollman, Nathaniel, and Gilbert W. Bonem. 1971. *The Outlook for Water: Quality, Quantity, and National Growth*. Resources for the Future. Baltimore.

Woytinsky, W. S., and E. S. Woytinsky. 1953. *World Population and Production: Trends and Outlook,* New York.

Zimmermann, Erich W. 1933. *World Resources and Industries: A Functional Appraisal of the Availability of Agricultural and Industrial Materials*. New York. Rev. ed., 1951.

4 Resource Use in Dry Places: Present Status and Potential Solutions

Douglas L. Johnson

Interest in dry environments is an ancient concern of humans, for it is out of such settings that the major civilizations spring. In northern China, the Indus, Tigris-Euphrates, and Nile valleys, central Mexico, and coastal Peru arose urban and agrarian civilizations that were seminal cultural hearths for present-day society and economy. Yet modern scientific concern for the wise use and management of the resources of drylands is a relatively recent phenomenon. Although warnings about the abuse of arid land resources appeared in the first half of the twentieth century (Bowman 1924; Stebbing 1937), concerted attention was not directed toward dryland resource problems until after the Second World War.

At that time the initiation of an arid lands research program under the auspices of UNESCO began to draw together researchers and institutions in a common endeavor (White 1956; McGinnies 1981). Initially there were few specialists in arid land resource management. For this reason the approach employed was to pull together researchers working on cognate problems in other environments and disciplines. Scholars with a focus on a similar, narrowly defined range of problems, such as salinity, climate change, hydrology, and plant ecology, were brought together in symposia in an effort to summarize existing knowledge and to identify the crucial needs for future research. While such projects were successful in promoting better interaction among individual scholars and producing valuable status reports, their overall effect on practical problems of dryland development was minimal (Baker 1979). Nonetheless, difficulties in implementing the knowledge generated by research and in applying technological advances appropriately should not obscure the fact that significant, albeit unevenly distributed, gains have been made. This chapter considers some of the salient characteristics of contemporary dryland resource management, the environmental problems that have emerged, and the strategies that might be applied.

Douglas L. Johnson is Associate Professor at the Graduate School of Geography, Clark University.

An assessment of the status and prospects for arid land research, resource management, and economic development is particularly appropriate in a volume dedicated to the seminal influence of Gilbert White. White has been involved in many aspects of arid land research and development, particularly in issues of water management, for the last quarter of a century. His 1960 inventory of knowledge (White, 1960a) about arid lands remains an essential point of departure for students of arid land management problems. Although outpaced in many of the specifics of data, this study remains remarkably perceptive about the generic nature of the problems that afflict dryland populations. Many of the issues noted in the 1950s and 1960s still await solution.

Four major characteristics of contemporary dryland resource management are identified here: recognition of the serious environmental constraints of arid and semiarid ecosystems; a sectoral approach to developing dryland resources; fascination with technological conquest of productivity limitations; and heightened concern about resource-base degradation. Each of these characteristics results in problems and issues found in many other ecosystems, but they assume unique manifestations in dryland settings.

Environments of Constraint

Sparseness and variability characterize the primary resources of drylands and impose serious constraints on resource-use systems. It is the paucity of water, both spatially and temporally, that gives drylands their distinctiveness. Heavy rainfall in one small district may be, and most frequently is, an isolated event, not replicated in nearby drainage basins. An entire year's rainfall may occur in just one or two storms of great intensity that produce rapid runoff, flash flooding, severe gullying, and erosion. Not only is annual variability great, but interannual extremes are typical. Such positive and negative pulses exhibit a tendency to persist (Hare 1977) on at least a biennial rhythm, and droughts of several years' duration are common. Faced with persistent extreme variations and climatic records of limited duration and number, calculation of average conditions is a frustrating and unrewarding activity. Instead, the existing vegetation, where not massively altered by human activity and indigenous livelihood patterns, probably constitutes the best measure of dryland productivity.

Dryland ecosystems cope with variability either through mechanisms that replicate the rhythm of the natural regime or by developing physiological devices for evading the worst effects of variability (Noy-Meir 1973). Thus, annual grasses complete their growth cycle and set seed quickly, the seed itself possessing the ability to resist long periods of desiccation until the next burst of moisture. Succulents rely on storage devices and transpiration-reducing surfaces that retain moisture, while other perennials develop deep-rooting structures that can reach groundwater. Rodents, gazelles, and camels all possess a degree of drought hardiness and water-use efficiency unknown in other environments (Noy-Meir 1973; Daly 1979). Indigenous livelihood systems exhibit similar adaptive ability, in that behavior patterns emphasizing spatial mobility, flexibility in diet, and maintenance of low population densities ensure resilience and enhance potential for survival (Noy-Meir 1979). Nomadic pastoralism is the classic example of a livelihood adaptation to widely dispersed, often ephemeral fodder and water resources that must be managed opportunistically if herd size is to be maintained and expanded. Only in spatially isolated situations where concentrated water resources are available, such as along allogenic streams (e.g. the Nile) or in oases, or where technology can be employed to bring water to otherwise deficit regions, are more sedentary land use practices useful.

Recent scientific research in dryland environments and their livelihood systems has followed a systems approach (Baker 1979) that pays particular attention to the constraints on use of arid lands. This flows from a more general concern within ecology for understanding the degree of stability and resilience that characterizes a particular ecosystem (Holling 1973). Proper management of an ecosystem that contains severe limiting factors makes identification of the system's stability boundaries important. Without a clear understanding of the tolerance of the system to human-induced changes, the possibility of exceeding limits is great. The result of pressure-induced changes can be a rapid shift toward a new equilibrium with very different and undesirable characteristics from the perspective of human use. Ecological analysis carried out in a wide variety of socioeconomic settings indicates that these stability limits are not easily identifiable, but that the least intensive management systems, ones that employ technology of limited power, are less likely to produce sudden and catastrophic change (Warren and Maizels 1977; Noy-Meir 1979) than are more intensive alternatives. As population growth in drylands puts increased pressure on both rural and urban resources, a shift toward more intensive and concentrated management seems inevitable. It is in this context that the struggle to maintain the resilience of dryland resource use systems must be carried out.

Two strategies offer potential productive alternative pathways for development. Both take the moisture constraints of the dry realm as their point of departure. The first emphasizes making the best use of existing, generally highly variable, moisture resources. The second attempts to overcome some of the aridity constraints of drylands by turning negative attributes into stimuli for growth.

The need for a more efficient use of existing water resources is evident in the drylands of both the industrialized and the developing world (White 1960a). In many parts of the American West and Southwest, water use is outstripping natural recharge, an overexploitation that threatens existing settlement patterns (Walsh 1980). In the Tucson basin, to cite but one example, there is disagreement about the extent to which current annual overdrafts of 124,000 acre-feet can be compensated for without drastic changes in water use practices to favor industrial and residential consumption at the expense of vested agricultural interests (Wilson 1977). In Phoenix, failure to recognize the constrained nature of groundwater supply has resulted in land subsidence (Laney, Raymond, and Winikka 1978). In substantial areas, subsidence has exceeded seven feet during the 1952–77 period as a result of overzealous mining of groundwater. At the other end of the development spectrum, rapid population growth in both Egypt and Sudan has spawned massive technological solutions, such as the Aswan High Dam and the Jonglei Canal, to the need to increase available water. Yet as the finite upper limit in total available moisture in the Nile drainage basin is approached, enhanced water use efficiency becomes increasingly crucial (Waterbury 1979).

Draconian controls on future growth are not what is required, for these would often preclude realization of the aspirations of dryland populations for improved standards of living and increased security of livelihood. Rather, one should emphasize a rational and controlled use of the resources as opposed to a restricted-use approach that in effect would stockpile the resource for future exploitation (White 1960b). Strategy should emphasize three salient principles: exploitation based on a conservative estimate of existing supply, employment of the most efficient water use and transmission technology available, and regeneration of water supplies depleted by development.

Accurate assessment of groundwater reserves and recharge rates is lacking in most dryland areas. Records are too limited in duration to assess recharge accurately in most areas, and many groundwater reservoirs are being tapped without an adequate inventory of existing capacity and recharge rates. In less industrialized countries, where comprehensive survey inventories are a relatively recent development, this lack of data is understandable, although rectification of the data gap is an essential ingredient in implementing development projects with prospects for long-term viability. In industrialized nations, where water resource inventories are an established activity, access to a more powerful technology encourages overuse in the expectation that alternative sources of water will be located or can be transported from surplus to deficit areas. In both political and economic settings the wisest course for development to take is to rely on locally available water, to employ the best available technology in assessing groundwater capacity, and to predicate development on a conservative estimate of existing water supply.

An efficient strategy must also insist on employment of the most efficient water use and transmission technology available (White 1955). Considerable efficiency differentials can be obtained in water consumption rates by employing careful management practices and technological innovations. Pipes rather than open canals, trickle rather than sprinkler irrigation, controlled environment rather than open-field agriculture, centralized and coordinated rather than individualized allocation systems can all contribute to more efficient water use. Selection of the technology to be used, however, must take account of the socioeconomic setting. The wealthier the economy and the more skilled its population, the more sophisticated and expensive can be the technology chosen.

Any effort to upgrade existing water delivery and use efficiency also must carefully consider the entire cultural and ecological system likely to be affected by any change. Thus, open-field, trickle irrigation systems that work well in experimental plots or in field trials, as they appear to be working in northwest India (Malhotra, Singh, and Sen 1977), may be much more difficult to disseminate to the bulk of the rural agricultural population. The new technology may well be beyond the economic means of most of the potential recipients. It is important that designers of technological systems pay careful attention to the entire life support system involved (White 1980). It could be false economy to line completely irrigation-water-delivery canals to reduce seepage losses, if recharge of groundwater exploited by domestic wells were to be threatened. In the Carson-Truckee (Townley 1977) irrigation district in northwest Nevada, the domestic water supply of farmsteads, ecologically significant wildfowl refuges, and native American water rights in Pyramid Lake would be jeopardized if the irrigation water delivery system were fully rationalized. Attention to the gains and losses associated with increased efficiency is essential if improved management practices and technology are to prove socially and ecologically acceptable and are to avoid unforeseen and undesirable effects.

Equally important is the regeneration of water supplies depleted by development. This is not categorically to rule out the treatment of groundwater as a "wasting resource" (Mandel 1977). In the interests of generating economic growth in marginal environments, some mining of groundwater supplies is probably tolerable and is certainly inevitable. But one must not mine groundwater in the absence of clear future alternatives to current water sources. No development project that necessitates substantial groundwater depletion should go ahead in blind faith that some future technological solution to increasingly scarce local water resources will be found, or that changing economic conditions will make previously uneconomic exploitation more feasible. Rather, all projects should contain within their basic rationale clearly demonstrable alternative supplies that will replace depleted stocks. The development of these alternatives should be integral to the infrastructure created during a period of depletion, not stopgap measures designed to bail out temporarily a faltering enterprise. In this sense, the water supplies exploited in this fashion are not wasted (although natural recharge may be far too lengthy to enter planning calculations), but instead represent productive investments in future livelihood capability. Failure to employ efficiency criteria of this type will condemn dryland development to the short-term boom and bust syndrome so characteristic of mineral exploitation.

Sectoral Approaches versus Holistic Solutions

Much of the contemporary approach to economic development and environmental management in arid lands is sectoral in design and implementation. In this respect, planning for arid lands has not differed significantly from development in other environments; examples of successfully implemented national land use plans are conspicuous by their absence. Almost invariably a reductionist approach to environmental management and enhanced productivity has been followed. This separates administrative responsibility into units defined by functions such as transportation, irrigation, animal husbandry, and industry. The result is increased bureaucratic accountability; another consequence, however, is an "administrative trap" (Baker 1976), because activities often compete for space and resources rather than being synergistically supportive. Ministry contends with ministry, foresters fight with peasant farmers, and cereal production vies with animal husbandry. Changes in land use are instituted in one sector without careful consideration of the full range of potential environmental impact. Thus, gains in dry farmed cereal crop yields are often realized through conversion of rangeland to mechanized cropland. The results are good for the ministry of agriculture, yet may seriously constrain the ability of the pastoral managers to maintain productivity without inflicting serious damage on the environment.

Gainers and Losers

One consequence of this sectoral approach is that gains in one area are frequently offset by losses elsewhere. A dramatic illustration of the general process is presented by Welcomme (1979) in his discussion of the ecology of floodplain rivers. In many dryland floodplain ecosystems, irrigated agriculture, pastoralism, forestry, and fishing possess overlapping ecologies. As long as the timing, spacing, and sequencing of these activities are sufficiently flexible, all can exist using the same floodplain resource base seasonally. Once one set of activities is favored over the others, however, adverse consequences may emerge in some or all areas. The areas favored by development are those least likely to exhibit adverse impact. This follows from the tendency to limit environmental impact assessments to a rigidly bounded development project. The benefits derived from development fall upon the project and its participants, while many of the costs, uncalculated, appear elsewhere.

Irrigation development is the most frequently employed technique for increasing the food production capacity of arid countries; yet its results are mixed (Worthington 1977). The doubling or tripling of agricultural output that results from converting floodplain farmland from annual to perennial inundation is threatened by salinization (White 1978), and other livelihoods and ecosystems too experience unforeseen complications. Fish yields decline when streamflow characteristics are altered; seasonal breeding grounds on the floodplain are eliminated; and swamps are reduced in size or drained (Welcomme 1979). Clearing of floodplain forests and bush savanna for agricultural expansion is costly but essential. However, this not only destroys an important source of fuel wood and construction material, but also can accelerate erosion, increase the turbidity of downstream water, and contribute to altered fish-stock composition. The net result is reduced fisheries production both in the artisanal fishery along the stream and in the commercial oceanic fishery offshore from the river mouth (George 1972), which often in turn results in diminished diet quality and economic losses for populations located far from the irrigation project itself.

Pastoral activities also are adversely affected by irrigation expansion. Conversion of seasonal fallow land to year-round arable eliminates a critically important source of dry season grazing for migratory pastoralists and can exacerbate both the long-term impact of grazing pressure on nearby rangeland used the year round and the socioeconomic impact of episodic drought (Beshah and Harbeson 1978). Traditional rights of access to dry season pasture in the interior delta of the Niger River were assured by a formal legal code (Gallais 1975), but customary rights protected pastoral grazing along most dryland floodplains. The dung deposit by the herds was exchanged for fallow fodder and stubble grazing. Not only has an expanding irrigation all but eliminated dry season fodder grazing, but continuous crop production removes stubble as a fodder source as well. Indeed, neither stubble nor floodplain fodder resources are sufficient for the animals that most irrigation project participants wish to keep. Most of the animals of sedentarists must be grazed outside the project boundaries where they compete for grazing with the herds of pastoralists. Since pastoralists are unlikely to reduce the size of their herds voluntarily, because the need to maintain herds sufficient to sustain the pastoral community remains, increased pressure on the rangeland is inevitable. Areas that formerly were grazed seasonally now receive year-round pressure. An expanding mechanized farming only increases this pressure by converting existing rangeland to cereal cultivation.

Interagency competition is compounded by the selective introduction of technology and change in one production sector without proper evaluation of the implications of the changes. Introduction of well technology and veterinarian services in pastoral areas has frequently produced its desired result: growth in the size and health of pastoral herds. Often this increase in animal numbers has occurred without adequate provision of market outlets, economic incentives, and effective local control of adjacent grazing and water resources. Severe overstocking around well sites is frequently reported (Heady 1972; Talbot 1972; Bernus 1977; Swift 1979) and is usually exacerbated during droughts. Much of the adverse ecological and social impact of the Sahelian drought in pastoral areas can be attributed to sectoral land management practices that promoted excessive concentrations of people and animals in ecologically vulnerable areas (Warren and Maizels 1977).

Selective introduction of individual items of technology is not the only feature of a sectoral development approach. Project-by-project evaluation of the environmental and socioeconomic impact of development is also a significant component of the problem. The linkage between development initiatives in one place and their impact on physical and human systems in adjacent areas is easy to ignore in sectoral development. For example, the decision to put development resources into the irrigation sector has serious environmental impact. Some of this falls within the floodplain itself and should be identifiable by traditional methods of benefit-cost analysis. Because populations affected often are not considered to be part of the project's target population, these adverse consequences receive minimal attention in the design process. Also, such costs are seldom counted in the project evaluation because they occur outside project boundaries and are difficult to calculate. Indeed, tracing the web of cause and effect from land-use change in one area to environmental impact in another is no easy task. The overgrazing that results in pastoral rangelands is most often regarded as a product of irrational traditional herders mismanaging a common property resource (Picardi 1976) rather than a consequence of forces external to the affected district and its denizens.

Although these problems are not unique to arid lands, they are particularly serious in dry regions precisely because the recovery time from environmental degradation is often longer than in more humid areas and because the constrained nature of the resource base makes any loss especially significant. Since these losses most frequently occur in the environments with highest potential, where use is most intense, their consequences are magnified.

Holistic Solutions

Awareness of the need for a holistic approach to the environment has grown in the last two decades (White 1980). Certainly a truly comprehensive and integrated planning design on a regional or national basis should go a long way toward avoiding the competitive and unbalanced approach to environmental management outlined above. Yet, despite lip service to comprehensive and holistic approaches, integrating physical and human components into a unified planning framework has proved difficult. Not only is much agricultural development research carried out in a narrowly technical framework, but it is applied to development through traditional sectoral bureaucracies that have little relation to the individual production unit of the farmer or herder (Baker 1979).

A tremendous gap exists between the production results and techniques generated on the individual research station and the conditions faced by the traditional producer. Communication across this gap is a fundamental problem in efforts to modernize the traditional sector of the economy. The effort to communicate is hampered by objectives, experience, and reward structures of administrators that may be at odds with local needs. The response of administrators may be rational in their own terms (Chambers 1979). Assignment to isolated rural areas may impose family hardships; language barriers may be difficult to overcome; cultural differences may be considerable; and the desire for immediate, quantifiable results upon which promotion can be based may necessitate a top-down development approach. All of these issues may be as relevant to the indigenous administrator from the capital city as they are to the expatriate technical adviser. Differences between traditional and modern patterns of conceptualizing the management and monitoring of resource use increase the severity of the problem. This frequently leads the modernizing sector to institute legal and administrative structures that bypass and undermine traditional management strategies. The environmental legislation of 1971 that was instituted in the Sudan, for example, was rational and progressive. But it failed to work because it destroyed local environmental management practices based on traditional leadership without replacing them with enough trained personnel to make the new system work effectively. Taken together, these factors add up to a considerable administrative barrier to the implementation of holistic development plans.

A more holistic treatment of arid land management would counter the reductionist tendencies typical of much of today's development planning and project implementation. This strategy could be viewed as a cultural-ecological unit strategy because it gives equal weight to physical and social parameters on the one hand and to goals of national and livelihood groups on the other. Analysis of the ecological units upon which local resource-use systems are based is a fundamental starting point. In this way the resources essential to each livelihood system can be identified. Attempts to change the current matrix of resource-use activities then can be based upon a clear understanding of actual and potential physical resources in a region and upon the ways they are exploited. Changes in one sector can proceed with an accurate awareness of where difficulties are likely to emerge in other components of the social and ecological system.

This has been attempted in planning development in the semiarid zones of western Sudan (Hunting Technical Surveys, 1976), where an effort has been made to integrate pastoral and agricultural activities in one management design. By structuring pastoral migration corridors between seasonal northern and southern grazing zones, it is possible to preserve the flexibility and mobility of the traditional pastoral livelihood system while reducing ecological pressures, reducing conflict over access to land resources, and enhancing development options in the agricultural sector. The same principle applied to integrated river basin development permits the allocation of water resources among competing uses on a more informed basis (National Research Council 1968). The result of such strategies in both industrialized and developing economies is to clarify the options available in a resource allocation and to identify the potential tradeoffs between competing uses.

But even where such plans exist, implementation remains difficult (Adams and Howell 1979). In government planning, much attention is given to large-scale initiatives in the modern sector that promise major contributions to national development goals. These goals are often quite different from those of the rural population, the bulk of whom are excluded from and remain untouched by the development process. As Adams and Howell point out (1979, p. 518), modern and traditional sectors are not necessarily competitive for labor and capital, although they are often competitors for access to land resources. The large capital investments required to initiate modern development ensure that central planners will give priority to large-scale schemes in the use of land resources. Viewing each scheme as a separate entity and minimizing the importance of the traditional sector and the adverse impacts on it fostered by much of contemporary development are ideal prescriptions for increasing adverse environmental impact. Because much of this social and physical degradation will occur outside the view and beyond the monitoring frameworks of most planning processes, considerable damage can occur before the problem comes to the attention of policymakers. In arid lands, where water is the limiting variable in development, insistence on strategies that examine options and effects in a holistic context is of the greatest importance.

The Search for the Technological Fix

Thus far, application of the holistic principle to dryland management has been flawed at best. In large part this is a consequence of countervailing tendencies in resource management practice. The most important of these tendencies is the search for primarily technological solutions to problems that contain a vitally important social component. The social impact of technological change is inadequately acounted for because the constraints on dryland productivity are often viewed as lying in the physical resource base. Technology applied to increase cash crop production and the scale of economic and social operations and to substitute new production and management practices for local systems is advocated. Other approaches can be used to enhance dryland productivity.

Cash Crops versus Subsistence Needs

Many arid areas are marginal, peripheral zones with limited productivity. Occasional pockets of higher productivity, such as along perennial streams, contrast starkly with the vast areas of lower potential. The zones of higher productivity receive a disproportionately large share of attention because they promise higher returns on investment. Indeed, one strategy frequently advocated (Dregne 1977) is to focus development resources primarily in such zones.

Almost invariably, this approach is linked to the production of crops that can earn foreign exchange in the international marketplace. Most of the irrigation schemes implemented in both industrialized and developing countries favor cash crops as the major agricultural activity. Often this involves project participants subscribing to rigid, centrally mandated production schedules and regimented crop allotments that run counter to the participants' perception of self-interest (Hoyle 1977; Sorbo 1977). Not only can this have adverse social consequences (Barnett 1977), but also ruthless environmental exploitation in the interest of immediate profits can have disastrous implications for future productivity (Thimm 1979). Similarly, overenthusiastic encouragement of cash cropping can seriously jeopardize the ability of traditional agricultural-pastoral systems to cope with severe drought when basic food grains and essential animal products are in short supply (Kates 1980). Much of this emphasis on cash crops is supported by the agricultural research establishment, which has been quite successful in developing crops and production techniques (Baker 1979) that meet narrowly defined goals in a high technology sector without taking into account the subsistence needs of a broad sector of the population. Much of the recent interest in dryland-adapted crop species in North America focuses on species such as guayule (*Parthenium argentatum*) and jojoba (*Simmondsia chinensis*) that have potentially important commercial applications (Maugh 1977; McGinnies 1979; Johnson and Hinman 1980). Attention is directed toward both large-scale commercial development, often using supplemental irrigation, and the harvesting of wild stands. The latter focus may have an adverse impact on the long-term reproductive viability of jojoba, the seeds of which are the source of a valuable lubricating oil, but present knowledge of the effects of continuous, substantial exploitation is too limited to predict or plan effectively (Foster 1980). Concern for the ethno-botany of indigenous cultures and its development possibilities is a relatively recent phenomenon. Careful consideration of the use of saguaro cactus (Crosswhite 1980; Fontana 1980) and the sunflower (Nabham 1979) in native American subsistence patterns suggests some of the development possibilities of the preadapted local flora and the ways in which security of local livelihood could be increased if greater efforts were made to enhance the productivity of traditional crops.

Without denigrating the significant technical work done in enhancing productivity in dryland agriculture, I would insist that emphasis on meeting basic subsistence needs of dryland populations is a long-overdue corrective to dominant tendencies. This approach complements rather than replaces cash cropping. It is based on the fundamental premise that, except in extraordinary circumstances, dry environments and their populations will at best be self-sustaining rather than a continual drain on national resources. Tempered by modest expectations of contributions to national productivity, this approach accepts drylands as possessing limited opportunities and anticipates few dramatic gains. By taking drylands on their own terms, a more subsistence-oriented development strategy would attempt to sustain basic needs by building on available local resources. It would also attempt to reduce the impact of humans on the sharp pulsations characteristic of dryland environments by concentrating on practices that anticipate and smooth out extreme variations.

Numerous approaches might be suggested. To exemplify the principle, a set of options that stress the role of trees in dryland livelihoods are chosen. This arboreal strategy emphasizes contributions to livelihood productivity and the mitigation of dryland hazard, particularly of the adverse consequences of drought, rather than forestation aimed at general environmental stabilization. Efforts such as the circum-Sahara green belt concept, spawned by the United Nations Conference on Desertification (1977a), are less likely to achieve a dramatic positive impact on dryland populations than those targeted directly toward such fundamental needs as fuel wood, increased food production, and drought fodder requirements.

Many traditional dryland livelihood systems make use of tree resources. For example, the collection of gum arabic (*Acacia senegal*) in the Sudanese savanna is an important element in the local economy. Pastoralists habitually lop the branches of acacia trees, particularly during drought periods, for animal fodder, and in at least some areas careful supervision of the process prevents degradation (Draz 1977). In semiarid Rajasthan, India, exploitation of *Prospis spicigera* by peasant farmers has reached a level of considerable sophistication. Valued for its multipurpose utility, *P. spicigera* is carefully preserved in fields for its edible pods, protein-rich leaves, bark (which, when mixed with vegetables, can be consumed as a food in times of famine), and valuable crops of fuel-producing branches.

Religious scruples reinforce the functional utility of *P. spicigera* and encourage its protection by the local population. Only its slow growth rate detracts from its otherwise estimable list of positive attributes and encourages some authorities to promote faster-growing introduced species (Mann 1981). Yet improvement of growth rate by genetic manipulation would seem justified given this shrub's superior adaptability to harsh local conditions. Development efforts that build on traditional uses of woody vegetation or attempt to introduce them into local production systems in an effort to diversify them have great potential.

The benefits of an arboreal strategy are maximized when forestation takes place within an integrated development plan. The guiding principle should be multipurpose use of the woody vegetation wherever possible to avoid conflicts between competing land uses and to integrate tree-related activities more fully into the local production system. Four main aspects of an arboreal strategy can be recognized: expansion of drought-resistant tree crops, enhancement of fuel wood resources, provision of fodder reserves, and increase in land stabilization and protection.

Tree Crops

Concentration of a substantial portion of dryland agricultural activity on commercially valuable tree crops is a major option in dryland development, but one that has not been systematically developed. Tree crops represent a high-value produce useful both in local diet and in external commercial exchange. Because tree crops are less susceptible to moisture shortfalls than are field crops, they provide increased stability to local livelihood systems. They remain, however, vulnerable to fluctuations in market price and in this sense suffer from the same liabilities as do all primary production systems in developing economies.

An important example of the use of tree crops in dryland development is found in the northern Negev of Israel. Here a variety of drought-tolerant tree crops are cultivated. The cultivation practices employed follow the dictum that traditional land management practices can provide valuable lessons for contemporary land use systems (Evenari 1977). In the northern Negev the ancient farming practices of the Nabateans provide the key to successful cultivation. Although the traditional systems can be shown to work successfully (Evenari, Shanan, and Tadmor 1971), the revenue produced is insufficient in general farming to satisfy the income requirements of Israeli citizens. Changes of scale in production make tree cropping possible. The operation emphasizes microcatchments for moisture collection around fruit trees. In the ancient system, water was collected from the lower-productivity hillsides of substantial drainage basins and channeled to terraced fields in the valley bottoms. Use of the same principles on a microscale makes possible commercially viable nut and fruit production. The ultimate success of commercial applications in Israel remains uncertain. But the system does appear to offer major opportunities for extension in less industrialized dryland locales. In particular, the role tree cropping might play both in filling local market demands and in diversifying the subsistence production base of dryland farming and herding populations makes its expansion an exciting development strategy.

The potential role such tree crops might make is foreshadowed by CAZRI's (Central Arid Zone Research Institute) experience with *ber* (*Zizyphus nummulaira*), and its improved cultivar, *Zizyphus mauritiana* (Pareek 1977; 1978). This dryland-adapted fruit tree possesses numerous advantages: it is drought-resistant; it flowers and fruits are coincident with the monsoon, so it uses available moisture efficiently; and it possesses a tough skin resistant to fruit-fly infestation, can be lopped for fuel, and has a leaf with good fodder potential. Moreover, a local market for the fruit exists at precisely the time when other fruits are in short supply, and the quality of the fruit delivered to the market can be improved by grafting *Z. mauritiana* onto locally adapted *Z. nummulaira* root stock.

Demand for *ber* seedlings exceeds CAZRI's ability to provide nursery stock (Malhotra, Singh, and Sen 1977). The receptivity of farmers is a product of two sets of factors. One is economic. With *ber* yielding 50 kilograms per tree under careful management and the fruit selling in 1978 for 3–4 rupees per kilogram, a farmer can make 150–200 rupees per tree each year. Fuel and fodder output accounts for perhaps an additional 8–10 rupees per tree. Although pruning, protection of young trees from grazing animals, reduction of crops due to pests and birds, and necessary land shaping diminish profits, relatively small inputs of labor and capital can produce a successful and profitable orchard.

Fitting with farmer objectives seems the additional factor in *ber*'s popularity. Provided one has land available, the trees require little labor once established, are good yielders for thirty years or more, place limited demands on skills, need no supplemental irrigation, produce substantial yields for limited ongoing input, can be dried to reduce storage problems, and represent an improved version of an already familiar item. For the farmer with a modest piece of low-productivity land, *ber* is a low-cost route of diversification, risk reduction, and substantial profits. Further efforts to upgrade productivity on existing *ber* plots are experimentally possible but run counter to farmers' objectives. Thus, spraying to reduce fruit-fly losses, or development of microcatchments, will undoubtedly increase yields but at the cost of additional capital and labor. Moreover, higher yields might well reduce prices by flooding the local market, a clear disincentive to increases in productivity when the market approaches saturation. But at present, *ber* cultivation is something of a success story because it fits neatly into local production strategies, a fortuitous coincidence not repeated everywhere.

Firewood, Fodder, and Stabilization

Traditional forestry management in drylands has tended to isolate forestation management from other aspects of land development (Kaul 1970). There is a tendency throughout the world to envisage forestation as a technical problem involving better management of nursery stock, improved planting techniques, and the like, rather than one that involves integration with and acceptance by indigenous resource users. Protected stands of timber are defended from encroachment by nearby livelihood groups at great cost and with indifferent success. It is the exceptional forestation project that has been designed to produce firewood, timber, fodder, and environmental enhancement as but one facet of an integrated assault on dryland degradation.

Recent progressive thinking has involved a shift away from a myopic focus on tree planting for purposes of general environment improvement and has emphasized the multipurpose role that trees and shrubs can play in local resource-use systems (Muthana and Shankarnarayan 1978; National Academy of Sciences 1979). A cautious approach to this multipurpose strategy is required for four reasons. First, local conditions, both social and environmental, are critical. What is successful in one area may turn out disastrously in another unless the local setting is carefully analyzed before new initiatives are implemented. For example, *Khaya senegalensis,* a valuable timber species in semiarid zones of northern Nigeria (Fishwick 1970, p. 75), can produce large specimens under precipitation regimes as low as 600 mm per year. However, when grouped in plantations, *Khaya* is highly susceptible to *Hypsispila,* which produces deformed shoots of minimal timber value. For this reason the most economical and efficient system for managing *Khaya* is least likely under local conditions to produce the desired result. Clearly either an alternative species or a different management practice is required. The introduction of new environmental management practices into social settings that are unprepared for or unresponsive to them is counterproductive. In the long run, the least costly, most effective approach is to identify the most pressing local need and attempt to build outward to other objectives while endeavoring to satisfy local demands. Effective management of the arboreal resource can only be attained—except at prohibitive cost and with the use of draconian measures—by successful articulation of planning objectives with local perceptions and practices.

Second, local trees and shrubs provide the best long-term opportunity for increases in productivity. These species are best suited by virtue of existing adaptations to local environmental conditions. Since their productive potential has seldom been investigated in concerted fashion, they represent an untapped resource of considerable economic potential. Well suited to their local setting, enmeshed in a web of predator-prey relations, indigenous species are less likely to become unforeseen, noxious pests than are exotic introductions.

Third, attention should focus on species that have more than one use. This makes it possible to maximize the productive potential of existing land resources by extracting several products from the same space. For example, quick-growing fuel wood trees that possess pods and leaves would seem to be ideal. Not only could they be planted as shelter belts around exposed agricultural fields, but they could also constitute an important source of dry-season and drought fodder for domestic animals.

Fourth, growth of trees for timber should receive a subsidiary priority to the effort to meet fuel, fodder, and protection needs. This prescription is based on the assumption that timber from outside the arid zone and alternative construction techniques that play down the importance of wood represent more feasible alterations than do timber plantations. This is not to underestimate the importance of pole-size timber for construction purposes but, rather, raises the question of whether even prolific but as yet unproven species such as languneo can hope to match the demands of burgeoning populations. Because most timber-producing species are slow-growing and because the moisture constraints of drylands are even more likely to retard rapid achievement of commercial size, lumber production appears competitive with and of lower productivity than other land uses. Moreover, food production needs inevitably will be accorded a higher priority than any form of arboreal management. This guarantees restriction of forestation to sites with poorer soil, slope, and moisture, again placing severe limits on productivity. Thus it is better to grow multipurpose trees on shorter rotations (5–8 years) in order to maximize growth potential, which is at its greatest during the first few years of an individual's growth, than to emphasize the longer rotations (in excess of 10 years) required if timber production is the objective.

The other major role for trees and shrubs in drylands is their specific integration into rangeland management systems. A recent study by the National Academy of Sciences (1979, p. 123) indicates that increasing costs of production coupled with continued population growth in dryland pastoral areas are giving shrubby forage plants a status they never enjoyed before. Both fashion and necessity make attention to the potential contribution of forage shrubs desirable. Until recently (McKell 1975), practice has emphasized the management of grass resources while relegating browse resources to secondary status. The failure of Western range science to recognize the role of shrub resources in range management has been one of the major obstacles to the sound development of integrated range management programs in Third World settings. For most traditional pastoralists, maintenance of herding activities would be impossible without the contribution of browsing animals such as the goat. In Syria, for example, substantial progress has been made in integrating shrub cultivation with pastoral activities. The development of pastoral cooperatives has made it possible for Syrian range managers effectively to control access to range resources (Draz 1977). This is an essential prerequisite to the introduction of managed *Atriplex numminaria* (Australian saltbush) plantations on cooperative land. *Atriplex*'s deep root structure and high salt tolerance mean that it can be raised in areas possessing water that is too salty to support crops. In less saline sites, *Atriplex* is interplanted with barley, a technique that enables the cultivator to realize a crop while the saltbush stand is maturing. Simultaneous grazing of *Atriplex*, with its high protein content, and barley, rich in carbohydrates, produces a balanced diet for sheep. Although *Atriplex* plantations on saline soils require rain to wash salt crystals from their leaves before grazing, they represent an important fodder reserve for consumption in the dry season or under drought conditions. Located on what would otherwise be wasteland, in areas inaccessible to and too risky for continuous cultivation, and integrated into production systems that effectively control and manage the resource base, shrub plantations have great potential for diversifying pastoral exploitation and increasing its security.

Scale and Technology

The scale at which a development project is designed has a crucial bearing on its ultimate success. Many projects fail to achieve their objectives because their management units are unwieldy and/or inefficient, or because ecological and social costs are exported beyond the project boundaries with eventual adverse impact on the project itself. Although it is possible to view scale problems as part of a conflict between small-scale and large-scale systems (Johnson 1979), this perspective is only partially correct. Indigenous systems in agriculture do tend to be smaller in scale than their counterparts in the industrialized world. This is especially noticeable if one is contrasting the size of landholding in each management unit with the number of individuals engaged in the enterprise. Tractor-assisted dryland agriculture is able to incorporate a much larger territory per person than is animal-powered cultivation. The opposite relationship is typical of dryland pastoral systems, for here the size of pastoral population is far larger (albeit still very small on a density scale), and the area exploited in a seasonal cycle more extensive, than is the case in an industrialized ranching scheme.

Differences in scale of operation relate both to fundamental objectives of the systems involved and to the linkages these systems have to other socioeconomic institutions. Smaller-scale systems tend to have the social and economic integrity of the local livelihood system as a paramount objective, whereas production for export to a national or international market is the dominating force motivating larger-scale systems. This can be viewed simplistically as a difference between a subsistence and a cash crop commercial orientation. This is not to say that cash cropping is absent in small-scale systems. Indeed, the production of specialty products for sale is a prominent feature of most small-scale systems. Yet the way in which technology is introduced into a system and its environmental and social impact can be profoundly dependent on the scale of the operation chosen and the degree to which local considerations enter into the decision-making process.

Technological Introduction and Environmental Impact

Technological innovations, often fostered and controlled by political and economic forces based outside the region, have a profound impact on local social and physical environments. Both large- and small-scale systems are affected, often in dramatic ways. The large-scale impact often attracts more attention, for its side effects influence more people, affect wider systems, and—because they represent large capital investments—evoke more concerted attempts by governments and international bodies to rescue the situation. The salinization of Egypt's Noubariya irrigation system (FAO 1978) and the decline of the Mediterranean fishery (George 1972) consequent upon the building of the Aswan Dam have attracted considerable international attention to the ecological change within the basin and to the political and management options available to decision makers (Waterbury 1979; Haynes and Whittington 1981). Less well recognized but equally profound in its impact, albeit at a more modest scale, is the introduction of gasoline-powered pumps into traditional Mauretanian oases. This has drastically altered local environmental and social ecology (Fauchon 1980). Not only have groundwater levels declined dramatically, but the social impact has been considerable also. Only those individuals with above-average access to capital can purchase larger pumps and deeper wells, and the effects of declining availability of groundwater have fallen inequitably upon those portions of the population least capable of coping with the change.

Careful consideration of the overlap areas between differing land use and livelihood systems, both in the need for adequate planning strategies and in the development of cultural-ecological resource management units, is a prerequisite for the successful integration of new technology and traditional expertise. This is exhibited by the systems under consideration for jojoba and guayule cultivation. These dryland-adapted plants of northern Mexico and the southwestern United States have potential as substitutes for rubber and industrial oils. Because they are suited to dry conditions, they could increase the incomes of marginal dryland populations by yielding commercial products in demand in nearby industrial economies. But any attempt to develop jojoba and guayule on a large scale can present serious socioeconomic problems. As Downing and Restrepo (1980) point out, other livelihood systems use the areas proposed for eventual inclusion in large-scale production schemes. The same marginal areas that are unsuited for present irrigation and dryland agricultural techniques sustain pastoral systems. Grazing animals rely on jojoba and guayule as part of their fodder supply. Development of production systems based on these plant resources could significantly reduce the fodder available to ranchers and peasant farmers. Since the use of the environment by these livelihood systems is extensive, that use is too light at any point in time and space to attract much attention. Gains in the commercial sector could, when developed, come at the expense of serious losses, often uncounted and not directly noticed, in other, less commercial sectors.

Fencing could be used to separate the grazer from the guayule collector (Downing and Restrepo 1980), but that is costly. Given the scattered natural distribution of jojoba and guayule, widespread use of fencing is probably impractical. Fencing to separate grazing from guayule collecting has the added disadvantage of imposing a rigid boundary upon a landscape that is constantly in flux. The resulting rigidity could have dire consequences for both types of livelihood when drought requires more flexible responses to existing ecological conditions.

Appropriate Scale in Dryland Management

For each livelihood system there is an appropriate scale of operation. At such a scale the system works efficiently, requires relatively little management and energy from outside the system, and can be absorbed into traditional approaches to resource use without massive disruption. Below this ideal level, operation is inefficient and produce seldom matches maintenance costs. Above this level, problems of giantism emerge, control of the operation becomes increasingly difficult, and unforeseen managerial and ecological problems limit social and economic gains.

Defining the most appropriate scale of operation is not easy, for it varies from livelihood to livelihood and from one socioeconomic milieu to another. Industrialized economies tend to favor very large-scale, technology-dependent systems as the bedrock of their development efforts. Thus, the Soviet Union proposes to divert southwest Siberian rivers that empty into the Arctic in order to provide water for irrigation in the central Asian republics and to replenish the Aral Sea (Gerasimov and Gindin 1977). Although objections to this plan have been raised, in terms of both the potential impact on the ecology of the Siberian environment and the possibility of global climate change due to alteration in Arctic Ocean water, determination to press ahead appears to be growing (Gustafson 1980). Equally grandiose schemes for interbasin water transfers characterize the arid land planning of nonsocialist industrial states, although greater sensitivity to escalating costs may retard their implementation.

In a attempt to accelerate the pace of economic development, many Third World countries have implemented technology-based projects that operate at a massive scale. Technocrats and politicians are often in league in this process. Both envisage the future in terms of larger- and larger-scale systems of which they are master and whose benefits they can claim as the product of their own genius and acumen. For the political decision maker, both in donor and in recipient country, visible, large-scale projects are tangible symbols of active intervention and achievement. Lured by messianic visions of attaining the promised land now, mesmerized by dreams of deserts converted to gardens, certain that quantum leaps forward can be undertaken, grasping for tomorrow's future yesterday, the central decision maker's drive to employ massive applications of technology is almost irresistible. The solutions, technologically perfect, gargantuan in scale, requiring major efforts at social engineering, symbolic of human mastery over nature, seldom work as intended. Large-scale projects such as the Aswan Dam (van der Schalie 1972; Worthington 1977), the Noubariya Canal (FAO 1978), and New Valley (Mackelein 1976), in addition to producing positive benefits (and often outweighing them), often result in ecological disruption, economic shortfalls, and social frustration. Yet the lure of the large-scale technological alternative remains strong and partially explains why the mistakes of the past are perpetuated.

Insistence on an appropriate scale of dryland project design does not imply that small-scale projects are desirable in and of themselves, as interpreters of the "small is beautiful" school have maintained. Rather, an appropriate strategy may insist that more numerous intermediate-sized projects will achieve the desired results more quickly and with less adverse impact than will single, giant-scale projects. Brief examples of scale considerations in three types of dryland agriculture follow.

Much interest is focused on dryland irrigated agriculture, its problems and its potentials. The 1976 symposium sponsored by the Committee of Water Research (COWAR) of the International Council of Scientific Unions (ICSU) indicated that nearly as much irrigated land was being lost to production due to salinization and related factors as was being brought into production by new projects (Worthington 1977; White 1978). These problems appear to be endemic to irrigation, since the enhanced evaportranspiration found in irrigated agriculture (as compared to natural arid land riverine systems) creates fundamental problems for the maintenance of productivity (Pillsbury 1981). Although there are many cases of deterioration in irrigation systems, one of the most important variables is the scale of the project. Bos's (1977) review of ninety-one projects indicated that projects of less than 1,000 hectares failed to make significant contributions to productivity. Larger projects commanding up to 100,000 hectares or more were characterized by serious managerial problems. Usually such projects were forced to rely on expensive expatriate managers with limited ability to interact with middle-level management or to relate to the skills and objectives of project participants. Projects that could be broken down into units of 2,000–6,000 hectares and could be grouped into managerial blocks of approximately 10,000 hectares, each with its own decentralized staff, yielded better results. More intimately scaled management units, particularly when the agricultural technology used is relatively close to that traditionally employed, can be more readily related to and absorbed by cultivators.

Downing and Restrepo (1980), in reporting Murrieta's analysis of optimal-size jojoba-collecting units in the Sonoran Desert of northern Mexico, indicate that farms of between 5 and 10 hectares are the most economic size. Below that size, production costs rise, while economies of scale fail to operate on units greater than 10 hectares. The favorable conditions that support 5–10-hectare jojoba-collecting operations are those that rely on intensive hand gathering of the jojoba seeds on rough terrain. Wherever land is flatter, and mechanized agricultural technology can be applied to harvesting operations, hand labor loses its comparative advantage, and machines (supported by low-cost labor, land, and water) are the most efficient system. If criteria of cost and profit are the dominant considerations, then larger units able to employ mechanized equipment are inevitable. This points out the importance of the socioeconomic setting in which decisions about scale must be made, as well as the role played by underlying objectives. Where social impact is a minor concern, or can be accounted for in other ways, larger-scale, technological solutions appear sensible, however reprehensible they might be from a humanistic standpoint. But where the desire to create greater employment opportunities is the fundamental objective, small-scale systems, using more modest technology designed to reduce the drudgery of daily chores, are more likely to stimulate substantial benefits.

In pastoral systems, an appropriate system must operate at several levels simultaneously. Generally, the argument in favor of encouraging nomadic sedentarization is that only by concentrating population into large agglomerations can adequate health, education, veterinary, and other social services be provided (and taxes collected). The extensive resource-use system and low density of pastoral populations are viewed as inimical to modernization (Khogali 1979). Sedentarization often leads to ecological deterioration around settlements when excessive numbers of stock are concentrated nearby (Heady 1972) and means abandonment of the most adaptive feature of nomadic pastoralism—its mobility and flexibility.

This problem can be overcome by encouraging pastoral systems that operate in decentralized fashion at two different scales for separate sets of objectives. The first level would contain a number of smaller-scale units. These might most appropriately be scaled at the level of the clan, so that an institution of control larger than the individual decision maker, but still conceptualizable in traditional terms, would exist. Management units of 10,000–20,000 hectares at the clan level could easily become the main focus for grazing decisions, control of stock numbers, and management of watering points. However, the population numbers of such clans are usually too small, and their place of residence too impermanent, for them to become the primary conduit for the introduction of social services into the pastoral community. These services can most efficiently be delivered through a larger-scale center that services a number of clan units. By reaching an area of 80,000–100,000 hectares, such second-order centers would contact a sufficiently large population base to be economically viable. They would represent a management unit within which pasture reserves could be reallocated during times of drought. This would make it possible to provide reasonable access to spatially scattered rainfall and pasture without transgressing the established boundaries of cultural-ecological units.

Monitoring Resource-Base Dynamics

Resource-base degradation has characterized the last several decades of the human use of dry environments. The growth of populations, especially in urban areas (Potter and Potter 1978), and the mistaken belief that semiarid environments can be made to produce higher levels of output that will not only sustain local populations but will also contribute to attaining national development objectives (Amiram 1966; 1977) has encouraged severe local degradation. Habitually this affects any site where people and animals are concentrated (Barth 1977; Warren and Maizels 1977). The problem is compounded whenever sectoral development favors the aggrandizement of one livelihood component at the expense of others. The spectacle of "successful" projects juxtaposed with a failed development process is depressingly familiar. Equally culpable is a myopic concentration on introduced technology and grandiose-scale solutions to the exclusion of any concern with building upon local knowledge and expertise in modestly and appropriately scaled activities. When an unbalanced development process is affected by a major environmental fluctuation such as the Sahelian Drought, the result is spectacularly catastrophic.

The obvious inability of transitional, developing dryland societies to cope with environmental fluctuation and inadequately managed resources has produced an outpouring of international concern for and interest in the future of dry environments and of the people who live in them. This has, perhaps, been reinforced by the recognition among researchers that even the most sophisticated and industrialized countries face daunting problems in managing their drylands and possess at best a checkered record. The United Nations Conference on Desertification, held in Nairobi in September 1977, was organized to address the process of degradation in dry environments.

Sparked by the human and ecological impact of the Sahelian drought of 1968–73 (Caldwell 1975; Glantz 1976; United Nations Conference on Desertification 1977b), the conference produced thirty-two official documents summarizing scientific knowledge, as well as a host of contributions from nongovernmental organizations (for example, Reining 1978a, 1978b). Interest in the issues raised at the conference was widespread and led to the appearance of numerous studies (Glantz 1977; D. L. Johnson 1977; Morales 1977; Mabbutt 1978; Wright 1978; Mabbutt 1979; Institute for Development Anthropology 1980; Walls 1980). This outpouring of research identified the basic causes of dryland degradation but failed to achieve a coherent perspective on how to cope with desertification. Although UNCOD produced a plan of action, failure to fund and staff a coordinated global effort to combat desertification under the auspices of the United Nations Environment Programme (UNEP) has concentrated efforts at the national scale. Even here, new initiatives to combat desertification have been slow to emerge.

Concern for dryland degradation in the UNCOD process began with the belief that existing knowledge about drylands and their management was adequate, and that the basic issue was how to apply that knowledge to the solution of management problems. UNCOD proved a valuable forum for this conviction and demonstrated that existing understanding of how to manage drylands was inadequate to the task. Knowledge of the complexities of dryland ecosystems was incomplete and insight into social systems deficient, providing no clear rationale for a global assault on dryland problems.

In part this was a consequence of inadequate longitudinal data about drylands. Few physical systems had been studied systematically over a long enough period of time for researchers to feel confident that the full range of fluctuations in dryland ecosystems was incorporated into existing management systems and development planning. Similarly, students of human systems seldom followed a livelihood system's progress over a long period. Identification of cause-and-effect relations and creation of a program of proposed solutions are difficult under such circumstances. Only by long-term study can basic processes of human-environment interaction be identified and the direction of change noted.

One fruitful direction for understanding dryland environments is to proceed on a more local and intimate, subnational scale to build up a network of archtypical environment-livelihood systems, the development of which can be monitored consistently. The need to establish such human-environment baselines is critical (Baker 1979). If one of its fundamental principles were the preservation of basic systems of life support (White 1980), an observational network could provide vital data on the status of life and livelihood. Critical indicators of negative change could be monitored to provide early indications of environmental disruption. An environmental monitoring system can combine both remotely sensed data and information from observers on the scene. The latter can be drawn from a cadre of local people who habitually monitor environmental indicators as part of their normal livelihood activities. Once adequate baselines have been established, information generated from such monitoring sites can be employed effectively in planning development to counteract the disruption of dryland environments and to offer thereby a greater number of viable future land use alternatives to the residents of drylands.

References

Adams, M. E. and J. Howell. 1979. "Developing the Traditional Sector in Sudan." *Economic Development and Cultural Change* 27, no. 3 (April): 505–18.

el-Afrifi, S. A. 1979. "Some Aspects of Local Government and Environmental Management in the Sudan." In Mabbutt 1979, pp. 36–39.

Amiran, D. H. K. 1966. "Man in Arid Lands: I—Endemic Cultures." In *Arid Lands: A Geographic Appraisal,* ed. E. S. Hills, pp. 219–37. London and Paris.

———. 1977. "Arid Zone Development: A Case of Limited Choices." In *Arid Zone Development: Potentialities and Problems,* ed. Y. Mundlak and S. F. Singer, pp. 3–17. Cambridge. MA.

Baker, R. 1976. "The Administrative Trap." *The Ecologist* 6, no. 7 (June): 247–51.

———. 1979. *Trends in Research and in the Application of Science and Technology for Arid Zone Development.* MAB Technical Notes 10. Paris.

Barnett, A. 1977. *The Gezira Scheme: An Illusion of Development*. London.

Barth, H. K. 1977. "Types of Semi-Arid Ecosystem: Stress and Stress Potential." *Applied Sciences and Development* 10: 59–80.

Bernus, E. 1977. *Case Study on Desertification: The Eghazer and Azawak Region, Niger*. Nairobi.

Beshah, T. W. and J. W. Harbeson. 1978. "Afar Pastoralists in Transition and the Ethiopian Revolution." *Journal of African Studies* 5, no. 3 (fall): 249–67.

Bos, M. G. 1977. "Some Influences of Project Management on Irrigation Efficiencies." In *Arid Land Irrigation in Developing Countries: Environmental Problems and Effects*, ed. E. B. Worthington, pp. 351–60. Oxford.

Bowman, I. 1924. *Desert Trials of Atacama*. New York.

Caldwell, J. C. 1975. *The Sahelian Drought and Its Demographic Implication*. Overseas Liaison Committee Paper no. 8. Washington, D.C.

Chambers, R. 1979. "Administrators: A Neglected Factor in Pastoral Development in East Africa." *Journal of Administration Overseas* 18, no. 2 (April): 84–94.

Crosswhite, F. S. 1980. "The Annual Saguaro Harvest and Crop Cycle of the Papago, with Reference to Ecology and Symbolism." *Desert Plants* 2, no. 1 (spring): 3–61.

Daly, M. 1979. "Of Libyan Jirds and Fat Sand Rats." *Natural History* 88, no. 2 (February): 64–70.

Downing, T. and I. Restrepo. 1980. "New Technology and Dry Land Agriculture." *Culture and Agriculture*, no. 7 (winter): 1–7.

Draz, O. 1977. *Report to the Government of the Syrian Arab Republic on the Development of the National Range Management and Fodder Production Program*. Rome.

Dregne, H. E. 1977. "Development Strategies for Arid Land Use." In *Arid Zone Development: Potentialities and Problems*, ed. Y. Mundlak and S. F. Singer, pp. 255–61. Cambridge, MA.

Evenari, M. 1977. "Ancient Desert Agriculture and Civilization: Do They Point the Way to the Future?" In *Arid Zone Development: Potentialities and Problems*, ed. Y. Mundlak and S. F. Singer, pp. 83–97. Cambridge, MA.

Evenari, M., L. Shanan, and N. Tadmor. 1971. *The Negev: The Challenge of a Desert*. Cambridge. MA.

FAO. 1978. *Control of Waterlogging and Salinity in the Areas West of the Noubaria Canal, Egypt: Land Drainage*. Technical Report 4. Rome.

Fauchon, J. 1980. "Oasis Agriculture in Mauvetania." *Ceres* 13, no. 4 (July-August): 41–45.

Fishwick, R. W. 1970. "Sahel and Sudan Zone of Northern Nigeria, North Cameroons and the Sudan." In *Afforestation in Arid Zones,* ed. R. N. Kaul, pp. 59–85. The Hague.

Fontana, B. L. 1980. "Ethnobotany of the Saguarao, An Annotated Bibliography." *Desert Plants* 2, no. 1 (spring): 63–78.

Foster, K. E. 1980. "Environmental Effects of Harvesting the Wild Desert Shrub Jojoba." *Desert Plants* 2, no. 2 (summer): 81–86.

Gallais, J. 1975. "Traditions pastorales et développement: Problèmes actuels dans la région de Mopti (Mali)." In *Pastoralism in Tropical Africa,* ed. T. Manod, pp. 354–68. London.

George, C. J. 1972. "The Role of the Aswan High Dam in Changing the Fisheries of the Southeastern Mediterranean." In *The Careless Technology: Ecology and International Development,* ed. M. T. Farvar and J. P. Milton, pp. 159–78. Garden City, NY.

Gerasimov, I. P. and A. M. Gindin. 1977. "The Problems of Transferring Runoff from Northern and Siberian Rivers to the Arid Regions of the European USSR, Soviet Central Asia, and Kazakhstan." In *Environmental Effects of Complex River Development,* ed. G. F. White, pp. 59–70. Boulder, CO.

Glantz, M. H., ed. 1976. *The Politics of Natural Disaster: The Case of the Sahel Drought*. New York.

———. 1977. *Desertification*. Boulder, CO.

Gustafson, T. 1980. "Technology Assessment, Soviet Style." *Science* 208, no. 4450 (20 June): 1343–48.

Hare, F. K. 1977. "Climate and Desertification." In United Nations Conference on Desertification 1977b. pp. 63–167. Oxford.

Haynes, K. E. and D. Whittington. 1981. "International Management of the Nile—Stage Three." *Geographical Review* 71, no. 1 (January): 17–32.

Heady, H. F. 1972. "Ecological Consequences of Bedouin Settlement in Saudi Arabia." In *The Careless Technology: Ecology and International Development,* ed. M. T. Farvar and J. P. Milton, pp. 694–711. Garden City, NY.

Holling, C. S. 1973. "Resilience and Stability of Ecological Systems." *Annual Review of Ecology and Systematics* 4: 1–23.

Hoyle, S. 1977. "The Khashm el-Girba Agricultural Scheme: An Example of an Attempt to Settle Nomads." In *Landuse and Development,* African Environment Special Report 5, International African Institute, ed. P. O'Keefe and B. Wisner, pp. 116–31. London.

Hunting Technical Surveys. 1976. *Savanna Development Project Phase II: Development Plan*. Khartoum.

Institute for Development Anthropology. 1980. *The Workshop on Pastoralism and African Livestock Development*. AID Program Evaluation Report no. 4. Washington, D.C.

Johnson, D. L. 1979. "Management Strategies for Drylands: Available Options and Unanswered Questions." In *Proceedings of the Khartoum Workshop on Arid Lands Management,* ed. J. A. Mabbutt, pp. 26–35. University of Khartoum and the United Nations University, 22–26 October 1978. Tokyo.

Johnson, D. L., ed. 1977. "The Human Face of Desertification." *Economic Geography* 53, no. 4 (October): 317–432.

Johnson, J. D. and C. W. Hinman. 1980. "Oils and Rubber from Arid Land Plants." *Science* 208, no. 4443 (May 2): 460–64.

Kates, R. W. 1980. *Drought Impact in the Sahelian-Sudanic Zone of West Africa: A Comparative Analysis of 1910–15 and 1968–74.* Office of Evaluation Working Paper no. 32. Washington, DC. Reprinted as Background Paper no. 2, Worcester, MA.

Kaul, R. N., ed. 1970. *Afforestation in Arid Zones*. The Hague.

Khogali, M. M. 1979. "Nomads and Their Sedentarization in the Sudan." In *Proceedings of the Khartoum Workshop on Arid Lands Management,* ed. J. A. Mabbutt, pp. 55–59. University of Khartoum and the United Nations University, 22–26 October 1978. Tokyo.

Laney, R. L., R. H. Raymond, and C. C. Winikka. 1978. *Maps Showing Water-Level Declines, Land Subsidence, and Earth Fissures in South-Central Arizona.* United States Geological Survey, Water Resources Investigation 78–83. Tucson, AZ.

Mabbutt, J. A. 1978. "The Impact of Desertification as Revealed by Mapping." *Environmental Conservation* 5, no. 1: 45–56.

Mabbutt, J. A., ed. 1979. *Proceedings of the Khartoum Workshop on Arid Lands Management.* University of Khartoum and the United Nations University, 22–26 October 1978. Tokyo.

McGinnies, W. G. 1979. "Rubber Production in the Desert: Guayule Bounces Back." *Desert Plants* 1, no. 2 (November): 52–57.

——— 1981. "UNESCO's Arid Zone Program: Looking Back, Looking Forward." *Arid Lands Newsletter,* no. 14 (July): 9–13.

McKell, C. M. 1975. "Shrubs—A Neglected Resource of Arid Lands." *Science* 187, no. 4179 (March 7): 803–9.

Malhotra, S. P., P. Singh, and M. L. A. Sen. 1977. *Operational Research Projects*. Jodhpur.

Mandel, S. 1977. "The Overexploitation of Groundwater Resources in Dry Regions." In *Arid Zone Development: Potentialities and Problems,* ed. Y. Mundlak and S. F. Singer, pp. 31–41. Cambridge, MA.

Mann, H. S. 1981. "Afforestation at the Village Level." *Arid Lands Newsletter,* no. 13: 11–15.

Maugh, T. M. II. 1977. "Guayule and Jojoba: Agriculture in Semi-Arid Regions." *Science* 196, no. 4295 (June 10): 1189–90.

Meckelein, W. 1976. "Desertification Caused by Land Reclamation in Deserts: The Example of the New Valley, Egypt." *Problems in the Development and Conservation of Desert and Semidesert Lands,* pp. 151–53. 23d International Geographical Congress, Working Group on Desertification in and around Arid Lands, Pre-Congress Symposium K 26, 20–26 July 1976. Ashkhabad, U.S.S.R.

Morales, C., ed. 1977. *Saharan Dust: Mobilization, Transport, Deposition.* Review and Recommendation from a Workshop held in Gothenburg, Sweden, 25–28 April 1977. Stockholm.

Muthana, K. D. and K. A. Shankarnarayan. 1978. "The Scope of Silvi-Pastoral Management in Arid Regions." In *Arid Zone Research in India (1952–1977),* pp. 84–89. Jodhpur.

Nabham, G. P. 1979. "Southwestern Indian Sunflowers." *Desert Plants* 1, no. 1 (August): 23–26.

National Academy of Sciences. 1979. *Tropical Legumes: Resources for the Future.* Washington, DC.

National Research Council. 1968. *Water and Choice in the Colorado Basin: An Example of Alternatives in Water Management.* Publication 1689. National Academy of Sciences. Washington, DC.

Noy-Meir, I. 1973. "Desert Ecosystems: Environment and Producers." *Annual Review of Ecology and Systematics* 4: 25–51.

———. 1979. "Stability in Arid Ecosystems and the Effects of Man on It." *Proceedings,* pp. 220–26. International Congress of Ecology. The Hague.

Pareek, O. P. 1977. "Arid Horticulture." In *Desertification and Its Control,* pp. 256–62. New Delhi.

———. 1978. "Arid Horticulture." In *Arid Zone Research in India (1952–1977).* Jodhpur.

Picardi, A. C. 1976. "Practical and Ethical Issues of Development in Traditional Societies: Insights from a System Dynamics Study of Pastoral West Africa." *Simulation,* January, pp. 1–9.

Pillsbury, A. F. 1981. "The Salinity of Rivers." *Scientific American* 245, no. 1 (July): 55–65.

Potter, R. B. and V. Potter. 1978. "Urban Development in the World Dryland Regions: Inventory and Prospects." *Geoforum* 9: 349–79.

Reining, P., ed. 1978a. *Desertification: Papers Prepared Before and as a Sequel to the Science Associations' National Seminar on Desertification.* Washington, DC.

———. 1978b. *Handbook on Desertification Indicators: Based on the Science Associations' Nairobi Seminar on Desertification.* Washington, DC.

Sorbo, G. M. 1977. "Nomads on the Scheme: A Study of Irrigation, Agriculture and Pastoralism in Eastern Sudan." In *Landuse and*

Development, ed. P. O'Keefe and B. Wisner, pp. 135–50. African Environment Special Report 5. London.
Stebbing, E. P. 1937. *The Forests of West Africa and the Sahara: A Study of Modern Condition.* London and Edinburgh.
Swift, J. 1979. "The Development of Livestock Trading in Nomad Pastoral Economy: The Somali Case." In *Pastoral Production and Society,* pp. 447–65. Proceedings of the International Meeting on Nomadic Pastoralism, 1–3 December 1976. Paris.
Talbot, L. M. 1972. "Ecological Consequences of Rangeland Development in Masailand, East Africa." In *The Careless Technology: Ecology and International Development,* ed. M. T. Farvar and J. P. Milton, pp. 694–711. Garden City, NY.
Thimm, H.-U. 1979. *Development Projects in the Sudan: An Analysis of Their Reports with Implications for Research and Training in Arid Land Management.* Tokyo.
Townley, J. M. 1977. *Turn This Water into Gold: The Story of the Newlands Project.* Reno, NV.
United Nations Conference on Desertification (UNCOD). 1977a. *Sahel Green Belt Transnational Project.* Item 5 of the Provisional Agenda, Action Plan to Combat Desertification. Nairobi, Kenya.
———. 1977b. *Desertification: Its Causes and Consequences.* Oxford.
van der Schalie, H. 1972. "World Health Organization Project Egypt 10: A Case History of a Schistosomiasis Control Project." In *The Careless Technology: Ecology and International Development,* ed. M. T. Farvar and J. P. Milton, pp. 116–36. Garden City, NY.
Walls, J. 1980. *Land, Man, and Sand: Desertification and Its Solution.* New York.
Walsh, J. 1980. "What to Do When the Well Runs Dry." *Science* 210, no. 4471 (November 14): 754–56.
Warren, A. and J. Maizels. 1977. "Ecological Change and Desertification." In UNCOD, *Desertification: Its Causes and Consequences,* pp. 169–260. Oxford.
Waterbury, J. 1979. *Hydropolitics of the Nile Valley.* Syracuse, NY.
Welcomme, R. C. 1979. *Fisheries Ecology of Floodplain Rivers.* London.
White, G. F. 1955. "Symposium on the Future of Arid Lands." *Geographical Review* 45, no. 3: 434–35.
———. 1956. "International Cooperation in Arid Zone Research." *Science* 123, no. 3196 (March 30): 537–38.
———. 1960a. *Science and the Future of Arid Lands.* Paris.
———. 1960b. "Alternative Uses of Limited Water Supplies." *Impact of Science on Society* 10, no. 4: 243–63.
———. 1980. "Environment." *Science* 209, no. 4452 (July 4): 183–90.

White, G. F., ed. 1978. *Environmental Effects of Arid Land Irrigation in Developing Countries*. MAB Technical Notes 8. Paris.
Wilson, A. W. 1977. "Technology, Regional Interdependence, and Population Growth: Tucson, Arizona." *Economic Geography* 53, no. 4 (October): 383–92.
Worthington, E. B. 1977. *Arid Land Irrigation in Developing Countries: Environmental Problems and Effects*. Oxford.
Wright, R. A., ed. 1978. *The Reclamation of Disturbed Arid Lands*. Albuquerque, NM.

5 River Basin Development

J. C. Day, Enzo Fano, T. R. Lee, Frank Quinn, and W. R. Derrick Sewell

The development of river basins has been a major focus of public policy throughout the world, particularly in the past two to three decades. Water is a key input into certain economic activities, and its development is viewed as a potential catalyst to economic and social improvement. As a consequence, schemes for the provision of domestic water supply, water disposal, irrigation, hydroelectric power, or navigation rank high on the list of projects competing for capital in various countries and are often a major component in annual budgets. In addition, water development has been given high priority by many international agencies, notably the United Nations Development Programme and the World Bank, in their technical and financial assistance schemes.

Geographers in many countries have taken an increasing interest in water management, particularly in the past decade. Some have undertaken important research on various aspects of river basin development, and others have acted as planners or advisers to government agencies or international bodies. Gilbert White has made outstanding contributions in each of these capacities.

White's introduction to the water resources field came when he was a doctoral student at the University of Chicago working with Harlan Barrows. Believing geographers had much to contribute to public service, Barrows encouraged White to join the Mississippi Valley Committee in 1934. During the subsequent seven years White also worked with the National Resources Planning Board. Following this he was a staff member of the Bureau of the Budget in the Executive Office of the President of the United States. Through these experiences and others, White was convinced not only of the need for economic and social improvement in various U.S. regions but

J. C. Day is Director of the Natural Resources Management Program and Professor in the Department of Geography, Simon Fraser University, British Columbia. Enzo Fano is Deputy Director and Chief, Water Resources Branch, Division of Natural Resources and Energy, Department of Technical Co-operation for Development, United Nations. T. R. Lee is Economic Affairs Officer, Water Resources Unit, Division of Natural Resources and Energy, Economic Commission for Latin America and the Caribbean, Santiago, Chile. Frank Quinn is Head of Social Studies, Inland Waters Directorate, Environment Canada, Ottawa. W. R. Derrick Sewell is Professor of Geography at the University of Victoria, British Columbia.

also of the potential contribution that might be made through the harnessing of water resources. He also became increasingly aware of deficiencies in contemporary approaches to water management. And ever since, through the scholarly literature and advice to governments and international agencies, White has helped pioneer major innovations in water resources policies.

White's contribution in this field can best be understood in the context of the changing economic and social circumstances of the United States in the past half-century. The 1930s and 1940s were a period of intense physical and economic problems. Improved land and water management were viewed as an effective mechanism to promote human welfare. Then, in the 1950s and 1960s, important shifts in social values led to a questioning of the efficacy of engineered water management solutions in the absence of nonstructural measures and a call for more sophisticated evaluations. The focus broadened from economic parameters to social and biophysical concerns, and the areal scope of water planning moved successively from local to regional, national, and global concerns.

The philosophy of river basin development changed dramatically in response to these shifts. An integrated approach emerged, which aimed to deal with several problems simultaneously. In this process the conceptual underpinnings enlarged to embrace the world experience (Teclaff 1967). And emphasis shifted from the needs of engineers and other practitioners to the implications for teaching, research, and advice to international government agencies such as the United Nations.

White was not only a keen observer of these trends but also a major contributor to their practical application. This is particularly evident in three benchmark treatises that he contributed to the literature in the 1950s and 1960s.

Theory and Practice

The first was his seminal article, "A Perspective of River Basin Development," which appeared in 1957. In it he traced the evolution of the components of the modern concept of integrated river basin development: namely, multiple-purpose storage projects, basinwide programs, and comprehensive regional development. Under this theory, "each major network of streams draining the land masses of the earth may be viewed as the backbone for a possible planned use of a unified system of multiple-purpose and related projects to promote regional growth. This view . . . has come, during the past sixty years, to be employed rather widely as a technical tool for achieving social change. . . . Like any tool, it is not inerently good. Its value must be judged in terms of the growth and changes it can effect and upon its flexibility and precision" (White 1957, p. 157).

He also observed that while multiple-purpose storage is now common in many parts of the world, and although unified basin plans have been drawn up for drainages covering several states, integrated river management has never been applied in an international basin. Moreover, unified administration has inspired more controversy than imitation. Beyond this, social and biophysical effects of existing river development experiences as a means of fostering social change have been largely ignored and undocumented (ibid., pp. 183–84).

The second major contribution, *Integrated River Basin Development* (United Nations 1958), was written by a United Nations panel with White as chairman. It proposes a model for advancing human welfare, primarily for developing countries, based on a synthesis of theoretical and empirical experience from around the world. It argues that river basin management can contribute to improved social conditions in national and international drainages based on integrated planning, construction, and operation of water-related activities throughout river systems. The approach is based on comprehensive assessments of physical parameters throughout the basin: geology, hydrology, soils, topography, sedimentation, vegetation, fish, wildlife, and natural hazards. Parallel socioeconomic assessments embrace population, income, transportation facilities, power, water supply, pollution, public health, and employment based on agriculture, forestry, fishing, trapping, manufacturing, and mining. Working from such a data base, one can define alternative strategies to promote economic development in directions that society chooses. In international settings, joint decision making is recommended by all countries involved because of the absence of a universally accepted set of legal principles to guide international river basin management. Ultimately, such a body should become a permanent international commission established by treaty (United Nations 1958). This document had an important impact on the design of projects throughout the world, notably those sponsored by the United Nations and the World Bank group.

In 1970, White prepared a preface to the second edition of *Integrated River Basin Development* (United Nations 1970), which reviewed new developments in water management. He reported a worldwide interest in multiple- as opposed to single-purpose projects, in basinwide development compared to smaller-areal units, and in improved training and scientific research in water disciplines; refinement of analytical methods for appraising the social consequences of developments; and a variety of other technical and institutional advances. But White noted that the United Nations had not yet created a unit to conciliate in international basins where conflict could develop (United Nations 1970, pp. ix-xiii), as recommended in the first edition.

The culmination of more than thirty years of practical experience and intellectual growth is expressed in a third major work, *Strategies of American Water Management*, published in 1969. This volume brings together a wide range of concepts and practical applications based largely, but not exclusively, on the American scene. It contains several new ideas: the importance of expanding the range of choice, the application of economic principles, the need to evaluate social and environmental impacts as well as traditional technological and economic considerations, the advantages of linking water planning to other needs, and the use of research as a management tool (White 1969).

These works provide an indication of White's emerging philosophy on water management. He has made several important contributions in connection with four specific dimensions of this philosophy, namely, assessment of experience, geographical analyses, the development of man-made lakes, and the use of river basin development as an instrument for promoting international peace.

Hindsight Assessment

Although much could be learned from past experience in water management, few formal attempts are made to do so. As White observed in 1957, "there has been a conspicuous lack of careful appraisal of the work accomplished. A tool capturing imaginative support, as this one does, deserves penetrating assessment, and such examination has been largely absent. For every hundred studies of what might or should be done with a river system, there is hardly one that deals with the results" (1957, pp. 183–84). In the absence of hindsight assessment there is a risk that mistakes are perpetuated and that more fruitful approaches are ignored.

White and others have recommended that hindsight reviews be introduced as an integral part of river basin planning and policymaking. Broadly, they would consist of assessments of biophysical and socioeconomic impacts of development in regions influenced by river basin development actions. Under White's influence, a number of studies of experience were carried out under the auspices of a United Nations seminar in Budapest in 1975. More recently he organized an international symposium under the sponsorship of the International Geographical Union's Commission on Man and the Environment. The papers presented at this meeting reviewed selected aspects of research on environmental effects of a sample of large-scale river management programs in different parts of the world (White 1977). They not only emphasized the value of hindsight assessment but also provided valuable guidelines on methods.

GeographicalAnalysis

The study of river basin development is of interest to engineering, geology, biology, economics, sociology, and political science. White has consistently drawn attention to the need to involve all these disciplines. He has also emphasized, however, the value of geographical analysis, particularly in connection with changing spatial distribution and human adjustments to differences in physical environment. Specifically, the preparation of background data and summaries (sometimes as atlases), the range of choice among alternative river management methods, outlines of improved estimates of resources and their uses, suggestions for the improvement of benefit-cost analysis, and the development of techniques for assessment of biophysical and socioeconomic impact are among potential contributions (White 1963a, pp. 35–36). White notes that although geographers might undertake some of this work, they should not be relied upon to do it all. Science, technology, social science, and education are mutually interdependent, and research on each is essential in promoting fruitful management of the earth's rivers (1964a, p. 37).

Man-made Lakes

A ubiquitous component of river basin development is the creation of impoundments by the construction of dams. In the process, a complex web of ecosystem changes is induced in human, biological, hydrologic, atmospheric, and geologic environments. While some of the changes are beneficial, occasionally highly disruptive environmental and social effects result.

In an effort to synthesize major world experience, White served on a committee to organize an international man-made lakes symposium in 1971. He also acted as chairman of the Scientific Committee on Problems of the Environment (SCOPE) which reviewed the symposium papers and other materials. Topics considered included major lake case studies, reservoirs as physical systems, limnology and biological systems, and reservoirs, man, and multiple-use management. The resulting volume represents a synthesis of the known global range of ecosystem effects that have been experienced, or may be anticipated, from the creation of man-made lakes (Ackerman et al. 1973, pp. 3–40).

River Development and World Peace

The most persistent theme in White's writing on river basin development is the potential role of this tool to enhance the possibility of international peace. An important illustration is his work as a consultant to the Lower Mekong Coordinating Committee in Cambodia (now Kampuchea), Laos, Thailand, and Viet Nam in an effort to apply principles of integrated river basin management in a resource-rich, comparatively undeveloped region where few development options had been precluded by earlier decisions (White et al. 1962). Not only did he provide valuable technical advice, but he also wrote extensively to educate and promote understanding of the potential of coordinated international action under the auspices of the United Nations to advance human welfare and avoid conflict in the lower Mekong Basin. Such an approach would be based on plans of the four nations involved and technical, scientific, and financial help from a broad spectrum of the world's nations (White 1963a, 1963b, 1964a, 1964b, 1967).

The foregoing review has outlined White's contributions to the development of principles for a comprehensive approach to river basin development. The extent to which these have been adopted in several parts of the world is reviewed in the following sections. Day notes the evolving situation in the United States and the Soviet Union, Quinn discusses the Canadian experience, Sewell describes western European conditions, Lee documents Latin American developments, and Fano deals with African and Asian circumstances.

The United States

The notion of a basinwide approach to water management appears to have evolved in the United States from observations by Powell and others in the late nineteenth century that were popularized by President Theodore Roosevelt early in the twentieth century. Initially the idea was applied to the small Miami Basin of Ohio in 1914, but interest in basinwide management subsequently turned to larger systems including the Columbia, Missouri, the upper Ohio, and the Tennessee in the late 1920s and later the Central Valley in California (White 1957, pp. 168–71). By 1960 it was widely accepted in the United States that basinwide development and management plans would be prepared for all major U.S. rivers and that these would be updated periodically as underpinnings for water resources decisions (Fox 1976, p. 2). A continuing fundamental role for comprehensive river basin planning in future U.S. water management actions was recommended in 1973 by the National Water Commission. It recommended that such plans be the basis for authorization and appropriation of funds for individual projects and programs within regions, that priorities for all river basin planning funds should be established by the National Water Council, and that states should consider the establishment of river basin authorities to plan and manage state and interstate river basins (United States, National Water Commission 1973, pp. 141, 145, 150).

Institutional Arrangements

Since the first application of basinwide planning there has been continual experimentation to design institutional arrangements and planning practices to promote sensitive water management. Federal agencies dominate water resources development with state governments generally playing a comparatively minor role. Often federal control is exercised through licensing and other regulatory actions (Fox 1976, p. 6). In a number of instances, however, its influence has extended far beyond this, notably through the construction and operation of facilities.

Rather than adopting a unified agency approach, the federal government created a number of water management agencies: notably the U.S. Army Corps of Engineers, the Department of the Interior Bureau of Reclamation, the Department of Agriculture Soil Conservation Service, the Tennessee Valley Authority, and the Environmental Protection Agency. A variety of bodies has been created in an effort to coordinate federal, state, and local interests. The National Resources Committee and the National Resources Planning Board were examples from the 1930s and early 1940s. These were replaced in 1943 by a series of committees that successfully resisted the need for formal integration of federal actions until the mid-1960s (White 1969, pp. 68–72).

More recently, the National Water Commission was established in 1968 under the 1965 Water Resources Planning Act. The two major goals of the act were to develop broad regional water and related land resource plans for the major U.S. river basins and to coordinate federal policies and programs and assess their adequacy to meet national water requirements (United States, National Water Commission, 1973, pp. 398–99). Seven river basin commissions were established under the Water Resources Planning Act for water planning: New England, Great Lakes, Ohio, Upper Mississippi, Souris-Red-Rainy, Missouri, and Pacific Northwest. In conjunction with federal-state compact commissions in the Delaware and Susquehanna basins and interagency committees in the Southeast Basins, Arkansas-White-Red Basins, and Pacific Southwest, these agencies embraced more than 75 percent of the area of the coterminous forty-eight states (ibid., p. 419).

In a 1973 evaluation of its work and experiences, the commission made recommendations concerning the comparative advantages of alternative approaches to river basin planning. Federal-state compacts were identified as the preferred institutional arrangement for water resource planning and management in multistate areas. River basin commissions were considered superior to ad hoc and interagency committees for planning in other circumstances (ibid., p. 418).

On other matters the commission observed that while some of its initiatives had been useful, the agency was ineffective in resolving conflicts among federal agencies. To remedy this problem it thought an independent full-time chairman on the White House staff was needed to report directly to the president. This would give the council a policy-making component and the ability to enforce decisions when consensus could not be reached (ibid., p. 403). The commission also urged the adoption of economic principles to improve the efficiency of water resources use based on the principle that the beneficiary pays. User charges to recover a major portion of water-based service costs were recommended as the primary vehicle to ensure efficient water use in municipal, industrial, and irrigation water supply; inland navigation; electric power; water-based recreation; municipal waste collection and treatment; flood control; drainage; shoreline, fish, and wildlife protection (ibid., pp. 169–75).

Recent U.S. Experience in River Basin Planning

The formulation of comprehensive plans for major basins based on detailed plans is being viewed increasingly as an expensive luxury. The original notion of a detailed comprehensive plan for each major river, updated periodically, now appears too costly. Instead, reconnaissance investigations for major tributaries using existing data as a basis for detailed studies appear more useful. These would indicate development potential, projected water demand, and the range of possible projects worthy of further study. Evaluative data are being presented in terms of national economic development, environmental quality, regional development, and criteria of and implications for social well-being (Fox 1976, pp. 10–13). Wengert (1980, p. 9) observed that, in the public mind, integrated river basin development "is being superseded by concerns for environmental improvement, protection of ecosystems, and the complex problems of water pollution control. At the professional and political level, the concept is being fragmented as more and more research and development effort is being put into specific issues of water control and management with increasing attention once again focused on site specific problems and solutions and analytic specialization." This observation was supported by a General Accounting Office (GAO) evaluation (United States, General Accounting Office 1981, pp. i–iv) which asserted that although the concept of river basin planning was sound, amendments to the Water Resources Planning Act were required to ensure that river basin commissions have sufficient support from, and cooperation with, relevant state and federal agencies to ensure that coordinated, comprehensive joint planning will occur.

Instead of adopting the GAO recommendation for restructuring, seven river basin commissions totally financed from federal funds were terminated by June 1982 as part of the Reagan administration's budgetary measures. Although disbanded federally, most of these bodies were either reconstituted with state support or their activities were transferred to preexisting agencies to continue regional cooperation on critical water and related land management issues. For example, the Great Lakes Commission took over the assets and management responsibilities of the Great Lakes Basin Commission. In summary, a major policy initiatives are underway in the United

States to shift the responsibility for river basin management from the federal level to state governments.

Canada

Two realities, one geographical and the other jurisdictional, have shaped the character of river basin management in Canada.

Canada has roughly 9 percent of the world's renewable surface runoff but only 1 percent of the world's population. Not surprisingly, this relative abundance of water has influenced Canadian attitudes toward its use and management. Since early times, water was less to be conserved than to be exploited for economic development. If immediate supplies became inadequate, industries and communities tapped the next convenient source or found another body of water to accept their wastes. The presumption that most new demands for water could be satisfied by exploiting new sources left Canada with the legacy that water management is largely synonymous with engineering measures of capture and delivery. This "supply-side" approach is evident in the many examples of major single-purpose projects that continue to overshadow more recent initiatives in basinwide and comprehensive regional water management. Of 514 large dams registered nationally in 1976, only 42 (8 percent) served multiple purposes; 348 (68 percent) were hydroelectric projects (International Commission on Large Dams 1977, p. 14). Numerous interbasin transfers have reinforced the dominance of hydroelectric development over other considerations north of the settlement fringe.

Jurisdictional division has also affected the course of river basin management. The primacy of Canada's ten provinces in resource matters goes well beyond that accorded subnational units in most other countries. Hence the variety of institutional experiences with water management from one province to another, the tendency to develop internal resources first, and the difficulties of developing rules of law or cooperative programs for interprovincial and provincial-territorial basins. Treaty arrangements have facilitated Canadian-U.S. cooperation successfully over several decades on international boundary water issues, but differences of national situation and objectives operate effectively against any tendency toward integrated basin management (Bruce and Quinn 1979, pp. 7–12).

The conservation authorities of southern Ontario, established from 1946 onward and now thirty-nine in number, constituted the first deliberate move by a province to effect soil and water conservation, flood control, wildlife protection, and recreation on a small watershed basis with a considerable degree of municipal participation. This pattern has not been duplicated, however, in the larger and less populated basins elsewhere in the country.

During the 1960s it became apparent that demands upon water resources were escalating, creating conflicts among uses and jurisdictions, and pollution was threatening both human health and the environment in general. Out of the growing debate on these issues, the federal minister responsible for coordinating water programs extended an offer to his provincial counterparts to finance a pilot exercise in river basin planning in each of the major regions of Canada. These exercises were to be undertaken jointly in an effort to overcome by federal-provincial cooperation the jurisdictional limitations of both levels of government; they were to be comprehensive in scope to give greater attention to social needs and environmental impact and to involve actively local residents; and they were to be basinwide in scale better to relate causes and effects of waterflow and quality modifications.

The first-generation basins selected in 1969 for this new approach were the Saint John in New Brunswick, the Qu'Appelle in Saskatchewan and Manitoba, and the Okanagan in British Columbia; Ontario and Quebec turned down the federal offer to finance a comprehensive study of the Ottawa Basin. The principles underlying the new approach were embodied in the Canada Water Act enacted in 1970.

A number of benefits have resulted from these exercises, owing to the shared nature of tasks and the consensus-forming nature of joint planning boards and committees. The provinces have benefited in the development of useful skills to apply to other regions of their respective jurisdictions or to subsequent problems, and the programs have encouraged them to change their laws and institutions. It is apparent that the programs have been a vehicle of change and of new perspectives in water management and will have a lasting impact on agencies and basin residents. Savings in capital costs were facilitated at times through the more comprehensive evaluations of solutions (Canada, Environment Canada 1983, pp. 58–60).

Not all results were positive, however. Detailed comprehensive investigations of relatively small basins such as the Okanagan and the Qu'Appelle have been costly in terms of both time and resources and were often insufficiently responsive to issues requiring more immediate attention. The zeal for planning often masked the necessity of providing a politically and institutionally sensitive, as well as explicit, strategy to implement the recommendations of the study. For these reasons and because of the unfocused collection of data, the inflexibility of comprehensive studies, and long delays before followup action was taken, interest in the comprehensive planning exercise began to wane. The focus of activities has shifted toward resolution of site-specific and immediate issues such as pollution, flooding, and project impact, which need not be addressed by full multiple-use and basinwide investigations, and, at the other end of the scale, toward larger regional framework plans which establish principles and priorities upon which subsequent development can be based. Examples of site- or river reach-specific planning agreements between the federal and provincial governments include water quality investigations on the St. Lawrence River and assessment of the impact of proposed hydroelectric developments on Lake Winnipeg and the Churchill and Nelson rivers. Examples of the framework variation include the Mackenzie and Yukon investigations in the sparsely settled north (table 5.1).

The idea of dividing a country into major basin regions and establishing permanent institutions and plans for each has never been considered seriously in Canada. Aside from jurisdictional concerns, governments do not have sufficient financial or technical resources to undertake more than a limited number of water exercises at one time. In interjurisdictional situations, these are negotiated sequentially according to priority.

Important changes in emphasis can be anticipated in future Canadian water management, out of strategic reviews currently underway and the results of recent planning activities. Two cases are cited here as evidence: allocation of water among users and apportionment among jurisdictions.

As the cost of major storage projects has risen, along with growing public opposition to the social and ecological disruption such structures often impose, governments have begun to explore other means of resolving use conflicts. The new emphasis is not on making more water available to satisfy growing demands, but on regulating the growth in demand through effecting greater efficiency in use. Opportunities abound in a country whose per capita water use rates are among the highest in the world. Waste is evident almost everywhere in the form of leaking municipal waterworks, excessive irrigation flooding, once-through industrial processes, and polluted watercourses. Research has shown that economic pricing, recycling, and other conservation methods can save impressive quantities of water, sometimes at a fraction of the cost of new supplies and with

Table 5.1 Federal-provincial river/river basin management activities, 1970–82

Preplanning	Planning	Implementation
Comprehensive	Okanagan Basin* (1969–74)	Okanagan Basin* (1976–82)
Souris Basin* (1972–73)	Qu'Appelle Basin (1970–73)	Qu'Appelle Basin (1975–84)
Lower Saskatchewan Basin (1974–79)	Saint John Basin* (1970–74)	Souris Basin (1979–)
Thompson Basin (1979–80)	Souris Basin* (1974–77)	
	Shubenacadie Basin (1975–79)	
Water quality	Lake Winnipeg (1977 deferred)	
	St. Lawrence River* (1973–77)	
	Ottawa River (1976–81)	
	English-Wabigoon River (1978–81)	
Environmental impact	Lake Winnipeg, Churchill-Nelson River (1971–75)	Northern (Manitoba) Flood (1977–)
	Peace-Athabasca Delta (1971–72)	
		Peace-Athabasca Delta (1974–75)
	James Bay Rivers (1971–77)	
	Churchill River (1973–75)	
Framework		
Mackenzie Basin (1976–78)	Mackenzie Basin (1978–81)	
Yukon Basin* (1978–79)	Yukon Basin* (1980–83)	
Other		
Winter River (1977. . .)	Ottawa Basin (1977–80)	
	Fraser Estuary (1979–81)	
	Waterford Basin (1980–85)	

* Excludes United States portion of basin. This table also excludes Canadian-U.S. water investigations undertaken through the International Joint Commission.

the advantage of avoiding environmental hazards such as the inundation of prime valley lands or biota transfers. Thus Alberta's management principles now include a requirement that the waters in each major basin be fully and efficiently used before interbasin augmentation could be considered. And the national Flood Damage Reduction Program, now in effect in seven provinces, is giving priority to identification of flood-risk areas and to discouragement of new development in these areas, with recourse to control structures only where essential to protect existing developments.

Related to the matter of efficiency of various water uses is that of apportionment or sharing of rights among neighboring jurisdictions to a water resource they hold in common. Apportionment agreements have been negotiated for a few Canadian-U.S. boundary and transboundary rivers and interprovincially for the east-flowing rivers of the Prairie Provinces. In each case, only water quantity has been apportioned. To minimize the likelihood of future conflicts as demands grow, there is interest in further apportionment agreements. Considerations have now expanded, however, beyond simple division of flow at boundary crossings to the seasonal distribution of that flow—so important for sensitive northern environments—and also to water quality. The Mackenzie River Basin Committee has recommended that the governments of Canada, Alberta, British Columbia, Saskatchewan, and the two territories reach without delay a permanent agreement including all these dimensions of apportionment. Such an accomplishment would be of historic proportions, incorporating as it would limitations on upstream provincial development options, especially hydroelectric developments, on the Liard, Peace, Athabasca, and Slave tributaries.

In summary, major single-purpose projects, especially for hydroelectricity, are still the most impressive symbols of Canadian water management. But increasing conflicts in more developed basins in the southern part of the country have led to a search for alternative means that can improve efficiency and equitable distribution of the water resource among users and jurisdictions. Integration of water concerns into larger regional development plans has barely been explored and in the case of international basins remains unlikely.

The European Experience

In Europe, as in North America, the approach to water management has broadened considerably in perspective in the past two decades. Stimulated by water shortages, declining water quality, conflicts in use, and the mounting costs of water development, there have been major changes in legislation, policies, and administrative structures. In several countries there has been a move from a single-purpose, single-means, locally oriented philosophy to one embracing several purposes, several means, and a regional and national perspective. In most instances the changes have been slow and incremental. In a few, however, wholesale reorganization and the introduction of bold new policies have taken place (Johnson and Brown 1976; Organization for Economic Co-operation and Development [OECD] 1976).

In several European countries there have been moves toward a more comprehensive approach. In certain instances this has been reflected in the establishment of agencies responsible for a wide variety of water management purposes, such as water supply, water quality control, fisheries management, or water-based recreation. In others it has been demonstrated in the preparation of long-term, basin-wide plans, or in the adoption of economic principles in the allocation of water among competing uses or in the prices charged for water-related services (National Water Council 1976). Adoption of the principle that the polluter pays is an important illustration (OECD 1976; McIntosh and Wilcox 1978). In yet other instances it has taken the form of a progressive widening of the range of alternatives considered in the selection of water management strategies. In two countries in particular, the United Kingdom and France, changes have been especially profound.

The United Kingdom

The United Kingdom has long prided itself as being a leader in water management. Not only does almost all of its population have access to high-quality potable water and effective sewage disposal, but also it has been a training ground for water engineers who have gone to many parts of the world to practice their craft. There were few concerns in the United Kingdom about water management until the years following the Second World War. Suddenly, economic expansion, population growth, and increasing affluence led to rapid growth of water demands and major quality deterioration in many of the nation's water bodies. The problems were exacerbated by a major shift of population toward drier parts of the country.

The central government responded by introducing new legislation and policies that led to a total reorganization of water management (Okun 1977; Porter 1978; Parker and Penning Rowsell 1980). The 1963 Water Resources Act established twenty-nine river authorities which were to be responsible for overall planning of water supply within their respective regions. The river basin was to be used as the basic unit for management. Periodic surveys, or long-term plans, were to be drawn up and revised from time to time, and charges were to be levied for water withdrawals, based on the volume, timing, and condition of water returned to the stream. A water resources board was established to provide a national perspective, furnish coordination, draw up a basic framework for long-term planning, and carry out research. While the new structure led to some important improvements, there remained major deficiencies (Rees 1976). Although there were moves toward rationalization in water supply, little progress was made in dealing with sewage disposal, and the condition of many major rivers seemed to be deteriorating rapidly. Plans for water supply continued to be prepared in isolation from water quality improvement programs. There was little attempt to base water charges on economic principles.

A decade after introduction of the new legislation, a 1973 Water Act was passed, aiming to deal with these problems (Sewell and Barr 1977; 1978). It called for establishment of ten regional water authorities, based on amalgamation of the existing river authorities. These were to have responsibility for all phases of water management except navigation and aspects of water-based recreation other than angling. They were to be "multifunctional bodies" which, with the exception of a few private water companies, would take over the role of all the existing water supply agencies, and all sewerage and sewage disposal agencies that hitherto had been lodged in local government. The new bodies were to integrate the two functions and institute direct charges to customers for services provided. Such charges were to be based on economic principles. The regional water authorities were to undertake long-range planning. A National Water Council was established, composed of the chairmen of the ten regional water authorities, members appointed by the government to represent various water management interests, and a chairman appointed by the government. Its major roles were to be the provision of information, a forum for the exchange of ideas, and a mechanism to improve training in the water industry.

For some observers, passage of the 1973 Water Act represented a bold and imaginative attack on existing problems (Okun 1977). In particular, the introduction of a corporate management approach and the integration of water management functions on a regional basis were lauded. There has been a major move toward implementation of a comprehensive perspective, but several difficulties remain (Rees 1976; National Water Council 1978, 1982; Parker and Penning Roswell 1980). The use of economic principles has been only partially accepted. For example, prices of water services are still based largely on average rather than marginal costs and on land values rather than values derived from use (Rees 1976; Herrington 1982). Moreover, there appears to be growing pressure for the introduction of uniform water prices not only within regions but also across them (National Water Council 1976). At the same time, the notion of effluent discharge fees has not been fully accepted in the United Kingdom. While some industries make payments to regional water authorities for the use of public sewers, there is not a comprehensive charging policy for waste disposal. Reliance continues to be placed mainly on permits and standards negotiated by regulatory agencies and polluters. Although the National Water Council and its various subcommittees provide an excellent forum for discussion of common problems, a mechanism to put forward a national water management perspective is absent. There is dissatisfaction in several of the regions about the extent to which the views of users of water services are consulted. There are also emerging anxieties about the ability of many authorities to replace aging pipes, treatment plants, and other facilities (National Water Council 1982). Some of the deficiencies have been recognized in reports of the National Water Council and in government white papers, and it is probable that further modification in the institutional framework will be made soon.

France

France faced water management problems in the postwar period similar to those prevailing in the United Kingdom. There was growing water deficiency in some regions and major deterioration of water quality throughout the country. The French government responded in 1964 with the introduction of a water law which called for the establishment of six river basin agencies embracing all of the country (LeFroy and Nicholazo-Grath 1979). Their purpose was to encourage and facilitate a broader approach to water management, principally through the formulation of basinwide plans and the introduction of charging schemes. The agencies are an interesting innovation in a country characterized by highly centralized government. Each has an executive board composed of a director, eight civil servants, and eight individuals elected by the basin committee. The latter is a controlling and advisory body, consisting of forty to sixty members, depending on the size of the region. One- third are drawn from central government agencies, one-third from water users, and one-third are elected officials: senators, deputies, mayors, and municipal council members. The committee acts, in a sense, like a water parliament. Its main functions are to oversee the planning of the basin's water resources and establish charges for water withdrawals or effluent disposal.

There is also a delegated basin mission, composed of twelve civil servants from water management and other agencies, and a secretary, who is director of the basin agency. Its function is to provide a link between the basin agency and the central government's Interministerial Mission on Water in connection with the formulation of long-range water management plans.

Fees collected by the agency are used to provide incentives, notably grants or subsidies, to industries and municipalities to construct water-supply or water-treatment facilities. The agencies pay only part of the cost. The remainder is provided in some measure by the central government, and partially by the polluters themselves.

The French approach has aroused considerable interest. It clearly recognized the value of a regional focus for water management as well as the merits of a system that combines standards and incentives. The 1977 Classified Establishment Act sets out a list of establishments which require licences, or permits, to discharge effluents. Subsequent decrees indicate various civil sanctions that may be imposed for noncompliance, including payment of fines, imprisonment, or the closing of an establishment. Policies instituted under the 1964 Water Law enable continued use of waters of a given river, but adopt the principle that the polluter shall pay for its use.

It is difficult to determine how successful the French effluent discharge fee has been (Harrison and Sewell 1980; Bower et al. 1981). Ideally, the fees should have been set at a level to force the polluter to consider whether to pay them or institute measures that would reduce the effluent load sufficiently to avoid such payments. The French authorities, however, believed it would be better to introduce low fee levels initially and gradually increase them once the idea was firmly established. At the same time, various industries brought pressure to bear to postpone payment or to keep it at low levels, arguing that the fees would severely impair their ability to survive. Export industries, notably pulp and paper, were especially vociferous. A partial compromise has been sought through a system of sectorial contracts or voluntary agreements in which major polluters agree to meet specified standards by a target date. In the interim, they are relieved of some part of the burden of effluent discharge fees but at the same time become eligible for grants and subsidies for pollution control equipment from the central government and the basin agency (Harrison and Sewell 1980).

Clearly, the 1964 Water Law was bold in intent and designed to attack water supply and waste disposal problems vigorously. The administrative machinery and the charging schemes are now firmly in place. There is an attempt to consult public views, notably through the circulation of drafts of proposed river basin plans, known as *Livres blancs*. But the French approach, like the British, falls short of being comprehensive. River basin agencies deal only with a limited range of purposes and functions. There remain considerable duplication of effort and confusion as to responsibilities. There have been reviews of the state of France's rivers, but so far only a few water-quality parameters have been measured. As a consequence, it is difficult to determine how far the discharge fee system has been effective.

Summary

The British and French experiences provide a useful measure of the extent of institutional innovation in water management in Europe over the past two or three decades. In terms of a broadened perspective and an integration of functions within a river basin context, England and Wales have probably advanced as far as any countries in the world. There and in France there has been a move toward the consultation of a broader range of choice in decision making with such measures as economic incentives and demand management being considered side by side with the traditional construction and regulation alternatives. In France, and to a lesser extent in England, an attempt has been made to treat water and water services like economic goods. Water planning in both instances has become increasingly sophisticated and increasingly comparable to overall economic and social planning.

These developments may be contrasted with the experience of several other European countries. Fragmented responsibilities are the rule in many. More than 25,000 local authorities and water associations deal with water matters in Germany. In the Netherlands there are over 600 water boards. River basin planning is weakly developed in the Scandinavian countries, and in many others focuses solely on the water resource.

Perhaps the greatest challenge is that of dealing with international rivers (Le Marquand 1977). Various attempts have been made to deal with divided jurisdictions, notably on the Rhine and the Danube, where commissions have been established to foster cooperation in navigation. Although there have been efforts to stimulate coordination in planning and management of such rivers, there remain important difficulties in dealing with the effects of upstream uses on downstream ones. Only a few of the principles set out in the United Nations report on Integrated River Basin Development (United Nations, Department of Economic and Social Affairs 1970) have been put rigorously into practice in the management of European international rivers. While progress has been made in the collection of data and planning of some jointly used facilities, a move toward fully comprehensive river basin development and management remains a distant prospect.

The Soviet Union

Before the Second World War, the construction of large hydroelectric power stations, ship canals forming a network of waterways between the Baltic, White, Caspian, and Azov seas, and large-scale irrigation and drainage works was undertaken in Soviet river basins. After the war, even larger multiple-purpose projects were completed on the Volga, Dneiper, Don, and other rivers until nearly all basins in the country were affected (Dunin-Barkovsky 1971, pp. 38–40).

As a consequence of rapidly increasing demands on water resources in the later 1940s and the 1950s, conflicts were commonly experienced among competing users. Consequently the Soviet Union established special bodies to improve water management. In 1965, the Ministry for Land Reclamation and Water Economy together with related ministries and committees in the union republics were empowered to supervise water use and conservation. The ministry's associated bodies include the Central Board for Integrated Water Resource Utilization, the State Inspectorate for Water Source Conservation, and a series of inspectorates for the main national river basins. All large-scale operations are formulated by the State Planning Commission of the Soviet Union, which examines long-term plans for national economic development, schemes for the integrated use of river basin resources, and the effects of major water projects (ibid., pp. 38–44).

In an effort to rationalize water management, a comprehensive national master plan has been developed to determine the main development trends in different water uses over the next thirty to forty years with emphasis on the first fifteen to twenty years.

The plan focuses on the nation's river basins and public administration needs. Principal topics covered in the master plan include the economic rate of return on investments, surface and groundwater resources, human impact on physical systems, current and projected water supply and disposal arrangements, irrigation, drainage, hydroelectricity, water transportation and floatage, fish farming, water quality, and preliminary management plans by basins. Although neither complete nor exhaustive, the master planning approach provides a valuable first approximation to the balancing of national water resources and needs, which can be modified to reflect experience, knowledge, and changing development priorities (Gangardt 1976, pp. 253–58).

Special attention is being devoted to the potential of interbasin water transfers to improve the national water resource balance. In particular, there is concern for the effects on the natural ecology of donor areas, receiving areas, regional climates, and social changes (Gerasimov and Grindin 1977, pp. 59–70; Vendrov and Avakyan 1977, pp. 34–37).

The Volga River experience exemplifies the sophisticated Soviet approach to river basin development. At all stages of the nation's development the Volga River Basin has been a locus of settlement, economic development, and culture. Containing the largest of the European rivers, the basin embraces forest, forest-steppe, semidesert, and desert biomes. Currently it is managed for multiple purposes by a variety of means.

Hydroelectric development was initiated during the late 1930s and intensified in the 1950s and 1960s. A cascade of nine hydroelectric stations has been built on the Volga and its northern tributary, the Kama; two more are under construction. Their total capacity is 12 million kilowatts. But as a consequence, the migration of certain fish species has been altered drastically. To compensate for losses, fish hatcheries and spawning farms have been constructed to maintain the Volga-Caspian biological resources. They account for 50 percent of the inland Soviet fishery and 90 percent of the sturgeon catch.

The Volga is also an integral component of the deep canal system that carries 75 percent of all domestic waterway traffic. It supplies large industrial water users and more than 0.5 million hectares of irrigated land. Water is also transferred for use in the Ural River to the east and the Don River and Azov Seas to the west. To compensate for these diversions water is transferrd from northern rivers to the Azov Sea.

Because of heavy industrial use, water pollution was increasing in the early 1970s. Following a series of measures taken since 1972, degradation has been arrested by waste treatment improvements, closed water supply systems, better choice of discharge sites, and burial of waste residues. Constant surveillance has been necessary to accommodate these multiple uses together with an intensive recreation industry as consumptive and nonconsumptive demands increase, changes in the hydrological systems multiply, and the potential for ecological change grows (Vendrov and Avakyan 1977, pp. 23–38).

In summary, since 1965 the Soviet Union has been experimenting with an approach to river basin management that relates national economic development planning to the integrated use of river basin resources and the effects of water projects. Special agencies have been established to enhance integrated water use, to promote conservation, and to manage the major river basins. A first version of a national water plan was created in an effort to predict future development opportunities and constraints. This innovative series of initiatives deserves careful analysis and evaluation to determine the lessons from the Soviet approach to resource management and conflict resolution in a large and complex environment.

Latin America

In recent years, water management techniques have been modernized in many Latin American countries. However, nowhere has a system developed that could be described as integrated river basin management. Great strides have been made in improving data collection concerning all aspects of the water environment. The modeling of flow regimes is becoming routine, and increasing concern is being shown, and action taken on, the ecological consequences of water management decisions. More tentatively, water demand is being seen as subject to management. Despite these and other management technology innovations, water management with certain exceptionsremains fundamentally use-directed and not integrated.

For example, less than 20 percent of the reservoir capacity behind large dams constructed or under construction in the 1970s could be classified as multipurpose (United Nations, Economic Commission for Latin America 1979). Similarly, basinwide water management programs even when undertaken, as in Brazil, Colombia, and Mexico, have a tendency to break down into a series of individual projects, often directed to a single use. Comprehensive regional development, although agreed upon as a laudable objective, has still to be applied on more than an experimental scale. The conclusions of a recent seminar on the environmental management of large water-resource projects called for the integration of such schemes into comprehensive regional development plans. Significantly, perhaps, the majority of the projects represented at the seminar were in practice multipurpose, although many were originally convened as single-purpose projects managed by sectorial agencies or specific agencies formed for the project independent of regional development plans and strategies (United Nations, Economic Commission for Latin America 1981).

Reasons for the persistence of use-directed or single-purpose management are manifold and vary from country to country. There do exist, however, some underlying general conditions: highly centralized decision-making, the physical configuration of the region, the relatively simple water-demand patterns prevalent until recently, and the general absence of regional integration that might lead to the development of international basin management agencies.

The Background of Water Management in Latin America

It is undoubtedly the high degree of centralization of decision-making authority that has played the heaviest role in obstructing the emergence of integrated management. Centralization militates against the successful establishment of regional institutions of any kind. The absence of decentralized government has aided the maintenance of strong, single-purpose management institutions responsible for irrigation, hydroelectricity, and water supply (United Nations, Economic Commission for Latin America 1979).

Mexico, Central America, and South America are divided into three major drainage areas: to the Pacific, to the internal drainage areas of the high plateaus, and to the Atlantic. The Pacific watershed is characterized by short, fast-flowing streams, the high plateaus by limited surface flow, and the Atlantic by large, complex river systems providing a superabundance of supply. This situation impedes the emergence of integrated management. Short rivers are normally dedicated to a single use, irregular and transient flows are not conducive to the elaboration of management activities, and excessive supply also weighs against integration in management.

Traditionally, individual Latin American countries have been outward-looking toward Europe and North America, and, despite the rhetoric, regional cooperation has been notable mainly by its absence. In South America, occupational of the continental interior, particularly in the Amazon Basin, is beginning to force international cooperation. Traditionally, the pressure of human activity on the water resource has been light in Latin America. Change in the simple patterns of past water use can be foreseen with the increasing industrial development upstream away from ports and increasing waterborne sewerage systems. In 1974, less than 40 percent of the urban population was served by a sewerage system (United Nations, Economic Commission for Latin America 1977). There is an increased threat of water-quality deterioration in many rivers and coastal areas. Such deterioration is already beginning to produce conflict with other uses, particularly drinking water supply and fishing, but also with irrigation and recreation. It leads, as well, to the destruction of the aesthetic quality of water as an environmental component.

Latin American Water Management Strategies

Water management strategies in Latin America are disparate. The simplest single-purpose approaches coexist with the most sophisticated, not only in the region as a whole, but also within the same country. Even inside individual countries there is a lack of homogeneity in water management practice. It is not suprising, therefore, to find that incorporation of particular considerations in water-management decision making is similarly heterogeneous. For example, the water management agencies are typically somewhat side-stepped in the environmental management process. Consequently, environmental considerations tend to be imposed from outside by international banks, ministries of health, or in the form of general codes of environmental conduct.

Management actions are characterized by the use of single-purpose and single-means strategies with isolated occurrences of multiple-purpose, multiple-means approaches. In most countries, one or more individual government agencies—sometimes ministries but more commonly autonomous corporations—dominate the water-resource management system. Thus there is an emphasis on individual water-use development projects. Machinery to resolve conflicts at the river basin level is rare. In water resources, as in most activities, integration tends to occur at only the highest, most centralized decision-making level.

Latin American countries can be grouped into three general types according to the nature of the water management system: (1) many agencies are active in water management, and there is no one dominant institution; (2) water management activities are concentrated in one institution but some significant functions are carried out elsewhere; and (3) water resource administration is conducted by one centralized institution.

1. The situation most commonly found is that the administration of the water resource is divided among various institutions. No one institution dominates the water management system even when coordinating institutions exist. Argentina, Bolivia, Chile, Colombia, Guatemala, Nicaragua, Paraguay, Uruguay, and Venezuela all fall into this category. Coordination of activities between institutions, or across sectors and uses, is important, and it is achieved through a great variety of interinstitutional bodies. These may be interministerial councils, specific coordination agencies, or ad hoc arrangements. There are cases, as in Argentina, where no formal means of coordination exist.

Within these systems, which might with justification be termed fragmented in terms of function, there is considerable variation in relation to centralization or decentralization of decision-making authority and of territorial units in which different institutions operate. Nearly everywhere, decentralization of specific functions to autonomous public agencies is common, particularly in hydroelectric power generation, public water supply, and irrigation. Territorial decentralization, outside the federal countries, Brazil and Argentina, is less common.

2. The system whereby one institution predominates over all others is found at the federal level in Brazil as well as in Costa Rica, El Salvador, Panama, and Peru. In Brazil, for example, the majority of water management activities are concentrated in the federal Ministry of the Interior. Coordination tends to be limited to the formation of interministerial committees to deal with one specific water use. In Peru, a unitary state, centralization of authority is even greater. The General Directorate of Waters and Irrigation within the Ministry of Agriculture has authority not only over irrigation water use but also over the management of the resource itself (Guerra 1974).

3. In four countries—Cuba, Ecuador, Honduras, and Mexico—and the majority of the Brazilian states, water resources adminis-

tration is centralized in a single institution. There are differences between these countries, but the important feature is consolidation of the water management system. The classic example is provided by Mexico. The basic water management institution, the Water Management Secretariat (Secretaría de Recursos Hidraulicos, or SRH), is wholly responsible for the development and conservation of the water resource. The SRH possesses the authority to define policies, plan uses, execute works, and conduct research into all areas of water resource use and conservation. The secretariat was recently incorporated into the Ministry of Agriculture.

A Summary Assessment

Throughout Latin America, water resources are progressively being used more intensively. Institutional change is more sluggish, but new ideas are penetrating, thanks to international agencies, the training of professionals overseas, and, more particularly, the education imparted in schools and universities of the continent. There has been considerable improvement in the data base, better assessment of ecological effects, and more demonstrable concern with the interactive effects of water management decisions.

It can hardly be claimed, however, that integrated river basin management is widely diffused in practice in Latin America. Examples can be found, but they are the exception rather than the rule. Even more rarely are the principles incorporated into a river basin management agency. This last institution is only commonly encountered in Colombia and Mexico, and even in these countries it operates under restrictive conditions. Elsewhere, whatever the form of management system, centralization continues to rule the day.

Africa

New impetus was provided for integrated basin development in the countries of Africa and Asia that achieved independence following the Second World War. The primary concern in many instances was to achieve as rapid an increase as possible in food production, which often hinged on the availability of basic inputs including irrigation water. The need to expand irrigation, to provide water to cities and rural areas, to increase hydroelectric potential, and to reduce flood losses encouraged planners increasingly to view basin development in an integrated context and within the framework of longer-term development plans with potential for international financing. This was done while still according priority to specific, short-term projects.

Difficulties centered on a shortage of financial resources, a lack of basic physical and economic data, a lack of trained technical and administrative personnel, and weak or outdated institutional structures. Obviously, the complexity of the required tasks varies greatly from basin to basin according to size. Included are parameters such as the level and type of activities; location, concentration, and number of inhabitants; rates of population growth or decline; and the available natural resource base.

Progress in African River Basin Development

Africa has more international river basins than any other continent, 57 of 215 worldwide. These make up 60 percent of the aggregated areas of all international river and lake basins. Not surprisingly, therefore, progress made in promoting various facets of integrated development has largely been international in scope, frequently entailing cooperation between neighboring states, particularly when the individual interests of different countries diverged.

The Nile River, with one of the largest basins in the world (2.9 million km^2) spread over nine nations and encompassing very different geographical and climatological zones, has a history of cooperation between Egypt and Sudan dating back to 1946. An agreement of November 1959 established a Joint Technical Commission for joint action by the two countries with the primary purpose of increasing the potential of the river to meet future water needs. Basically, the Nile Valley Plan envisaged eventual control of the Nile and its tributaries for irrigation and possibly hydroelectric development, especially in Egypt and Sudan. For this purpose, comprehensive simulation models were developed in one of the earliest attempts to use this technique for integrated development in the upper reaches of the basin (Mageed 1976, pp. 16–20). The agreement between Egypt and Sudan recognized the riparian rights of other basin states. In 1960, Ethiopia initiated a large-scale study of the Blue Nile Basin and its tributaries, leading to the establishment of a hydrometeorological network for basic data collection. A number of hydroelectric and irrigation projects were identified, and these activities, added to the 1959 Egyptian-Sudanese agreement, prompted the formation of a Nile Water Committee among the East African countries. This led to further collaborative efforts between Egypt and Sudan on the one hand and the newly created East Africa Committee on the other. These joint ventures resulted in the initiation of hydrometeorological surveys of the catchments of lakes Victoria, Kioga, and Mobutu Sese Seko (Lake Albert) with the assistance of the United Nations Development Programme in order to assess the water balance of the lakes, and in the establishment of a technical committee to act as an intergovernmental cooperating agency to assist in planning and conservation of upper basin reaches. The project is nearing completion in the countries of the Nile River Basin: Burundi, Egypt, Kenya, Rwanda, Sudan, Tanzania, Uganda, and Zaire. A model of the hydrological regime of the Upper Nile catchment has been developed and transferred to the participating governments for use in planning water resource development. A separate project, with assistance from the United Nations Environment Programme, is designed to obtain information on water quality in the Upper Nile lakes and to develop a model of the related ecological processes for use in the future environmental management of these water resources. This project, which involves the above countries, was completed in 1984.

The Volta River Basin experience is aptly described by Futa (1976, pp. 220–27). He recalls that Ghana, which started the initial basin development, received its independence earlier than its French-speaking neighbors (Togo, Benin, Ivory Coast, and Upper Volta). Apart from the specific energy requirements that had to be met for the development and operation of bauxite mines and of a large smelter, the decision to build a large hydroelectric scheme at Akosombo was motivated by the desire to provide energy for the industrial and domestic requirements of the country and to develop a fishery and commercial transportation facilities on Volta Lake. The Volta River Authority, established in 1961, also undertook to deal with health care and resettlement made necessary by the creation of the lake which led to the evacuation of eight thousand people, and to carry out research in association with other agencies closely related to lake development in such areas as hydrobiology, public health, and shoreline agriculture. The river basin represents about 67 percent of the Ghanaian surface area. According to Futa, most known suitable Volta River dam sites are in Ghana or on its borders.

In the late 1960s, the Ghanaian authorities began to turn from purely national interests to a concern with the multinational character of the river. In 1967, negotiations were initiated with Togo and

Benin for the provision of energy from the Akosombo scheme to the two other countries. The Canadian government assisted in discussions that led to the construction of a transmission line in 1970. With the initial provision of energy in 1972, the Electrical Community of Benin was created by the two recipient countries to deal with the Volta River Authority in Ghana on all matters pertaining to the agreement. Subsequent negotiations, beginning in the early 1970s, were entertained between Ghana and the governments of Upper Volta and Ivory Coast, also entailing development and provision of hydroelectric resources. Detailed feasibility studies of potential new dam sites on the Black Volta were conducted by Ghana and Upper Volta, while the possibility was explored by Ghana and Ivory Coast of having an intertie between the Ghana and Ivory Coast grids, with a possible further link to hydrodevelopment on the Black Volta. The possibility of having the United Nations Development Programme and the United Nations Department of Technical Cooperation for Development provide assistance for the feasibility study on the Black Volta has been under consideration by the government of Upper Volta. These international contacts between five neighbors reflect the desire by each to view basin development in an integrated context that will go beyond the initial considerations of energy to encompass agriculture, fisheries, industry, community water supply, health, and transportation.

Interest in the development of the Senegal River Basin dates as far back in colonial times as 1918. The idea of integrated river basin development, however, was conceived only after independence in 1963, when Guinea, Mali, Mauritania, and Senegal established an Interstate Committee for the Development of the Senegal Basin to promote agriculture, energy, industrial development, and navigation improvement within an integrated development focus. The interstate committee was later replaced by the Organization of the Senegal River States (OERS). In 1970, the OERS Council of Ministers decided that on the basis of a basic stream flow of 300 cubic meters per second, the following activities would be included in a development plan: a hydroelectric regulatory dam at Manantali, a delta irrigation dam, development of a river and sea port at St. Louis, and improvement of port facilities at various locations and of riverbed sills to facilitate navigation. This decision was confirmed in 1972 by the newly established Organization for the Development of the Senegal River (OMVS), which replaced OERS, with Mali, Mauritania, and Senegal as members.

In 1973, the principle of a midterm action program was accepted, and shorty thereafter came the establishment of national coordinating committees for river development. The midterm plan was to determine the methods and approaches for resource exploitation, to indicate criteria for creating the supporting institutions and administrative mechanisms at national and regional levels, and to ensure a satisfactory development rate.

Subsequently, through the mechanism of OMVS, a common approach for integrated basin development was formulated, including the principle that each participating country contributes an agreed percentage share of the costs of the joint projects, and the corresponding principle that participating countries undertake financial liability vis- à-vis sources of funding in respect to loans obtained for the execution of joint river basin development projects. These far-reaching principles provide the indispensable legal framework for implementing the program of integrated development of the Senegal River Basin and are quite innovative in that they pioneer a truly regional approach to the development, conservation, and use of a river basin shared by more than two countries. It is on the basis of these principles, formalized in agreements and council of ministers' decisions, that funding for the two priority projects—a multipurpose dam at Manantali in Mali and a dam for the control of saltwater intrusion at Diama on the river estuary—could be obtained from various financial sources, bilateral and multilateral. Construction of the two dams is now underway. The United Nations Development Programme, the Food and Agriculture Organization of the United Nations, and the United Nations Department of Technical Co-operation for Development served as a catalyst in these endeavors, which are continuing with enthusiastic support by the OMVS member countries.

An overriding concern in the Sahelian zone is the limited availability of water. In the Senegal River Basin an essential consideration is the desire to improve the basin environment by offsetting limitations imposed by the scarcity of this vital resource.

Activities relating to investigations preparatory to integrated basin management were conducted in a number of other important basins in Africa over the last few years. These include such aspects as hydrological surveys, damsite investigations, and socioeconomic and financial analyses. Among the significant basins studied are the Gambia, the Logone, Lake Chad, the Niger, and the Kagera.

Asia

A unique example of interstate cooperation for integrated basin development in Asia is the Mekong River Basin. Although the first international actions to cooperate in Mekong River use occurred in 1926 when France and Thailand signed an agreement that neither would impede river navigation, the real stimulus for international cooperation came from the establishment of the Economic Commission for Asia and the Far East (ECAFE). Evolution of the approach to planning on the Lower Mekong has been described by Sewell and White (1966). They noted that in 1949, ECAFE established a Bureau of Flood Control to advise and assist governments in the ECAFE region on flood control and related river problems. Two years later, in response to an ECAFE request for a review of technical problems of developing international rivers, the bureau suggested a study of the Lower Mekong. The governments of the four riparian countries agreed, and a preliminary investigation was carried out by the ECAFE staff concerning flooding and other technical water management problems in 1952. A more detailed study was undertaken by a team of experts working with the four basin riparian nations: Cambodia (now Kampuchea), Laos, South Vietnam, and Thailand. The group's report concluded that it was possible to develop the region for a wide variety of purposes and benefit the entire area. It also provided a conceptual framework for planning. At a 1957 meeting, A Statute of the Committee for Coordination of Investigations of the Lower Mekong Basin was unanimously adopted by the four countries.

The first decision endorsed by the committee at this meeting concerned conservation of the regime of the Mekong mainstem during low-flow periods, allocation of water for irrigation and other consumptive uses, and establishment of priorities for project investigations in the four countries. This involved mainstem structures at Sambor and Pa Mong and a major project at Tonle Sap in Cambodia (Sewell and White 1966, p. 19).

The Mekong experience clearly illustrates that a sophisticated approach to river basin planning is possible even in regions with limited technical and financial resources and where political disharmony has been characteristic for more than a century. A series of well-organized steps was taken to gather the basic data not only on the supply side but also with respect to the services to be provided if the plans were implemented. Beyond this, there was a conscious attempt to provide training for those who would occupy leadership roles either in river management or in providing and using specific water and related services. Another remarkable aspect of the planning was that it continued despite the constant hostility, especially in Vietnam. Further, the planners recognized that they must gain experience by starting with relatively small projects before embarking on large ones. Accordingly, several small tributary projects were built, each involving a considerable degree of international cooperation. An important checkpoint was reached with the presentation of an indicative plan in 1975 setting out a range of options from which member countries could select.

The Secretariat of the Mekong Committee recently described the evolving situation. So far, available financial resources have tended to be channeled toward undertakings primarily of interest to one of the member states as opposed to projects of the mainstream. Twelve tributary projects have been completed for hydroelectric generation, irrigation, flood control, fisheries, and navigation purposes. In addition, however, twelve years of study at a cost of some $20 million have been devoted to feasibility studies of the Pa Mong dam, on the mainstream, for multiple purposes. What is believed would be one of the largest multipurpose dams in the world is regarded as the centerpiece of the entire integrated development program of the Lower Mekong Basin, which also includes watershed management, water quality control, and self-sustaining agricultural development combined with protection of the complex environmental conditions of the basin. All these activities have continued in spite of the problem created by the temporary absence of Kampuchea from the committee since 1975. In January 1978, the other three members of the committee established an Interim Committee for the Coordination of Investigations of the Lower Mekong Basin, which is implementing the basin development program in the three member countries and is continuing basic data collection and long-term planning activities.

Emerging Trends

The contemporary approach to comprehensive river basin management is based on several basic concepts. These include integrating planning for water and water-derived services, increasing the range of alternatives considered, creating effective institutions, evaluating the full range of environmental impacts, using research as a management tool, and recognizing water demand management and conservation as potential strategies.

The highest level of achievement appears to be developing in the United States as expressed in the National Water Commission's seven water management themes and in subsequent legislation and policy directives (United States, National Water Commission 1973, pp. 6–10). Significant progress has been made in other countries as well. River basin planning incorporating several purposes is undertaken not only in economically advanced countries but also in developing countries. There has been a search for new forms of administrative structure, including strategies to involve several governmental levels. Water commissions or councils at the national, state, or provincial levels have been created. Examples may be found in the United States, Canada, and the United Kingdom. Another device is the river basin agency; it can perform a variety of functions ranging from planning to implementation. The United States Delaware River Commission, the Tennessee Valley Authority, regional authorities in the United Kingdom, and the French river basin agencies are a third kind of institution related to improvement of water quality. Although not comprehensive in terms of management authority and interests, these institutions progressively embrace a wider range of functions within one or more levels of jurisdiction. There are also examples of these in the United States, Canada, and western Europe.

It is becoming increasingly important to evaluate the impact of water development, in part because of the rising cost of water projects and in part because of an increasing concern about potential adverse social and environmental effects of the construction of water-control facilities. Recent advances in techniques for the identification and measurement of impact have been incorporated into improved analytical frameworks. In some countries, notably the United States, there are legislative requirements for impact assessment at various administrative levels. In others, there are policy directives that call for varying degrees of evaluation. For example, it is common practice for the Canadian federal government under the Environmental Assessment Review Process to require such an assessment. In the United Kingdom, Australia, West Germany, and New Zealand, similar policies ensure consideration of environmental parameters.

In contrast to the progress reported above, there have been far fewer attempts to implement three other contemporary water management themes. In most instances there continues to be an emphasis upon construction alternatives in the search for the solution to water problems. Although demand management and conservation, for example, offer considerable promise for the improvement of water use efficiency, they are used relatively infrequently. Moreover, while economists argue that consideration of alternative and marginal analysis would lead to such improvements, it has been only in the past few years and only in a few countries that the use of economic analysis has been formalized or that pricing has been soundly based on economic principles.

In summary, Gilbert White began his contribution to public service during the 1930s in the field of river basin management. Over four decades he contributed to its developing philosophy as well as to its application. The seminal questions he raised then concerning the efficacy of the emerging approach in this field and promising potential initiatives charted the course for rewarding research directions and innovation in the following years. As the world with its rapidly increasing population confronts the need for more food-growing soil and the concomitant fact of forest degradation, as well as the need for global energy growth, it will be necessary to experiment widely with available policy options to manage the global freshwater resource. River basin management, guided by a variety of institutional arrangements, will be used to satisfy diverse social needs for water and related resources.

References

Ackerman, W. C., G. F. White, and E. B. Worthington, ed., 1973. *Man-Made Lakes: Their Problems and Environmental Effects*. American Geophysical Union. Geophysical Monograph 17. Washington, DC.

Bower, B. T., B. Remi, J. Kuhner, and C. S. Russell. 1981. *Incentives in Water Quality Management: France and the Ruhr Area*. Washington, DC.

Bruce, J. P. and F. J. Quinn. 1979. "What Difference Do Boundaries Make?" *Canadian Water Resources Journal* 4, no. 3: 4–14.

Canada, Alberta, British Columbia, Saskatchewan, Northwest Territories, and Yukon Territory. Mackenzie River Basin Committee. 1981. *Mackenzie River Basin Study Report*. Regina, Sask.

Canada, Environment Canada. 1983. *Canada Water Year Book 1981–1982*. Ottawa.

Dunin-Barkovsky, L. V. 1971. "Planning of Integrated River Basin Development." In *River Basin Management*, pp. 38–44. Proceedings of the Seminar by the United Nations, Economic Commission for Europe, Committee on Water Problems. New York.

Fox, I. K. 1976. *Water Resources Planning in the United States and Some Lessons for Canada in U.S. Experience*. Summary of the Environment Canada, Inland Waters Directorate Seminar on Water Resources Planning Policy, Environment Canada, Western Northern Region.

Futa, A. 1976. "Volta River Development: A Case Study." In *River Basin Development: Policies and Planning* 2, pp. 220–27. United Nations Natural Resources Water Series no. 6. New York.

Gangardt, G. 1976. "Methodological Guidelines for the Master Plan for Integrated Water Management in the U.S.S.R." In *River Basin Development: Policies and Planning* 2, pp. 253–58. United Nations Natural Resources Water Series no. 6. New York.

Gerasimov, I. P. and A. M. Grindin. 1977. "The Problems of Transferring Runoff from Northern and Siberian Regions of the European USSR, Soviet Central Asia, and Kazakhstan." In *Environmental Effects of Complex River Development*, ed. G. F. White, pp. 59–70. Boulder, CO.

Guerra, J. 1974. "Legislacion de Aquas en el Peru." Paper presented at the Seminar on Legal and Institutional Aspects of Water Resources Development, Merida, Venezuela, May 1974.

Harrison, P. and W. R. D. Sewell. 1980. "Water Pollution Control by Agreement: the French System of Contracts." *National Resource Journal* 20 (October): pp. 765–86.

Herrington, P. 1982. "Water, A Consideration of Conservation." *Journal of the Royal Society of Arts*, 332–46.

International Commission on Large Dams, Canadian National Committee. 1977 n.d. *1976 Register of Dams in Canada*, n.p.

International Joint Commission, Canada and the United States. 1979. *Great Lakes Water Quality: 6th Annual Report*. Ottawa.

Johnson, R. A. and G. R. Brown. 1976. *Cleaning up Europe's Waters: Economics, Management and Policies*. New York.

LeFroy, C. and Nicholazo-Grath, P. 1979. *Les Agences financières de Bassin*. Paris.

Le Marquand, D. G. 1977. *International Rivers: The Politics of Cooperation*. Vancouver.

McIntosh, P. and J. Wilcox. 1978. "Water Pollution Charging Systems and the E.E.C." *Water* 23:2–6.

Mageed, Y. A. 1976. "Problems Encountered in Integrated River Basin Development: Case Study of the River Nile." In *River Basin Development: Policies and Planning* 2, pp. 16–20. United Nations Natural Resources Water Series no. 6. New York.

Mitchell, B. and J. Gardner, ed. 1983. *River Basin Management: Canadian Experiences*. University of Waterloo Department of Geography Monograph no. 18.

National Water Council. 1976. *Paying for Water*.London.

———. 1978. *Water Industry Review, 1978*. London.
———. 1982. *Water Industry Review, 1982*. London.
Okun, D. A. 1977. *Regionalization and Water Management*. London.
Organization for Economic Co-operation and Development (OECD). 1976. *Pollution Charges: An Assessment*. Paris.
———. 1976. *Study on Economic Land Policy Instruments for Water Management*. Paris.
———. 1977. *Water Management Policies and Instruments*. Paris.
Organization of American States, Secretary General. 1978. *Environmental Quality and River Basin Development: A Model for Integrated Analysis and Planning*. Washington, DC.
Parker, D. J. and E. Penning Rowsell. 1980. *Water Planning in Britain*. London.
Porter, E. A. 1978. *Water Management in England and Wales*. Cambridge.
Rees, J. A. 1976. "Re-thinking Our Approach to Water Supply Provision." *Geography* 61, no. 4: 232–45.
Science. 1978a. "Water Projects: President Facing a Defiant Congress" 201, no. 4360 (September 15): 996.
Science. 1978b. "Devastating Blow Dealt Water Projects Pork Barrel" 202, no. 4366 (October 17): 408.
Sewell, W. R. D. 1978. "Environmental Improvement and Water Resources Planning: The Mekong Experience." In *Science and the Environment,* Occasional Paper 32, UNESCO, Canadian Commission, pp. 11–19. Ottawa.
Sewell, W. R. D. and L. R. Barr. 1977. "Evolution in the British Institutional Framework for Water Management." *National Resources Journal* 17 (July): 395–413.
———. 1978. "Water Administration in England and Wales: Impacts of Re-Organization." *Water Resources Bulletin* 14, no. 2 (April): 337–47.
Sewell, W. R. D. and G. F. White. 1966. "The Lower Mekong." *International Conciliation,* no. 558. New York.
Teclaff, L. A. 1967. *The River Basin in History and Law*. The Hague.
United Nations, Department of Economic and Social Affairs. 1958. *Integrated River Basin Development*. New York.
———. 1970. *Integrated River Basin Development*. Revised edition. New York.
United Nations, Economic Commission for Latin America. 1977. *The Water Resources of Latin America*. Santiago, Chile.
———. 1979. *Water Management in Latin America*. Oxford.
———. 1981. *Report on the Regional Seminar on Environmental Management and Large Water Resources Projects*. Santiago, Chile.

United States, General Accounting Office, Comptroller General. 1981. *River Basin Commissions Have Been Helpful But Changes Are Needed*. Washington, DC.

United States, National Water Commission. 1973. *Water Policies for the Future*. Washington, DC.

Vendrov, S. L. and A. B. Avakyan. 1977. "The Volga River." In *Environmental Effects of Complex River Development,* ed. G. F. White, pp. 23–28. Boulder, CO.

Wengert, N. 1980. "A Critical Review of the River Basin as a Focus for Resources Planning, Development, and Management." In *Unified River Basin Management,* ed. R. M. North et al, pp. 9–27. Minneapolis, MN.

White, G. F. 1957. "A Perspective of River Basin Development." *Law and Contemporary Problems* 22, no. 2: 157–87.

———. 1963a. "Contributions of Geographical Analysis to River Basin Development." *Geographical Journal* 129: 412–36.

———. 1963b. "The Mekong River Plan." *Scientific American* 208, no. 4: 49–59.

———. 1964a. "Rivers of International Accord." *UNESCO Courier* 17, no. 7–8: 32–37.

———. 1964b. "Vietnam: The Fourth Course." *Bulletin of the Atomic Scientists* 20, no. 10: 6–10.

———. 1967. "River Basin Planning and Peace: The Lower Mekong." In *Problems and Trends in American Geography,* ed. S. B. Cohen, pp. 187–99. New York.

———. 1969. *Strategies of American Water Management*. Ann Arbor, MI.

White, G. F., ed. 1977. *Environmental Effects of Complex River Development*. Boulder, CO.

White, G. F., et al. 1962. "Economic and Social Aspects of Lower Mekong Development." *A Report to the Committee for Coordination of Investigations of the Lower Mekong Basin*. Bangkok, Thailand.

6 Decision Making in Hazard and Resource Management

Howard Kunreuther and Paul Slovic

Problems of decision making have been a focal point of research in the social and behavioral sciences over the past twenty-five years. These topics were introduced to geographers by Gilbert White, whose work subsequently sensitized economists, psychologists, and sociologists to problems in hazard and resource management that were amenable to formal and behavioral anaylsis.

White has emphasized the importance of understanding how individuals and groups make decisions about alternative programs for coping with hazards. Specifically, he has sought to demonstrate how empirical study of decision processes can aid the development and selection of public policy alternatives. His concern with linking descriptive models of choice to prescriptions for policy is summarized in the preface of his book *Strategies of American Water Management* (White 1969, p. viii): "The theme of this volume is that by examining how people make their choices in managing water from place to place and time to time we can deepen our understanding of the process of water management and thereby aid in finding more suitable ends and means of manipulating the natural water system."

The present paper highlights this theme by selectively surveying research on decision making and describing the implications of these results for hazard and resource policy. We will indicate what has been learned from this research, its influence on public policy, and promising directions for future study. Figure 6.1 provides a schematic model that will guide our approach to decision making. Decision makers collect and process information on the basis of their perceptions of the environment and the available options. Their final choice reflects numerous constraints imposed by their limited ability to collect and process information. As we will see below, White's

Howard Kunreuther is Professor of Decision Sciences and Director of the Wharton Risk and Decision Processes Center, University of Pennsylvania. He is coauthor of *Disaster Insurance Protection: Public Policy Lessons* and *Risk Analysis and Decision Processes*. Paul Slovic is Research Associate at Decision Research / A Branch of Perceptronics, Inc., Eugene, Oregon, and Adjunct Professor in the Psychology Department of the University of Oregon. He has coauthored or coedited *Disaster Insurance Protection: Public Policy Lessons; Judgment under Uncertainty;* and *Acceptable Risk*.

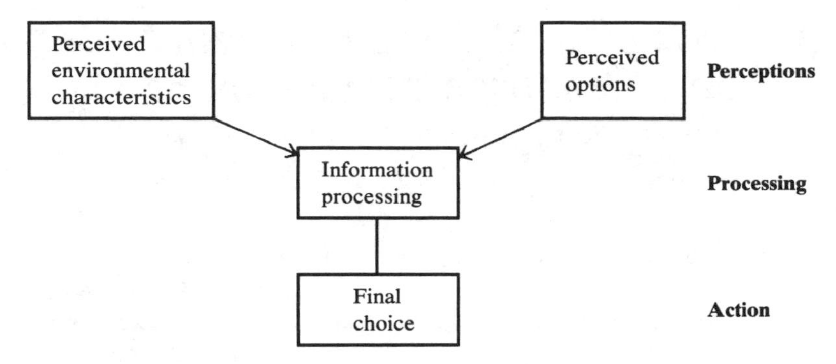

Figure 6.1 The decision making process

empirical analyses have deepened our understanding of the limitations of individual and societal decision making. His work reflects a concern with the question, What should we do differently now that we have learned more about human behavior? Underlying this concern is a philosophy that policies and programs should be based on the realities of the environment and human behavior rather than on unproven theoretical models. We share this perspective.

White has been concerned with decisions made by both individual managers and the public. We will survey the research in these two broad areas, concentrating on problems of hazard and resource management. The concluding section of the paper provides guidelines for future policy-related research.

Individual Decision Making

What protective actions do individuals undertake to deal with hazards that they face? What actions should they undertake? Some hazards, such as floods, occur rarely but may produce severe damage. Other hazards, such as hail, may occur more frequently but result in relatively little damage. Recognizing that each type of hazard requires a special set of protective adjustments, White made an important contribution to the literature of resource management by developing a framework for structuring the analysis of adjustment decisions. In particular, he distinguished between the theoretical and practical ranges of choice: "The theoretical range of choice open to any resource manager is set by the physical environment at a given stage of technology. The practical range of choice is set by the culture and institutions which permit, prohibit, or discourage a given choice" (White 1961, p. 29).

Consider the options open to homeowners residing in the floodplain. Individuals would have an opportunity to reduce flood losses by elevating their structures or adopting floodproofing measures. They could deal with the financial consequences of disasters by purchasing insurance, relying on federal aid, or bearing the loss entirely themselves. The practical range of choice open to any homeowner may be smaller than the above set either because of a blocked option or because of limited knowledge. For example, until 1953, the federal government did not have a systematic program of disaster relief. Flood insurance was not available to homeowners until 1968, when the National Flood Insurance Program was initiated. Techniques for reducing flood damage to residential structures in the floodplain have been effective in recent years, but many residents are still unaware of these possibilities.

Table 6.1 depicts the practical range of choice considered by a typical homeowner on the floodplain (Ms. Waterman) and the estimated consequences to her personal wealth under three different states of nature; no flood, mild flooding, and severe flooding—with estimated probabilities of 0.90, 0.09, and 0.01, respectively.

If there were no flooding, she would be better off not purchasing insurance or elevating her house. Minor or severe flooding justifies

Table 6.1 Practical range of adjustments and consequences to Ms. Waterman

	State of nature		
	No flooding	Minor flooding	Severe flooding
Loss (dollars)	0	−6,000	−20,000
Probability	0.90	0.09	0.01
Adjustment (dollars)			
Bear loss herself	0	−6,000	−20,000
Flood insurance[a]	−50	−250	−250
Federal relief[b]	0	−4,000	−18,000
Elevating house[c]	−2,000	−2,000	−6,000

[a] We are assuming $20,000 coverage at 2.50 per $1,000 and the following deductible schedule—maximum ($200, 2 percent of loss).

[b] For illustration purposes, we are assuming a $2,000 forgiveness grant and no low-interest loans.

[c] We assume it would cost $2,000 to elevate the structure, and the damage from severe floods would be an additional $4,000.

both of these options in comparison with the other two. Ms. Waterman would also conclude that it would never be appropriate for her to bear the loss herself if federal relief were easily available. If, however, there were considerable red tape involved in obtaining disaster relief of if she opposed relief from the government, then she might decide to bear the loss.

The analysis of Ms. Waterman's problem can be structured in a number of different ways. Decision analysis is the most sophisticated of the methods as it forces the decision maker to evaluate systematically each alternative. Behavioral approaches, which are of more recent vintage, incorporate the limitations of individuals in processing information. We will survey these two broad approaches in the context of Ms. Waterman's problem.

Decision Analysis

Structuring the problem to determine relevant adjustments, events, probabilities, and consequences (as in table 6.1) is a crucial first step, still as much an art as a science. We will assume the structure presented in table 6.1 and proceed to the calculations involved in comparing the various options. The method for determining the best solution "requires that preferences for consequences be numerically scaled in terms of utility values and that judgments about uncertainties be numerically scaled in terms of probabilities" (Raiffa 1968, p. x). A principal argument for using such an approach is that it is based upon a reasonable set of assumptions regarding behavior and choice. These assumptions imply that the consistent decision maker assigns probabilities to different states of nature (e.g. chances of a severe flood), assigns numerical utilities or disutilities to the possible results of each course of action (e.g. the disutility of a severe flood with no insurance protection), and then chooses the action yielding the highest expected utility. In other words, the theory provides a rational means for making decisions by prescribing the course of action that conforms most fully to the decision maker's own goals, expectations, and values.

An integral part of decision analysis is constructing a utility curve that reflects the value of different outcomes to the decision maker. In the case of Ms. Waterman, assume that she is averse to risk so that a gain of $100 is worth proportionately less to her than a loss of $100. One way of representing this attitude toward money is to convert dollars into utilities by presenting Ms. Waterman with a specific lottery or gamble and asking her to specify a dollar value

A that reflects an indifference to receiving this amount with certainty or playing the lottery. For example, since she is averse to risk, she might specify A to be $40 when presented with a lottery consisting of a coin flip to determine whether she has won $100 (heads) or received nothing (tails). By undertaking a series of such comparisons between lotteries and certainty equivalents, we can draw Ms. Waterman's utility curve, which enables us to evaluate different alternatives such as the ones in table 6.1. Such a curve has been drawn in figure 6.2, where we have arbitrarily specified the relevant end points of $0 to have a utility of 0 and $20,000 to have a utility of −100.[1]

We are now ready to specify an optimal choice for Ms. Waterman. The analysis is graphically depicted in figure 6.3 by a decision tree: at the end of each path there is a disutility associated with a particular decision and a specific event. For example, the expected utility associated with bearing the loss herself would be −2.8. The optimal decision for this example would be to purchase insurance, since the expected utility is −0.067.

White (1966) has pointed out that the resource manager may often want to choose a combination of adjustments to deal with a particular problem. Decision analysis also enables one to undertake such alternatives. In the previous example, if Ms. Waterman were able to obtain reduced insurance premiums for elevating her house (because it was now less prone to flooding than before), she might have con-

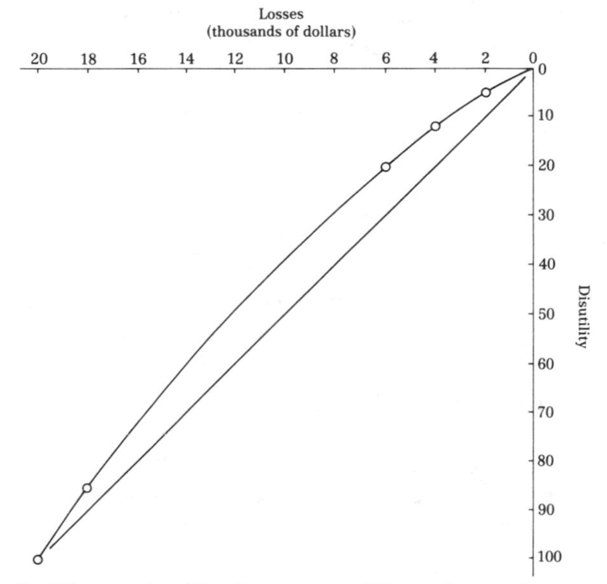

Figure 6.2 Ms. Waterman's utility function for different losses of money.

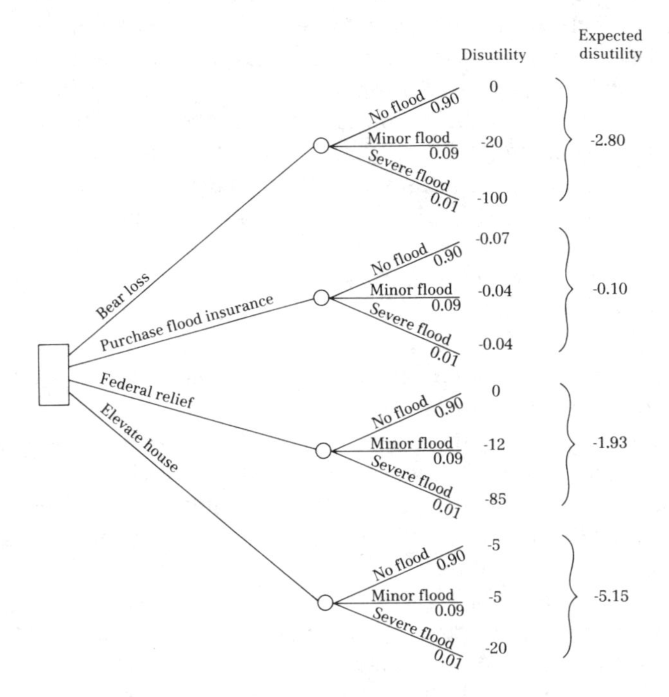

Figure 6.3 Decision tree for Ms. Waterman's adjustment

sidered adopting both of these options. The decision tree would then have been expanded to include a fifth option, "Elevate house and purchase flood insurance," and the expected utility computed in the same manner as outlined above.

In summary, there are three interesting factors that jointly determine the optimal choice using decision analysis: the shape of the utility curve, the estimate of probabilities of different states of nature, and the estimates of the consequences associated with each alternative, given a specific state of nature. In the above example, if Ms. Waterman had assumed that minor or severe flooding in her area would produce little damage to her home, then she might not have found insurance attractive. Similarly, if she had felt that any sort of flooding in her area were more probable, then a protective measure, such as elevating her house, would have been more attractive to her. If she were a risk taker rather than being averse to risk, her utility function would have had a different shape (convex rather than concave), and she would be uninterested in insurance or other mitigating measures unless the cost were subsidized.

Recent Extensions of Decision Analysis

Recent extensions of decision analysis have focused on three general topics: incorporating processes of data collection, assessing uncertainty of unknown parameters, and expressing preferences. We will consider each of these developments in turn, by extending the previous example.

Data Collection Processes:

The importance of costs of obtaining data under conditions of uncertainty have led to the development of search models. These models purport to explain how individuals behave when they have imperfect or incomplete market information. The objective is to specify the optimal number of price quotations if there is a fee associated with collecting information from each seller. The fee can be interpreted as the time and effort required to obtain this estimate.

In the previous example, suppose that Ms. Waterman were considering elevating her house on stilts but did not know how much this would cost. Suppose she obtained an estimate of $2,500. She now has to decide whether it is worthwhile to obtain another estimate, which may be higher or lower than $2,500. If it is lower, then she will have improved her position (assuming that the quality of the job was the same). If it is higher, she will have wasted time (although she will have gained some information about the nature of prices in the market). The basic question addressed by these search models is, How much search should be undertaken if the decision maker's objective is to maximize expected utility and there are benefits and costs associated with search?[2] In the case of Ms. Waterman, she would have to assess the likelihood of obtaining an estimate lower than $2,500 and balance this potential benefit against the costs of collecting these data.

Assessing Uncertainty

In recent years there has been considerable work that formally incorporates the cost and value of information as a part of the decision process. Through Bayesian analysis one can revise prior estimates of key quantities on the basis of new data. Furthermore, one can determine whether or not it is worthwhile to collect further information. To illustrate this approach, suppose that Ms. Waterman believes there is a direct relationship between the height of flooding of the river in her community and the magnitude of damage to her house. When asked what the height of the river is likely to be during the flood season, she gives three estimates with respective possibilities, as shown in table 6.2. These three estimates refer to the three states of nature (no flooding, minor flooding, or severe flooding) that enabled Ms. Waterman to evaluate the alternative adjustments in table 6.1.

Suppose she now decided to consult historical records to obtain a distribution of flood heights of the river over the past fifty years. By combining this new information with her initial subjective prior estimates, she could arrive at an updated or posterior estimate. These revised estimates would depend on the distribution of flood height and the confidence that Ms. Waterman places in the data.

As should be clear, there is a direct connection between the amount of search one undertakes and the updating of information from the search. If, for example, Ms. Waterman were convinced that the historical data accurately reflected current distribution of flood losses, she would have little incentive to incur additional search costs. However, if she were under the impression that there had been structural changes in the river flow in recent years, she might want to explore this matter further.

There is a growing literature of the relationship between search costs, the updating of prior information, and the choice of a final alternative (see e.g. Howard, Matheson, and Miller 1976).

Expressing Preferences

Individuals may also be concerned with tradeoffs between more than one attribute when expressing their preferences. For example, suppose that in choosing between the adjustments in table 6.1, Ms. Waterman is concerned not only with the financial expenditures both before and after a flood, but also with the time required to undertake each adjustment. Models have been developed to incorporate multiple dimensions of concern. For example, the utility associated with different time delays can be evaluated in much the same manner as the utility of money. The alternatives can then be evaluated by constructing a multiobjective value function that reflects Ms. Waterman's tradeoffs between time and cost.

Table 6.2 Ms. Waterman's prior estimates of river height

	Probability	Event
Less than 35 feet	0.90	No flooding
35–45 feet	0.09	Minor flooding
More than 45 feet	0.01	Severe flooding

As one adds additional attributes and considers more adjustments, the data collection and computational process become more burdensome to the individual. Keeney and Raiffa (1976) discuss ways of simplifying this process, but point out that there may be, nevertheless, substantial costs in obtaining the relevant data in complex decision problems.

Behavioral Approaches

The concept of bounded rationality developed by Herbert Simon (1959) forms the basis for behavioral approaches to individual decision making under conditions of uncertainty. In contrast to decision analysis, this approach seeks to understand what factors influence decision processes. The underlying philosophy is that the time and energy required to collect information, coupled with the decision maker's cognitive limitations, lead a person to construct a simplified model of the world that differs in important ways from the models employed by decision theorists.

An important feature of this simplified world is a strong tendency toward maintaining the status quo unless there is sufficient motivation for change. Rather than making the tradeoffs between the costs and benefits of searching for new options, a person is likely to *avoid* addressing the question, What else can I do? unless the current position is believed to be unsatisfactory.

In cases where alternative options are presented and the decision maker has to make a choice, simplifying strategies are likely to be used. One such behavioral strategy is elimination by aspects (Tversky 1972). Each alternative is viewed as a set of aspects or attributes, and each aspect is weighted according to its relative importance in relation to the others in the set. The higher the weight, the more likely the aspect will be selected for consideration. All the alternatives that do not contain the particular aspect are eliminated from consideration. The process continues until only one alternative remains.

To illustrate elimination by aspects, consider the alternatives listed in table 6.1 plus the alternative of floodproofing whereby the base of the structure is protected from minor flooding by a retaining wall or siding material. Table 6.3 depicts the four aspects of choice: the nature of the activity, the amount of time required to obtain information, predisaster and postdisaster costs, and their characteristics for each of the five adjustments. The first two modify the vulnera-

Table 6.3 Four aspects of choice for evaluating flood adjustments

Adjustments	Nature of activity 0.3*	Time to get information 0.1*	Predisaster cost 0.4*	Postdisaster cost 0.2*
Floodproofing	Modify vulnerability	High	Medium	Medium
Elevate house	Modify vulnerability	Medium	High	Low
Insurance	Distribute losses	Medium	Low	Low
Federal relief	Distribute losses	Low	Low	High
Desired state	Modify vulnerability	Low–medium	Low–medium	Low

* Importance weight.

bility of the event (i.e. reduce the potential damage) while the other two distribute losses (i.e. relieve the financial burden after a disaster).[3]

For the other three aspects or attributes we have assigned high, medium, or low values. We have also assigned importance weights that sum to 1, to indicate how critical each aspect is to the decision maker. At the bottom of each column we have listed the desired state for each aspect. An individual would thus prefer an alternative that modified the event, had a low predisaster cost and a medium–low postdisaster cost, and required low–medium time to be adopted. Obviously no alternative to this set satisfies all four of these aspects. Hence, the order in which one selects the attributes becomes critically important in the final selection process. Figure 6.4 illustrates two sequences of selecting the attributes and the resulting differences in final choice. In process 1, the nature of the activity is selected first and then predisaster cost. In process 2, the two aspects selected are predisaster cost and postdisaster cost, so the final choice is now different.

It should be obvious from the above example that elimination by aspects does not always maximize expected utility. Its major weakness is that it may eliminate at an intermediate stage in the process alternatives with an overall quality greater than those options remaining. It is used by decision makers because it is easy to state, defend, and apply.

Tversky and Sattath (1979) have extended elimination by aspects by suggesting that for many problems there is a natural order associated with attributes that enables one to build preference trees rather than selecting an attribute at random. For example, an individual may first focus on predisaster cost being medium or low

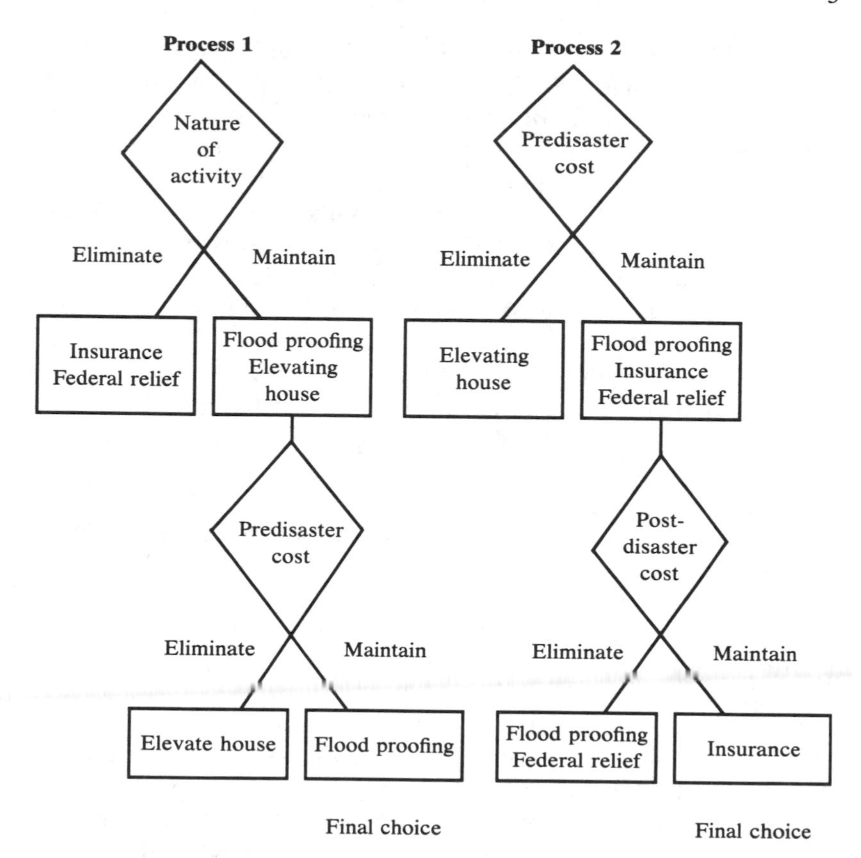

Figure 6.4 Two processes using elimination by aspects

because he or she is constrained by short-run budget limitations. Only later would the other three elements of the problem be considered. To illustrate this approach, process 2 in figure 6.4 would represent one of the possible preference trees while process 1 would not (because it focuses first on the nature of the activity). This approach can be looked at as setting an agenda for the decision maker. As the agenda or order of introducing different attributes varies, the choice will also vary. We will return to this point again in our discussion of public decision making.

If there are a number of alternatives and relatively few aspects, it is likely that at the conclusion of the process, several alternatives will remain. In this case, the decision maker may want to use more sophisticated approaches, such as decision analysis, to choose between options. Payne (1976) observed a two-stage process of this sort among students choosing among apartments in experimental studies: those unsatisfactory apartments were first eliminated on the basis of certain criteria, and then a more thorough evaluation was undertaken for choosing between the reduced set of alternatives.

A similar process was observed by White, Bradley, and White (1972) in their study of the choice of water sources by East African households. In many of the communities studied, the women rarely considered more than five alternative sources. Rather than searching for a particular "best" source, the women discarded options that were unsuitable by one or more criteria. For example, some sources were eliminated because the perceived quality of the water was deemed unsuitable or because the energy cost associated with obtaining the water was viewed as too high. Other factors considered important were the technological means associated with drawing the water and a concern with meeting or avoiding certain individuals who frequented a particular source. Even after a set of alternative sources was rejected, more than one source perhaps remained. At this stage, the women were more likely to trade off one attribute against another (e.g. quality of water with energy cost) in making their final decisions.

According to White, Bradley, and White, this decision process was lexicographic in nature. Each attribute was ranked in the order of its importance, and a prespecified standard was set; all alternatives not meeting a given standard were deemed unsatisfactory. In contrast to elimination by aspects, the lexicographic model of choice assumes that there is a fixed prior ordering of attributes so that the choice process is deterministic once this set of priorities is known. To illustrate, assume that the East African household ranks quality, technological difficulties, and energy costs in that order. This rule would choose the source that had minimum energy costs, but met minimum quality standards and technological constraints. If no alternative met the minimum quality constraint, then the criterion would have to be lowered until at least one option satisfied it. If no source met the technological constraints, then the one that came closest to doing so would be chosen. Individuals may decide to order attributes in different ways depending on their preferences. Thus, an individual who wants to avoid certain people at all costs might choose social relationships as his first attribute and only consider sources that were not frequented by those people.

Recent Extensions of Behavioral Approaches

In recent years, descriptive models have been investigated through field surveys and laboratory experiments. This work can be categorized under the same three general types of problem that have guided recent research in extending decision analysis. These areas, processes of data collection, assessment of uncertainty, and the expression of preferences, are considered below.

Data Collection Processes

Research in data collection has focused on the factors that influence the decision to collect data and the sources of information. Laboratory experiments on insurance decisions (Slovic et al. 1977) have shown that people are often unwilling to protect themselves against events with a low probability of occurrence (e.g. 1 in 1,000) even though the potential loss from the event would be relatively high (e.g. $1,000) and insurance was actuarially fair (e.g. $1) or even subsidized ($0.90). These results suggest that in some situations people are not inclined to worry about the potential losses from a future disaster if they perceive its probability to be below some threshold. The threshold concept assumes that there are only so many things in life an individual can worry about. People are forced to restrict their attention to events that they feel are sufficiently probable to warrant protective action.

Most low-probability events involve the type of transactions where, in addition to price, forces such as media exposure or advice from friends and neighbors influence an individual's behavior. Some of these features have been captured by a sequential model of choice that has been examined using data from a field survey of 3,000 homeowners residing in hazard-prone areas (Kunreuther et al. 1978a). According to this model, a person is reluctant to collect data on protection against hazards unless motivated to do so by some external event such as a recent disaster. Even then, the person may seek information only from easily accessible sources. Such a sequential model of choice suggests that individuals fail to protect themselves because of limited knowledge rather than as a result of their own benefit-cost calculations. The sequential model describes how different environmental events and the behavior of other people affect individual action.

Evidence from the social sciences forms the basis for this sequential model of data collection. A series of cross-cultural field surveys summarized by White (1974) and Burton, Kates, and White (1978) reveals the limited ability of individuals to deal with information about natural hazards. In the latter book the three geographers characterize individual behavior related to hazard adjustments by postulating that the choice process does not begin unless a first threshold of awareness of actual or anticipated loss is reached. The authors suggest the importance of past experience in triggering this awareness. A large empirical literature on the diffusion of innovations (see Rogers and Shoemaker 1971) consistently shows that most individuals are first made aware of a product or of a protective mechanism through the mass media. Before purchasing the item they are likely to turn to friends or neighbors for additional data that may not have been available to them from initial sources of knowledge.

Personal communication may also be particularly important because there is a tendency to trust implicitly the judgment of a friend or colleague. In addition, accepting the judgment or advice of friends develops and strengthens social relationships that are viewed as desirable ends in themselves. After discussing a new product with someone who has adopted it, one is likely to feel that this person has carefully evaluated the information on which to base a decision. By making such an assumption, which may not necessarily be correct, an individual considering the purchase of a new product can justify not having to collect detailed information.

Further light has been shed on the accuracy of people's perception of risk by Kunreuther and others (1978a). When homeowners participating in the field survey were asked to estimate the chances of a severe flood or earthquake damaging their property in the next year, 15 percent of the respondents in flood areas and 8 percent of those in earthquake areas were unable to provide any sort of estimate. Of those who did respond, some thought the probability of a disaster hitting them in the following year was quite large—at least once chance in ten—yet said they had purchased no disaster insurance even though they knew it was available. Others believed the chance of a disaster affecting them was minuscule—1 in 100,000—yet they *had* purchased disaster insurance. These findings raise the question as to how well individuals understand the concept of probability or know how to incorporate it into their decisions. It also suggests that there may be other factors influencing choices that are not evident to researchers.

Assessing Uncertainty

Efficient adjustment to natural hazards demands an understanding of the probabilistic character of natural events and a desire to think in probabilistic terms. Because of the importance of probabilistic reasoning to decision making in general, a great deal of recent experimental effort has been devoted to understanding how people perceive, process, and evaluate the probabilities of uncertain events. Although no systematic theory about the psychology of uncertainty has emerged from this descriptive work, several empirical generalizations have been established. Perhaps the most widespread conclusion is that people do not follow the principles of probability theory in judging the likelihood of uncertain events. When estimating probabilities, people rely on mental strategies (heuristics) that sometimes produce good estimates, but all too often yield serious biases (Slovic, Kunreuther, and White 1974; Kahneman, Slovic, and Tversky 1982).

Some of the most dramatic demonstrations of these sorts of biases come from a series of laboratory experiments conducted by Tversky and Kahneman (1974). One heuristic documented in these studies is that of availability, according to which one judges the probability of an event (e.g. a severe flood) by the ease with which relevant instances are imagined or by the number of such instances that are readily retrieved from memory. Any factor that makes a hazard highly memorable or imaginable—such as a recent disaster or a vivid film or lecture—could increase the perceived risk of the hazard. According to this bias one would expect that personal experience with misfortune would play a key role in an individual's estimate of the probability of a future disaster.

There are extensive field data indicating that the risks of natural and other hazards are misjudged in ways predictable from laboratory research. For example, the biasing effects of availability are evident in the observations of Kates (1962, p. 140): "A major limitation to human ability to use improved flood hazard information is a basic reliance on experience. Men on flood plains appear to be very much prisoners of their experience. . . . Recently experienced floods appear to set an upward bound to the size of loss with which managers believe they ought to be concerned." Kates further attributes much of the difficulty in achieving better flood control to the "inability of individuals to conceptualize floods that have never occurred" (p. 92). He observes that, in making forecasts of future flood potential, individuals "are strongly conditioned by their immediate past and limit their extrapolation to simplified constructs, seeing the future as a mirror of that past'' (p. 88). The purchase of earthquake insurance increases sharply after a quake, but decreases steadily thereafter as the memories become less vivid (Steinbrugge, McClure, and Snow 1969).

Some hazards may be inherently more memorable than others. For example, one would expect drought, with its gradual onset and offset, to be much less memorable, and thus less accurately perceived, than flooding. Kirkby (1972) provides some evidence for this hypothesis in her study of Oaxacan farmers. She found also that memory of salient natural events seems to begin with an extreme event, which effectively blots out recall of earlier events and acts as a fixed point against which to calibrate later points. A similar result was obtained by Parra (1971), studying farmers in the Yucatan. Parra found that awareness of a lesser drought was obscured if it had been followed by a more severe drought. He observed also that droughts were perceived as greater in severity if they were recent and thus easier to remember.

Additional demonstrations of availability bias come from studies by Lichtenstein and others (1978) that showed that frequencies of dramatic causes of death such as accidents, homicides, botulism, and tornados, all of which get heavy media coverage, tend to be greatly overestimated. In contrast, the frequencies of death from unspectacular events, which claim one victim at a time and are common in nonfatal form (e.g. asthma, emphysema, diabetes) are greatly underestimated. A followup study by Combs and Slovic (1979) showed that these biases in judgment were closely related to the amount of coverage given to the various causes of death by the news media.

One would expect that since people have a great deal of difficulty thinking about uncertainty, they would tend to view the world as more certain than it is. Evidence supporting this hypothesis comes from the work of Kates (1962), who found that floodplain dwellers used a number of mechanisms for dispelling uncertainty, such as denying the risks from flooding or perceiving floods as repetitive or cyclical phenomena.

Expressing Preferences

Once the consequences of a decision have been enumerated and their uncertainty assessed, some value must be attached to them. When it comes to tradeoffs between such issues as deaths today versus deaths in the future or between economic development and possible catastrophic natural disasters, we have little choice but to ask people for their opinions. Unfortunately, for such unfamiliar and complex issues, people may not have well-defined preferences. Fischhoff, Slovic, and Lichtenstein (1980) show that values may often be incoherent, not thought through. For example, in thinking about risk we may be unfamiliar with the terms in which issues are formulated (e.g social discount rates, minuscule probabilities, or megadeaths). We may have contradictory values (e.g. a strong aversion to catastrophic losses of life and a realization that we are not more moved by a plane crash with 500 fatalities than one with 300). We may occupy different roles in life (parents, workers, children), which produce clear-cut but inconsistent values. We may vacillate between incompatible, but strongly held, positions (e.g. bicycles are an important mode of transportation, but are too dangerous to be allowed on most streets). We may not even know how to begin thinking about some issues (e.g. the appropriate tradeoff between the opportunity to dye one's hair and a vague, minute increase in the probability of cancer twenty years from now). Our views may undergo changes over time (say, as we near the hour of decision or the consequence itself) and we may not know which view should form the basis of our decision.

In such situations, where we do not know what we want, the values we express may be highly labile. Subtle changes in the way issues are posed, questions phrased, and responses elicited can have marked effects on our expressed preferences. The particular question posed may evoke a central concern or a peripheral one; it may help clarify the respondent's opinion or irreversibly shape it; it may even create an opinion where none existed before.

Three features of these shifting judgments are important. First, people are typically unaware of the potency of such shifts in their perspective. Second, they often have no guidelines as to which perspective is the appropriate one. Third, even when there are guidelines, people may not want to give up their own inconsistency, creating an impasse (Lichtenstein and Slovic 1973; Tversky and Kahneman 1979).

Summary

Behavioral research depicts the process of choice as based on incomplete and often biased information and simplistic rules for evaluating alternatives. Studies have shown that one's decisions are typically guided by past experience or discussions with other people rather than by a detailed comparison of the costs and benefits of different alternatives.

From a policy perspective, decision analysis is netural with regard to the optimal locus of decision making. In contrast, those who have been influenced by behavioral analysis contend that people may not act in their own best interest because of limited or biased information. They argue that if major societal costs are incurred because of "poor" decisions by individuals, some form of regulatory control may be necessary. Thus they would favor land use regulations if it were found that individuals were not sensitive to the hazards they faced. They would also support some form of required insurance if empirical data revealed that individuals were not willing to protect themselves voluntarily (e.g. the Flood Disaster Protection Act of 1973, which required all homeowners to purchase flood insurance as a condition for federally financed mortgage). We will elaborate on these policy implications in the concluding section, where we discuss guidelines for future research.

Public Decision Making

The broad area of hazard and resource management frequently requires investments in projects such as flood control dams that affect large numbers of individuals. These public goods have two principal characteristics that differentiate them from products offered on the market such as insurance, floodproofing materials, or wooden pilings for elevating one's house. The costs are so high and the individual gets such a small share of the benefits that he or she has no incentive to invest in such projects. Furthermore, an individual will benefit when others provide the goods, so there is no incentive to pay for them. For example, if half of the community paid for a flood control dam, the other half would still be given the same protection. One of the main justifications for the existence of government programs in hazard and resource management is to provide citizens with such public goods as highways, national defense, and flood control projects that would not otherwise be provided by the public sector.

In this section we will examine alternative approaches for dealing with the allocation of limited governmental funds among competing projects. Until recently, benefit-cost analysis was the principal tool employed in this process, and we will review its concepts first. White has been one of the leaders in stressing the importance of including multiple objectives explicitly in the analysis, and we will indicate the types of models of choice that come under the broad heading of multiobjective planning. Finally, there has also been considerable interest by White and others in such behavioral questions as the way existing institutions as well as disasters and crises affect public decision making. We will conclude this section by touching on work in this area.

Benefit-Cost Analysis

Benefit-cost analysis systematically incorporates tangible and intangible benefits and costs of different projects in much the same way as decision analysis does. The decision maker lists a set of alternative options and then determines possible outcomes under different states of nature. The alternative that produces the greatest net benefit (total benefit minus total costs) is considered the most desirable. The tool was originally used in evaluating water resource projects in the 1930s, but it has had widespread use only since the Second World War.

To illustrate the application of benefit-cost analysis, consider a decision by the Army Corps of Engineers as to whether it should invest in flood control project A or B on a given river basin. In order to determine the expected benefits from each project, it has computed the expected annual savings in flood damage (i.e. probabilities times damages avoided from building the dam) to the community for the next fifty years appropriately discounted to the present.[4] Similarly, the costs of each project are discounted to the present year so an appropriate comparison can be made.[5] A summary of the relevant figures appears in table 6.4 As we can see from these values, project B has a higher net benefit to society but assists primarily the upper-income group. Project A protects primarily low-income individuals who live in the floodplain.

The above example does not distinguish between the benefits from each of the projects accruing to the federal government and to the local community and hence ignores issues of cost sharing. The importance of making this distinction can be illustrated by two contrasting examples of disaster programs. Suppose federal disaster relief or subsidized flood insurance provides recovery funds for a substantial portion of flood losses; then the construction of a project will mean a reduction in federal disaster expenditures. If, in contrast, flood victims in a community rely on their own resources for recovery, then the federal government stands to gain little financially from a flood control project. In the latter case, potential

Table 6.4 Evaluation of two projects using benefit-cost analysis

Sector	Discounted expected annual savings in flood damage ($000)	Project cost ($000)	Net benefits ($000)
Project A			
Low-income	500		
Middle-income	100		
High-income	80	—	—
	680	600	80
Project B			
Low-income	120		
Middle-income	130		
High-income	400	—	—
	650	500	150

victims may want to place pressure on their local government to help build the project with their own funds. The issue of cost sharing between federal and local governments thus has important implications for efficient resource development. Until recently, agencies such as the Corps of Engineers required relatively little cost sharing by local governments for large flood projects, but specified higher percentages for other techniques such as channel improvements, levees, and diversion channels. Marshall (1970), who analyzed data for thirty-four corps projects authorized by the 1968 Flood Control Act, found that the local cost sharing ranged from 0 percent for large reservoirs to approximately 50 percent for some levees and channel improvements. There was considerable variation in the cost-sharing amounts for the same type of projects in different regions of the country.[6]

One of the important current issues in benefit-cost analysis is the proportion of costs that local interests should be required to absorb for specific projects. Marshall (1973) has shown that an association rule, whereby local beneficiaries are charged a percentage of the cost equal to the ratio of marginal local benefits to marginal national benefits computed at the nationally efficient scale of output, induces local interests to select the nationally efficient project design. Furthermore, he points out that there needs to be consistency between the cost-sharing practices of different agencies as well as between different techniques such as structural and nonstructural measures.

On a theoretical level, this type of analysis makes excellent sense. The challenge from the point of view of evaluating alternative cost-sharing rules is determining what the national and local benefits are likely to be for different types of projects and understanding how local constituencies decide on whether they will sanction a particular project, given a fixed percentage of cost they must assume. There is thus a need to understand the decision processes of governmental units and to collect detailed information on the effects of different actions before deciding on a particular course of action.

Multiobjective Planning

One of the principal criticisms leveled at benefit-cost analysis is that it focuses almost entirely on criteria of economic efficiency without concern for other objectives such as income redistribution (Maass 1966). One way of coping with equity considerations is to use other means such as transfer payments to low-income residents rather than explicitly incorporating other objectives into the analysis. In the above example, project B would still be deemed most desirable, and special grants from taxpayers' money could be given to low-income flood victims.

An alternative approach is explicitly to incorporate income distribution and other goals of a particular project as part of a multi-criterion objective function. White and his colleagues on the Committee on Water (1968) went to great lengths to highlight the diverse objectives that must be taken into account when planning for water management of the Colorado River. Aside from the standard goal of national economic efficiency, four other aims were suggested: income redistribution, political equity, controlling the natural environment, and preservation and aesthetics. The report indicated the nature of these different objectives, but left it to the policymaker to determine how tradeoffs between them should be made.

This concern with incorporating multiple objectives into the analysis of resource management projects was expressed by the U.S. Water Resources Council (WRC) which in 1973 adopted principles and standards that indicated that the beneficial and adverse effects of projects be assessed under four general accounts: national economic development (NED), environmental quality (EQ), regional economic development (RED), and social well-being (SWB). The 1979 WRC "Principles and Standards for Planning Water and Related Land Resources" emphasized the importance of evaluation of NED and EQ and then provided detailed guidelines for quantitatively measuring the beneficial and adverse effects of each of these two objectives. The document, however, notes that "the statement of the objectives and specification of their components to these standards is without implication concerning priorities to be given to them in the process of plan formulation and evaluation" (U.S. Water Resources Council 1979, pp. 72).

Over the past twenty years considerable research has been done on ways that a multicriterion objective function can be evaluated by a policymaker in a systematic manner. After stating the objectives of the policy proposal, one has to define attributes that can measure how well each alternative meets specific objectives. For example, one attribute measuring NED might be "number of new jobs created." There are likely to be several attributes that map onto each objective. In the case of qualitative objectives, it may be more difficult to define a set of attributes. For example, how does one measure "environmental quality" in a quantitative manner? One way to get a handle on this objective is to divide it into subobjectives (e.g. creating recreational opportunities, preserving wildlife), which may then suggest specific attributes (e.g. number of visitor days in a park).

If there are several attributes describing a given objective, then appropriate weights must be given to them. If the attributes are independent of each other, one estimates the utility function for each attribute separately in a manner similar to that described in the previous section (see figure 6.2). One then determines scaling constants to specify the appropriate weights in the overall objective function. The process is somewhat more complex if attributes are dependent.[7] It is questionable how well the process is likely to work in practice. It requires a sophisticated policymaker and does not explicitly incorporate the decision processes of different individuals and stakeholders.

A related approach is goal programming, whereby the policymakers set desired goals for particular objectives and a penalty associated with deviating from these goals. It is then possible to develop a formal model for evaluating the impact of different alternatives on the multicriterion objective function. In essence, this approach is a hybrid between a lexicographic model and a multiattribute utility model. Acceptable levels are set as in a lexicographic approach, but deviations below this level and simultaneous consideration of alternatives take place as in a multiattribute utility model. How different weights should be determined for the different goals and whether such a multiobjective function captures the decision process remain open questions.[8]

An alternative theory of measurement has recently been proposed by Saaty (1977). His method consists of decomposing the problem into relevant attributes and then combining them to make an overall choice. Rather than asking individuals to estimate utility functions for each attribute, he requires them only to make pairwise comparisons that can be combined in a way that reflects the decision-makers' preferences. For example, consider the five objectives for the Colorado River project investigated by White and his Committee on Water. The policy analyst would have to make ten pairwise comparisons across those objectives, reflecting the relative importance he attached to each one in relation to another. For example, the analyst would be asked to specify the importance of national economic efficiency relative to income redistribution, and income redistribution relative to preservation and aesthetics. If each objective were subdivided into a set of attributes, a similar set of comparisons would have to be made at this level.

Multiattribute utility models, goal programming, and Saaty's scaling method assume that there is a single decision maker who must determine a course of action based on his or her estimates of appropriate weights and utilities. If there are several interest groups represented (e.g., the floodplain resident, the general taxpayer, representatives from different state, federal, and local agencies) then members of each group are likely to have their own rankings with respect to the importance of different attributes and objectives. Either some type of weighting scheme has to be assigned to the preferences of each of these different interest groups, or some type of consensual procedure such as a nominal group or delphi process[9] must be employed.

Understanding Public Decision Making

On the descriptive side there has been an interest in understanding the impact of different institutional arrangements and specific events such as disasters on formulation of public policy. White (1969) indicates three principal ways that state and federal agencies can gauge people's attitudes toward possible solutions to their problems. The public hearing is most democratic, permitting citizens' groups, industries, and other special interests to state their points of view. A second method is congressional committee hearings held as part of budget agency recommendations or for determining the need for special legislation. White thinks that by far the most important sources of public preferences are the informal comments from different citizens' groups, lobbyists, and other interested parties. The role of personal contact and informal networks in influencing strategies and final courses of action has a parallel in the studies of the individual decision-making process, where friends and neighbors play a key role in influencing choice.

Although these three insititutional arrangements provide insight into how information is elicited by public agencies, they do not indicate how the choice is made. One of the most interesting studies on this question is a description of the decision-making process used by the Delaware River Basin Commission in its analysis of water quality on the Delaware River (Kneese and Bower 1968; Haefele 1973). A system of advisory committees provided estimates of costs and benefits of five different alternatives with respect to water quality. These five alternatives (one of which was the status quo) were then presented at a public hearing, where different groups were able to voice their concerns and state their preferences. Finally, the choice was made by an interstate compact with three state commissioners, each having one vote. After considerable discussion, the commission chose water-quality standards falling somewhere between the two choices favored by almost everyone at the public hearings.

The commission's study on water quality used existing institutional mechanisms to evaluate the benefits and costs of a set of alternatives. Furthermore, there was general agreement by all parties as to the preferred alternatives. One reason for this agreement is that the benefits of improved water quality were restricted solely to recreation.

When there are a number of projects to be considered and a number of different attributes are relevant, the decision-making process may be somewhat more complicated than the one followed by this commission.[10] In that case, the ordering of different attributes or, more generally, the construction of an agenda may play a key role in the final decision. We have already discussed this point in the section on individual decision making by indicating how descriptive techniques, such as elimination by aspects or preference trees, will yield different choices, depending on which aspects are chosen first. Similar results have been shown to hold for group decision processes, where changes in the agenda have influenced outcome.

One of the most interesting recent studies to explore the effects of agenda was by Plott and Levine (1978), who demonstrated that one could change the probability that certain types of planes would be purchased by a flying club by altering the order in which aspects pertinent to the decision were considered. They also replicated their results in a series of laboratory experiments. A simple example in the water resources area illustrates the effect of agenda on decisions. Suppose that a committee deciding which water resources projects to fund, is considering two regions (Florida and Louisiana), and is deciding whether project costs should exceed a certain number of dollars. One agenda might determine the set of available projects by focusing first on cost and then on location. Another agenda would reverse the order. The group choice may yield a different solution depending upon which question is presented first.[11]

Frequently the agenda is ordered by external events such as disasters or crises, which focus attention on specific remedies. White and Haas (1975) point out that most federal legislation on natural hazards follows within a few months or a year of a major disaster. The history of the flood insurance program illustrates this point.[12] Severe flooding in the northeastern states in 1955 created a clamor among victims for a government-backed insurance program. As a result, Congress passed the Flood Insurance Act of 1956, which provided for government subsidized rates, but refused to appropriate the funds for the program because there were serious questions raised both within government and by outsiders as to the potentially harmful effects on floodplain development.

Two fortuitous events helped to launch the National Flood Insurance Program. The Bureau of the Budget appointed the Task Force on Federal Flood Control Policy, which explicitly recognized the need for a different type of flood insurance program and indicated how such coverage could be related to other types of adjustments such as regulation of land use. At approximately the same time, Hurricane Betsy devastated a large portion of the Gulf Coast, including New Orleans. As a result, a congressional task force was authorized to undertake a study on the feasibility of some form of federal flood insurance. The results of this study, coupled with the task force report, culminated in the National Flood Insurance Act of 1968.

The most recent change in the flood insurance program was triggered by Tropical Storm Agnes in 1972. Many communities struck by these disasters had not entered the flood insurance program, and hence residents could not buy coverage. In other communities, where flood insurance was available (including Wilkes Barre), few residents had voluntarily purchased coverage. This lack of voluntary interest in the program on the part of homeowners and communities induced Congress to pass the Flood Disaster Protection Act of 1973, which required insurance as a condition for any mortgage or home loan partially or fully financed by the federal government.

This brief summary of the impact of flood disasters on insurance policy suggests a behavioral model of choice for public decision making that has similar features to the sequential model for individuals. Stage 1 brings on an awareness of the problem. This is frequently triggered by a disaster with its resulting inequities or by some concern by Congress in reviewing the performance of a given program. Stage 2 consists of examination of feasible alternatives. Through a task force report and/or public hearings, a set of options is outlined for possible adoption. In stage 3 an option is chosen. Either the proposed program is rejected because its costs are likely to exceed its benefits (Federal Flood Insurance Act of 1956) or the program is adopted (National Flood Insurance Act of 1968). Stage 4 sees a reevaluation of the choice. Should the program be unsuccessful in meeting its objectives, then it will be reexamined. Policymakers are then made aware of a problem and have reentered stage 1. The reexamination of the Flood Insurance Program after Tropical Storm Agnes illustrates this phase of the process.

Directions for Future Research

Our survey illustrates the motivating forces behind the alternative approaches to decision making under uncertainty. Tools such as decision analysis and benefit-cost analysis are primarily concerned with ways to improve behavior and hence are prescriptive in nature. Behavioral analyses look at how the world actually works, and thus are concerned with institutional arrangements and decision processes; these approaches are descriptive in nature. White would like to see policies designed with sensitivity to both prescriptive and descriptive considerations. It is in this spirit that we will offer a few suggestions for future research in the two broad areas surveyed above.

Individual Decision Making

In the area of individual decision making, we need to develop techniques for structuring the decision. The logic of decision analysis cannot be applied until the alternatives, critical events, and outcomes are specified. We need algorithms for accomplishing this and for simplifying the large, complex decision trees that may result. Crises, where stakes are high, time is short, and the alternatives and information are continually changing, pose particularly difficult problems of structuring.

Subjective judgments of probability and value are essential to decision analyses. We still do not know the best ways to elicit these judgments. Now that we understand many of the biases to which people are susceptible, we need to develop "debiasing" techniques to minimize their destructive effects. Simply warning a judge about a bias may prove ineffective. Like perceptual illusions, many biases do not disappear upon being identified. It may be necessary to restructure the judgment task in ways that circumvent the bias, use several different methods allowing opposing biases to cancel one another, or correct the judgments externally, on the basis of an estimate of the direction and strength of the bias.

Much progress has been made recently toward understanding judgmental and decision-making processes. We need to continue this pursuit of basic knowledge. Simon (1965, p. 92), outlining the historical development of writing, the number system, calculus, and other major aids to thought, indicates the importance of synthesizing descriptive and prescriptive approaches:

> All of these aids to human thinking, and many others, were devised without understanding the process they aided—the thought process itself. The prospect before us is that we shall understand that process. We shall be able to diagnose the difficulties of a . . . decision maker . . . and we shall be able to help him modify his problem solving strategies in specific ways. We have no experience yet that would allow us to judge what improvement in human decision making we might expect from the application of this new and growing knowledge. . . . Nonetheless, we have reason, I think, to be sanguine at this prospect.

Public Decision Making

In the document *Water and Choice in the Colorado Basin,* White and his colleagues (Committee on Water 1968) detailed a set of objectives and alternatives for managing the Colorado River basin. The report offers prescriptive policy suggestions while recognizing the constraints imposed by existing institutional arrangements. Thus, it proposes that a set of plans, which involve structural and nonstructural methods, be developed that explicitly recognizes both the objectives of different parties and institutional or cultural constraints that limit the range of choice.

The study of the Colorado River basin provides guidance to policymakers as to how one might develop a strategy for evaluating alternative programs. It does not, however, address the question of what is likely to emerge from this activity. The Federal Flood Loss Reduction Program, the result of the 1966 Bureau of the Budget task force, which White chaired, offers a blueprint for a plan of action. Its principal recommendation is that both structural measures (e.g. dams and protection works) and nonstructural means (e.g. land use and building regulations, warnings, and flood insurance) be considered in coping with flood problems.

Despite the public commitment to this program, there have been severe problems in implementing a multiple means strategy. The many agencies involved in the flood problem, conflicting objectives, and limited data have made it extremely difficult to coordinate programs in floodprone communities. On the positive side, a direct outcome of this task force report was the National Flood Insurance Program with its emphasis on land use regulations and building codes as a condition for subsidized insurance.

One way to facilitate communication among agencies with a common data base is to develop some form of decision support system for analyzing resource management problems. The term "decision support system" implies the use of computers to assist managers in their decision processes; support, rather than replace, managerial judgment; and improve the effectiveness of decision-making.[13] The key feature of a decision support system is that it enables policymakers and interested users to experiment with alternative programs in the confines of their office or agency. The computer plays a key role in facilitating data analysis, standardizing the data bases so that different agencies have common points of communication, and permitting relatively easy comparisons of costs and benefits of different programs.

At a descriptive level, a group at the University of Pennsylvania has developed an interactive decision support system for analysis of disaster policy in the hope that it will facilitate the decision and choice process of resource managers (see Kunreuther et al. 1978b). The modeling system can deal with sets of individual homeowners and businesses. This feature enables the user to construct representations of hazard-prone communities and examine the impact of mitigation and recovery programs on inhabitants as well as on external sectors such as federal, state, and local governments.

To illustrate the use of decision support systems in the context of a specific problem, let us consider the evaluation of alternative floodplain management problems. Any adjustment or combination of adjustments will have effects on a number of different stakeholders. Not only are the residents and businesses of the floodplain directly or indirectly affected, but so also are the general taxpayers who have to pay part of the disaster bill. Businesses and industrial concerns such as the insurance industry, financial institutions, and the construction and real estate industries are also affected by hazard mitigation and recovery programs.

Figure 6.5 illustrates the interaction among alternatives and stakeholders affected by particular measures. The first four items represent simple adjustments for dealing with floodplain management; the remaining alternatives would be combinations of several adjustments. We have listed a representative set of stakeholders affected by each of the strategies. The cells in the matrix can be used to indicate costs and benefits of any strategy. For example, a strategy of subsidized flood insurance would involve costs to floodplain res-

Strategies \ Stakeholders	Flood plain residents and businesses	Community	General taxpayers	Private sector groups	Government agencies
A. Floodproofing					
B. Subsidized insurance					
C. Land use regulations					
D. Flood control works					
E. , , Combinations of , different types , of above , approaches ,					

Figure 6.5 Strategy-stakeholder matrix

idents and businesses in the form of premiums and provide benefits in the form of claims following a disaster; the general taxpayer would incur the costs of subsidizing premiums but would benefit by having to pay for less disaster relief. The private insurance agents would have administrative costs of operating the program but would receive commissions for their efforts. Similarly, governmental agencies such as the Federal Emergency Management Agency (FEMA) would incur program costs but would help fulfill their responsibility of reducing future flood losses.

The challenge in developing a meaningful program of floodplain management is to evaluate data entered in the various cells in the matrix shown in figure 6.5 and to use criteria for selecting among them. A flexible decision support system enables policymakers to investigate the relative performance of alternative strategies in various situations—such as the 100-year flood. Sensitivity analysis can be performed to determine the impact of different socioeconomic and physical characteristics of the floodprone area or the nature of flooding on the performance of different alternatives. The computer facilitates data analysis, standardizes data bases so that policymakers can communicate with each other, and contrasts the relative performance of different strategies on different constituencies.

The computer cannot decide which policy or set of strategies is most desirable. Rather it is a tool along with benefit-cost analysis and multiobjective planning for enabling policymakers to weigh the tradeoffs among strategies and to arrive at solutions. Policymakers will still have to make relevant value judgments in determining a final course of action.

Decision support systems can also provide users with insights into the impact of other decision makers on their own activities. One of the most interesting recent experiments is the design of a system for allocating public goods among individuals. Ferejohn, Forsythe, and Noll (1977) developed an interactive computer model that enabled public broadcasting stations to allocate their budget to different programs based on the actions of others. The more stations that selected the program, the lower the cost to each individual station. After an initial set of program selections, price information on the various programs was disseminated to each individual station, and they had an opportunity to revise their choices. Within a relatively small number of iterations a stable solution was found.

The important lesson of this experiment for our purposes is the opportunity of providing decentralized information to resource managers who have to allocate a budget among a number of activities. A similar mechanism for eliciting preferences through prices may lead to more efficient allocation of scarce resources and better coordination among federal, state, and local agencies facing similar problems. In contrast to the program budgeting system where a budgetary decision must be made at regular intervals, no specific deadlines force coordination in the area of resource management. By developing an interactive system for communication, there may be opportunities for sharing data and bringing groups facing the same problems to make their decisions on budgetary allocation more systematically.

The use of decision support systems for policymaking is only as good as the assumptions made by users. In the case of resource management problems that involve a number of interested parties, each having its own objectives, detailed analyses of the impact of different programs have to be made. For such tools to be useful, there must be an explicit recognition of the criteria on which policies must be judged, as well as the constraints under which one is operating. These are the basic ingredients for any choice model, as Gilbert White has stressed in his papers on the subject and in his public service activities. He has been instrumental in awakening public and private decision makers to the need for systematically evaluating different alternatives. The extent to which we can reap the benefits of his efforts rests with our future endeavors.

Acknowledgements

We would like to acknowledge the helpful discussion and comments on an earlier draft of this paper by Mike Eleey, Ed Haefele, John Jackson, Paul Kleindorfer, Allen Kneese, Harold Marshall, Jerry Milliman, Clifford Russell, William Wallace, and William White. Support for this paper comes, in part, from the Technology Assessment and Risk Analysis Program of the National Science Foundation under Grant PRA79-11934 to Clark University under subcontract to Perceptronics Inc., and the Bundesministerium für Forschung and Technologie, FRG Contract No. 321175911 RGB 8001. Any opinions, findings, and conclusions or recommendations expressed are those of the authors and do not necessarily reflect the views of these sponsors.

Notes

1. Utility curves are unique up to a linear transformation. Hence, two end points can be arbitrarily specified and the other points on the curve estimated in relation to these two. For an excellent introduction to the properties of utility functions and ways to assess them, see Keeney and Raiffa (1976, chapter 5).

2. The seminal work in this area is by Stigler (1961). Extensions of the analysis and a comprehensive set of references on the subject appear in Rothschild (1974).

3. The nature of the activity is based on the classification scheme described by White and Haas (1975). They categorize (p. 57) different measures as modifying the causes of the hazard (e.g. cloud seeding of a hurricane), modifying vulnerability to the event (e.g. floodproofing), or distributing losses (e.g. insurance).

4. A more extensive discussion of the selection of a discount rate appears in the essay by Platt, this volume, chap. 2.

5. For purposes of this review, we will not dwell on the detailed calculations of benefits and costs. Water resources have been the subject of a number of excellent analyses using this technique. See e.g. Krutilla and Eckstein (1958); Hirshleifer, DeHaven, and Milliman (1960); and Haveman (1965). A comprehensive summary of the benefit-cost method can be found in Herfindahl and Kneese (1974).

6. For example, local cost sharing on levees ranged from 0 to 49.7 percent and channel improvements from 7.8 to 54.3 percent.

7. A more detailed discussion of this process is found in Keeney and Raiffa (1976).

8. Programming approaches for structuring and solving these problems have been developed in the literature (see Dyer 1972) and have been proposed for solving specific problems in resource management (e.g. Charnes et al. 1979).

9. For a description of these group techniques for program planning, see Delbecq, Van de Ven, and Gustafson (1975).

10. Russell (1979) contains a set of papers that describe empirical tests of alternative theories of public decision making.

11. A more detailed discussion of the impact that ordering the items has on choice can be found in Plott and Levine (1978).

12. A more detailed description of the Flood Insurance Program and its changes appears in the essay by Platt, this volume, chap. 2. The discussion here supplements Platt's historical review by calling special attention to the relationship between crisis and legislation.

13. This definition is taken from Keen and Scott-Morton's (1978) book on the subject (p. 1).

References

Burton, I., R. W. Kates, and G. White. 1978. *The Environment as Hazard*. New York.

Charnes, A., W. Cooper, K. Karwan, and W. Wallace. 1979. "A Chance-Constrained Goal Programming Model to Evaluate Response Resources for Marine Pollution Disasters." *Journal of Environmental Economics and Management* 6: 244–74.

Combs, B. and P. Slovic. 1979. " Newspaper Coverage of Causes of Death." *Journalism Quarterly* 56, no. 4: 837–43, 849.

Committee on Water. 1968. *Water and Choice in the Colorado Basin: An Example of Alternatives in Water Management*. Washington, D.C.

Delbecq A., A. Van de Ven, and D. Gustafson. 1975. *Group Techniques for Program Planning*. Glenview, Il.

Dyer, J. 1972. "Interactive Goal Programming." *Management Science* 19: 62–70.

Ferejohn, J., R. Forsythe, and R. Noll. 1977. *An Experimental Analysis of Decision Making Procedures for Discrete Public Goods: A Case Study of a Problem in Institutional Design*. California Institute of Technology, Social Science Working Paper 155. Pasadena.

Fischhoff, B., P. Slovic, and S. Lichtenstein. 1980. "Knowing What You Want: Measuring Labile Values." In ed. T. Wallsten, *Cognitive Processes in Choice and Decision Behavior*, pp. 117–41. Hillsdale, NJ.

Haefele, E. 1973. *Representative Government and Environmental Management. Baltimore. Haveman, H. 1965. Water Resource Investment in the Public Interest*. Nashville TN.

Herfindahl, C. and V. Kneese. 1974. *Economic Theory of Natural Resources*. Columbus, OH.

Hirshleifer, J., J. C. DeHaven, and J. W. Milliman. 1960. *Water Supply*. Chicago.

Howard, R. A., J. E. Matheson, and K. L. Miller. 1976. *Readings in Decision Analysis*. Menlo Park, CA.

Kahneman, D., P. Slovic, and A. Tversky. 1982. *Judgment under Uncertainty: Heuristics and Biases*. Cambridge, England.

Kates, R. 1962. *Hazard and Choice Perception in Flood Plain Management*. Research Paper no. 78, University of Chicago Department of Geography.

Keen, P. and M. Scott-Morton. 1978. *Decision Support Systems: An Organizational Perspective*. Reading, MA.

Keeney, R., and H. Raiffa. 1976. *Decisions with Multiple Objectives*. New York.

Kirkby, A. V. 1972. "Perceptions of Rainfall Variability and Agricultural and Social Adaptation to Hazard by Peasant Cultivators in the Valley of Oaxaca, Mexico." Paper presented at the 22d International Geographical Congress, Calgary, Alberta, Canada.

Kneese, A. V. and B. T. Bower. 1968. *Managing Water Quality: Economics, Technology, Institutions*. Baltimore.

Krutilla, J. V. and O. Eckstein. 1958. *Multiple Purpose River Development*. Baltimore.

Kunreuther, H., R. Ginsberg, L. Miller, P. Sagi, P. Slovic, B. Borkan, and N. Katz. 1978a. *Disaster Insurance Protection: Public Policy Lessons*. New York.

Kunreuther, H., J. Lepore, L. Miller, J. Vinso, J. Wilson, B. Borkan, B. Duffy, and N. Katz. 1978b. *An Interactive Modeling System for Disaster Policy Analysis*. Boulder, Co.

Lichtenstein, S. and P. Slovic. 1973. "Response-induced Reversals of Preference in Gambling: An Extended Replication in Las Vegas." *Journal of Experimental Psychology* 101: 16–20.

Lichtenstein, S., P. Slovic, B. Fischhoff, M. Layman, and B. Combs. 1978. "Judged Frequency of Lethal Events." *Journal of Experimental Psychology: Human Learning and Memory* 4:551–78.

Maass, A. 1966. "Benefit-cost analysis: Its Relevance to Public Investment Decisions." *Quarterly Journal of Economics* 80: 208–26.

Marshall, H. 1970. "Economic Efficiency Implications of Federal-Local Cost Sharing in Water Resource Development." *Water Resources Research* 6:673–82.

———. 1973. "Cost Sharing and Multiobjectives in Water Resource Development." *Water Resources Research* 9:1–10.

Parra, C. G. 1971. "Perception of Past Droughts in Ticul, Yucatan." In *Proceedings of the Great Plains-Rocky Mountain Meeting of the American Association of Geographers*. Colorado Springs.

Payne, J. W. 1976. "Task Complexity and Contingent Processing in Decision Making: An Information Search and Protocol Analysis." *Organizational Behavior and Human Performance* 16: 366–87.

Plott, C., and M. Levine. 1978. "A Model of Agenda Influence on Committee Decisions." *American Economic Review* 68: 146–60.

Raiffa, H. 1968. *Decision Analysis*. Reading, MA.

Rogers, E., and F. Shoemaker. 1971. *Communication of Innovations*. New York.

Rothschild, M. 1974. "Searching for the Lowest Price when the Distribution of Prices is Unknown." *Journal of Political Economy* 82: 689–711.

Russell, C., ed. 1979. *Applying Public Choice Theory: What Are the Prospects?* Washington, DC.

Saaty, T. 1977. "A Scaling Method for Priorities and Hierarchical Structures." *Journal of Mathematical Psychology* 15: 223–81.
Simon, H. A. 1959. "Theories of Decision Making in Economics and Behavioral Science." *American Economic Review* 49: 253–83.
———. 1965. *The Shape of Automation for Man and Management*. New York.
Slovic, P., B. Fischhoff, S. Lichtenstein, B. Corrigan, and B. Combs. 1977. "Preference for Insuring against Probable Small Losses: Implications for the Theory and Practice of Insurance." *Journal of Risk and Insurance* 44: 237–58.
Slovic, P., H. Kunreuther; and G. White. 1974. "Decision Processes, Rationality, and Adjustment to Natural Hazards." In *Natural Hazards: Local, National and Global*, ed. G. F. White. New York.
Steinbrugge, K. V., F. E. McClure, and A. J. Snow. 1969. *Studies in Seismicity and Earthquake Damage Statistics*. U. S. Department of Commerce Report (Appendix A), COM 71–0053. Washington, DC.
Stigler, G. 1961. "The Economics of Information." *Journal of Political Economy*. 69: 213–25.
Tversky, A. 1972. "Elimination by Aspects: A Theory of Choice." *Psychological Review* 79: 281–99.
Tversky, A., and D. Kahneman. 1974. "Judgment under Uncertainty: Heuristics and Biases." *Science* 185: 1124–31.
———. 1979. "The Framing of Decisions and the Rationality of Choice." *Science*.
Tversky, A., and S. Sattath. 1979. "Preference Trees." *Psychological Review* 86: 542–73.
U.S. Water Resources Council. 1979. "Principles and Standards for Planning Water and Related Land Resources." *Federal Register* 44 (242): 72892–990.
White, G. F. 1961. "The Choice of Use in Resources Management." *Natural Resources Journal* 1 (March): 23–40.
———. 1966. "Optimal Flood Damage Management Retrospect and Prospect." In *Water Research* ed. A. V. Kneese and S. C. Smith, pp. 251–69. Baltimore.
———. 1969. *Strategies of American Water Management*. Ann Arbor, MI.
White, G. F., ed. 1974. *Natural Hazards: Local, National and Global*. New York.
White, G. F., D. Bradley, and A. White. 1972. *Drawers of Water*. Chicago.
White, G. F. and J. Haas. 1975. *Assessment of Research on Natural Hazards*. Cambridge, MA.

7 The Citizen-Scholar: Education and Public Affairs

Elliot J. Feldman

In his classic essay "Science as a Vocation," Max Weber posed the dilemma of all socially responsible scholars: where is the line to be drawn between scholarship and citizenship? In his plea for an objective social science, Weber argued the importance of value-free inquiry. He emphasized the independence of science; yet he also insisted that the scholar could not escape the obligations of citizenship.

The quest for an objective social science, on the one hand, and for "relevance," on the other, has been a continuing dilemma in western scholarship. The Marxian response to Weber was to deny the possibility of value-free scholarship, and so to abandon the quest. There could be no knowledge for knowledge's sake, Marxists said; learning must serve the people. But this commitment denied disagreement over discovery of the public interest. If the followers of Weber tended toward irrelevance, the Marxian scholars tended toward dogmatism.

Every social science discipline has undergone demands for relevance since World War II. The challenge to represent an international power summoned many Americans into government service. The civil rights movement and the Vietnam War asked every scholar to become engaged in socially responsible activity. At first, those scholars called to Washington to serve in government were respected, admired, and generally rewarded when they returned to their universities. They rendered service to their country, applying the talent and knowledge of the academy to the needs of the nation. They rarely doubted the wisdom, or the rectitude, of answering Washington's call, and it is not without reason that many were later bewildered by the accusatory tone in revelations of covert research in intelligence and other areas. Later, however, when the American

Elliot J. Feldman is Research Associate Professor at Tufts University and Research Director at the University Consortium for Research on North America, Harvard University. He is the author of *Concorde and Dissent: Explaining High Technology Project Failures in Britain and France* and *A Practical Guide to the Conduct of Field Research in the Social Sciences*, and has coauthored or coedited five other books.

academy became identified with the policies of war and destruction in Indochina, the cry for relevance changed. Colleagues and administrators generally continued to admire and reward scholars serving and advising in Washington, but students organized against such activity and belittled the academy's strong identification with government. Responsible scholarship was to be at once humanitarian and critical, many students said, and it was to be conscious of the social consequences of public policy.

The pressures of change in the definition of service have touched everyone in American universities. Physical and applied scientists, dependent on government contracts and government financing for expensive equipment and experimentation, persevere in close relations with government expectations and objectives, but many have organized into associations of concerned scientists exploring the ethical propriety of certain research. Some social scientists became wary of government-sponsored research institutes such as the Rand Corporation; organizations such as the Association of Concerned Asian Scholars appealed for socially responsible scholarship divorced from the influence of government.

These developments served to emphasize the dilemma. How does a scholar guarantee the integrity of research and inquiry without ascending an ivory tower or closing eyes and ears to the concerns of society? There are at least four possibilities. Weber's solution was to separate scholarship from citizenship, to live in several spheres but to recognize the full scope of obligation to both. The Marxian solution was to integrate social responsibility and scholarship, to premise all intellectual inquiry on the Marxist conception of the state, and to pursue only those areas of science which promised to serve society.

A third possibility, one chosen by Gilbert White during those years of escalated contact between the American academy and government, combines Weber and Marx and adds a new dimension. Here, the scholar chooses his or her research concerns on the basis of perceived social need but attempts to conduct the inquiry free of influence from outside the inquiry itself. The scholar is committed to social and political action as follow-up to the research, based on the findings and the quality of the inquiry. But political activity is, during the course of the scholarship, kept separate. Moreover, political and social activity may be pursued outside the field of scholarship, but not under the guise of scholarship. Finally, in the new dimension added by White, the citizen-scholar teaches public affairs both within a curriculum and by example, letting his or her own activities demonstrate good citizenship and good scholarship without being didactic.

Developing the Citizen-Scholar Alternative

In developing an alternative vision of citizen-scholar, Gilbert White contributed to education in public affairs in at least five distinct career tracks. In three, the presidency of Haverford, the presidency of the Association of American Geographers (AAG), and the chairmanship of the University of Chicago Program in Public Affairs, White intentionally sought to develop a public affairs curriculum. In a fourth track, his own research, his commitment to subjects of great public consequence and his concern for reform have provided a model for students and scholars and cast White in the role of public educator, teaching a Senate committee or a government department or agency what may be required in pursuit of public responsibility. The fifth track, White's work in the American Friends Service Committee, has been personal, and White has kept it separate from his administrative and scholarly functions in the academy. In this track White has been the responsible citizen envisioned by Weber, championing causes because of deep religious and humanitarian feelings without recourse to specific research or scholarship.

Liberal Arts and Public Affairs

When Gilbert White accepted the presidency of Haverford College in 1946, there were eighty students at the Quaker institution catering to medical school aspirants through a curriculum conservative in content and style and organized around rote learning. Psychology was taught within the Department of Philosophy; political science, economics, and sociology were taught hardly at all. White had not been a university teacher and had been chosen for the administrative skill he had demonstrated in the American Friends Service Committee. The faculty in place at Haverford was skeptical about his appointment.

When White left Haverford after ten years, it had 450 students (at which level it has remained), was financially stable, and had a radically new curriculum. He recruited a young faculty which included later deans of the Fletcher School of Law and Diplomacy and of Princeton, a president of Johns Hopkins, a president of the

American Psychological Association. The Haverford junior faculty moved on to senior teaching posts at Stanford, Cornell, Berkeley, and Harvard, and programs were established which gave the school a reputation as one of the outstanding small colleges in the United States.

Haverford's program under White was built around a commitment to education for excellence and public service. However, because of the commitment to the liberal arts, "public service" was defined very broadly. On the annual "campus day," for example, the entire college community, led by the president, worked together on some project affecting the physical environment, whether clearing a path through the woods or painting an institutional building. Weekend work camps exposed students to poverty first hand. Classrooms were reorganized to stimulate participation. All faculty committees were interdepartmental in order to emphasize the community of scholars. In short, a community was created with a humanitarian conscience; the learning environment was informed by participation and mutual responsibility. The whole educational enterprise was built around the concept of public service.

Specific programs encouraging an appreciation of public service also took hold. A specially endowed visitors program brought prominent public officials from many countries to spend three weeks or more in close contact with the college community. Students interned with public service agencies, especially the American Friends Service Committee. Student government became a vital force; an honor system replaced adult supervision. Self-government and a constant awareness of government in the United States and abroad were emphasized.

Haverford remains, twenty-five years after White's presidency, the home of a public service–oriented student body in an educational program devoted to the interdisciplinary implications of government. Public service is not taught, but it is learned—through example and through individual and community experience.

Public Affairs and the Professional Geographer

White sought to develop further his citizen-scholar alternative in his capacity as a professional geographer. Because the professional geographer would have the greatest impact on public policy, White worked to stimulate recognition of the geographer's special skills. Formally, he converted a scholarly society into a professional organization, the Association of American Geographers. Membership has grown sixfold since his presidency in the early 1960s. More significantly, however, he set educational and professional standards by coaxing the discipline into ever stronger commitments to policy-relevant research.

As chairman of the University of Chicago Department of Geography, White engaged graduate students as apprentices on projects with public consequence. The educational process was featured by specific research tasks, fieldwork, and partnership among students and faculty. He thrust research results into the public arena, always setting an example for students of the urgency and importance of their work. Other essays in this volume testify to the extent to which the educational objectives were accomplished and geography has emerged as a policy-relevant profession.

Geography and Public Affairs for the Nonprofessional

For the professional's education, citizenship and scholarship are integrated because the experience focuses on research which is chosen for its public significance. As one descends the scale of higher education, this integration is less pronounced.

The college geography instructor, as White saw it, should "stress the method of sorting out and analyzing the minimum number of facts with the maximum use of explanatory models" (White 1965, p. 17). College geography, he argued, should elaborate a new high school curriculum and relate geographic knowledge to other disciplines: "Majors which require or permit undergraduate students to devote the greater part of their time to geography as now taught are a disservice to liberal arts students" (pp. 21–22). Apart from "carrying the student to more advanced understanding of problems and applications," White held that "further work would seem justified only as a service to other disciplines or as a sinewy framework for study in related fields." The "intellectual growth of students rather than imperial growth of departments" held sway in White's vision of the undergraduate program (p. 22).

The preparation of college students to understand concepts and methods, particularly in geography, depends upon their preparation in high school. As president of the AAG, White launched the High School Geography Project (HSGP), whose objective was to develop a new curriculum. After nearly a decade of work, in 1970 a package of materials entitled "Geography in an Urban Age" became available to high school programs, generating important changes for the teachers, administrators, and scholars who participated and signaling the probability of notable changes for the high school students who gradually would be introduced to the new curriculum. By 1978, some 5 percent of high school students taking geography courses in the United States (up to perhaps 250,000 students per year) were using these new materials, and the impact reached into colleges and into foreign countries, especially in Australia, Canada, and the Netherlands. Japanese and West German schools have expressed growing interest in adoption.

As White himself (1970a, p. 1) has described the final curriculum package, "there is little in the course materials that smacks of rote learning or of description of earth features for its own sake. There is much that speaks of geographic concepts and skills. The result is a product which differs significantly from most other attempts to assist geographic thinking at the high school level. There is concern for the attitudes which students develop toward their fellow members of the human race and their common habitat. Description of parts of the world is secondary." Indeed, the whole package is infused with White's essential educational philosophy: an emphasis on concepts and "regularities", a concern for process; a commitment to undermine ethnocentrism;[1] an appreciation of the limits of technology; an enunciation of general rules helping prediction—particularly of the environmental consequences of human behavior; a devotion to learning in service to mankind. White defined the HSGP from the outset with these values, and he guided it to the completion of its first instructional materials.

If the HSGP appeared to concentrate on the narrow concerns of one discipline's approach to one level of instruction, White's own vision was far more global:

> The opportunity lying ahead is to help the young people of the world recognize in similar ways the processes that account for diversity and order on the earth's surface. From such common inquiry might come a more sensitive understanding of the images which the human family shares of its domain, as well as of its own capacity to live together in peace while modifying that habitat for the human good. The time is not far off when through some international channel social and natural scientists will work together to cultivate appreciation of the different images of the globe perceived by the human race. (White 1970b, p. 71)

As the new curriculum won ever wider adoption through the 1970s, the AAG tested and revised the materials with a growing world view. In 1979 new materials emerged with greater emphasis on environmental issues and problems—indeed, with greater emphasis on the very intellectual concerns which have animated Gilbert White's career as a professional geographer.

White's approach to a special curriculum in public affairs was similar to his educational approach for the nonprofessional geographer. The undergraduate experience was intellectual and academic—not professional and not vocational. In the Public Affairs program at the University of Chicago, each student tailored an individual curriculum with faculty advisers drawn from sociology, psychology, education, economics, political science, law, geography, and anthropology. All students joined together for an introductory seminar and a senior seminar. Emphasis was placed on interdisciplinary theses requiring fieldwork, often outside the United States, and the introductory seminar involved multiple approaches to policy problems in which students interacted with prominent public officials. Credit was given for off-campus research but not for internships. Analytical skills, not the accumulation of information, dominated the program. It was the first undergraduate public policy major in the United States and, as we shall see, perhaps the only one to sustain a commitment to strictly intellectual objectives.

The Development of Public Policy Programs

Because of Gilbert White's prominence as a professional geographer, his influence on education in the discipline was widespread. His pioneering interdisciplinary contribution in public affairs, however, was not disseminated as a model for others because formal training in this field, despite its tendency to be interdisciplinary, was dominated by historians, economists, and political scientists—not geographers. Public policy education in the United States defined the fourth approach to the citizen-scholar dilemma.

In the 1970s, programs of public policy proliferated in the United States, first at the graduate level and subsequently in the undergraduate curriculum. The graduate programs derived initially from two established educational traditions—foreign service and public administration. Some new programs, however, were stimulated by scholars pursuing the field of public policy more as a domain for analysis than as an avenue to power.

The foreign service–oriented graduate programs at the Fletcher School of Law and Diplomacy, Georgetown's Foreign Service School, Johns Hopkins's School of Advanced International Studies, and Princeton's Woodrow Wilson School all grew out of educational philosophies emphasizing the role of the mandarin: students were to acquire a keen sense of history and a world view; they were trained chiefly in programs awarding master's, not doctoral, degrees; and they were prepared to move into public service agencies and government offices more than into educational institutions. The schools focused on international relations and diplomatic history and on area studies; they did not teach the developing analytical skills of social science.

Public administration schools at the University of Pennsylvania, Harvard, and elsewhere, offered curricula as ethnocentric as the foreign service programs were international. They concentrated on certain skills, such as accounting and budgeting, personnel relations, and organizational behavior and management, but they moved less quickly into the interdisciplinary areas of political economy which were winning attention in Washington policy circles. They were essentially vocational programs training students for public responsibility.

In the 1960s and 1970s, graduate programs grew in essentially two directions. Business schools introduced public administration programs of their own, reasoning that the public sector was growing large enough to justify training students with business skills that would serve the public as well as the private sector. This development was informed by a philosophy which encouraged government to be treated as a business. "Efficiency," "cost-effectiveness," "streamlining," and other terms from the business world became the jargon of government in the late 1960s and into the 1970s, spurred by the infusion of businessmen (such as Robert McNamara, who led the government-management revolution) into positions of government authority.

The second policy direction at the graduate level employed many of the skills and concerns of the business schools but with greater emphasis on social science and the academy. Jack Walker (1976), in explaining the conversion of the nation's oldest school of public administration at the University of Michigan into the first graduate program in public policy analysis, observed that "graduate programs in public policy analysis were born in the late 1960s out of frustration both with the staggering ineptitude of governments struggling to launch new social programs and with the growing intellectual irrelevance of traditional courses in public administration" (p. 90). New graduate programs specifically for instruction in public policy thus emerged. Aaron Wildavsky (1979), the first dean of the Graduate School of Public Policy at Berkeley, said, "Look at it this way: after World War II, the United States, aware of other countries' difficulties, established numerous centers to study foreign areas; by the late sixties people realized the United States had problems, too; thus they started schools of public policy" (p. 408).

The Berkeley school, like the John F. Kennedy School of Government at Harvard, Michigan's Institute of Public Policy Studies, and others, grew out of a program in public administration and was designed to prepare students for public service. Government was growing faster than American higher education was training appropriate personnel; by the late 1960s, fewer than a thousand students were graduating annually with master's degrees in public administration. Government was therefore being staffed mostly with lawyers and graduates of business schools (Walker 1976, p. 91). Walker has described the objectives of Michigan's Institute in light of the prevailing critique of public administration:

> We aim to produce the administrative generalists of the future—modern public managers and policy planners who combine the latest tools of problem solving and quantitative analysis with a subtle understanding of the principal social, political and economic processes at work in the public sector. We would like to produce graduates who will be able to approach the daily routines of government with a creative, inquisitive, questioning spirit. (Ibid., p. 92)

The programs thus were professional and action-oriented, as Wildavsky (1979, p. 413) has explained: "Whereas scholars in the liberal arts tradition want to understand their material as fully as possible (whether or not anybody can do anything about the situation), a policy analyst would rather figure out who can change what with how much effort."

Wildavsky's definition of the policy analyst was shared broadly in the development of graduate curricula. Emphasis was placed in all programs on the master's degree, following the format of the foreign studies centers and schools of public administration. Education was defined as training for positions of action and decision making. The Berkeley program is formally and officially devoted to American domestic public policy. Not until 1978–79 did Michigan's Master's of Public Policy program introduce a track with international emphases, and the Kennedy School, while vigorously recruiting future leaders from around the world, concentrates its curriculum almost exclusively on American cases, American methods, and American experience.

The graduate approach to policy, then, has been professional, skill-oriented, and ethnocentric. Undergraduate programs tended to be imitative. The Ford Foundation was crucial in launching the graduate programs; the Alfred P. Sloan Foundation underwrote most of the undergraduate initiatives (Chicago's was a notable exception in developing prior to Sloan's interest, but later benefited briefly from Sloan support). Like Ford, Sloan had little sense at the outset of how a program should be designed, and encouraged creative and diverse approaches. Faculty representing some twenty-five programs from different universities and colleges were invited by Sloan in 1976 to confer and to share experiences; although each program subsequently reflected the peculiar needs and capabilities of the respective institution, they all acquired certain common characteristics.[2]

First, all the programs shared a concern for a structured curriculum. The master's programs had organized orderly courses, but the very few doctoral candidates in public policy at, for example, Berkeley and Carnegie-Mellon were invited to shape their own programs around interdisciplinary interests. Gilbert White was unique in applying this principle to undergraduates.

The Chicago Program in Public Affairs was unique, too, in rejecting internships. Every other undergraduate program has emphasized the internship experience; MIT made it a program centerpiece, Princeton the pinnacle of an undergraduate education. Duke, which formed a department in public policy studies six years after Chicago's major was launched (and three years after Chicago graduated its first students with degrees in Public Affairs), called internships "policy-oriented field experience," a wholly different conception from White's field research requirement in which students worked independently in the development of an honors thesis.

The first undergraduate public policy programs (apart from Chicago's) developed in relation to graduate programs already in place. Duke's program, Sloan's model for predominantly liberal arts schools, emerged out of the graduate program of the Institute of Policy Sciences and Public Affairs. MIT, which offered Sloan a complementary model for an essentially technical school, sold its undergraduate major in relation to the Technology and Policy Program established in 1975 and already funded by the Sloan Foundation. Almost inevitably and despite their publicly stated preferences (see Smith 1974),[3] these programs began to promise undergraduates "enough analytical competence and knowledge of specific problems to perform effectively in public policy-related employment," or preparation for "careers in government, in work affecting public affairs, or for graduate work in law, public policy, management or social sciences."[4]

Other schools have tended to promise less and to build more on existing liberal arts commitments. Williams College, for example, used Sloan monies to enhance its thirty-year-old program in political economy. Oberlin launched, with Sloan's help, a concentration, not a major, the "Public Service Studies Program," in order to "train students in the application of analytical problem-solving methods to public concerns and to further multi-disciplinary research on contemporary public policy issues." Similarly, Dartmouth developed a major utilizing faculty in engineering, business, medicine, chemistry, and philosophy: "The intention is not to produce "policy specialists," but to give students with a wide variety of career interests an opportunity to develop and test an integrated, multidisciplinary approach to problem-solving."[5] And Lawrence University in Appleton, Wisconsin, perhaps the most ambitious and innovative of all, joined an undergraduate emphasis in political science and economics to the natural sciences in order to stimulate studies of the environment, health, energy, and other multidisciplinary problems.

The Sloan Foundation gave its biggest grants in the early 1980s to an ethnocentric interdisciplinary training program at Stanford and a vocational problem-solving program operating out of the Woodrow Wilson School at Princeton. Nevertheless, grants were renewed at Lawrence, Oberlin, Swarthmore, and Williams, and fledgling programs oriented to science and environmental policy emerged at Smith and Carleton Colleges. They reflected, more than the newest and most expensive efforts, the commitments underwriting Chicago's pioneer project of 1966. Despite considerable change in the Chicago program, a decade after its founding it still described itself in terms consistent with its original mission:

> "The Public Affairs program is concerned with the study of public policy, both as a theoretical concept and as theories of policy are manifested in specific programs and jurisdic-

tions . . . the specific topics and curriculum chosen by students should relate policy and program to the larger questions growing out of a concern for social change and public policy."[6]

Hence, Gilbert White's creation eschewed professional or vocational training for undergraduates, just as his concept of undergraduate education for geographers had done. It emphasized concepts and theories, but required the inculcation of abilities to apply theory.[7]

Where Is Public Affairs Education Going, and Do We Want to Go There?

Consideration in detail of university programs in public affairs and public policy reveals two contradictory trends, and the stronger trend offers an approach which rejects the central principles of education promoted by Gilbert White. It is important to appreciate these developments because, whatever White's impact in geography, future key policymakers in all fields will be the products of the prestigious public policy schools.

Public policy programs grew rapidly into the early 1980s, despite doubts about field definition, study content, or even lasting value (see Smith 1974). The initial inspiration for them, outside the traditional foreign service orientation, came from a belief in the capacity of government to succeed where the private sector had failed—in civil rights and in the reduction of inequality and poverty. They got underway as government began to fail, at home in the War on Poverty, and abroad in the war in Southeast Asia. The idea thus began with a commitment to learn the requisite skills, to appreciate the inescapable complexities, and then to make government work. Programs developed in the continuing belief that government could do the job but that the occupants of administrative and legislative responsibility and power were incompetent. More and better-trained talent might succeed. The programs thus developed a continuing rationale.

The social and political changes of the 1970s did not reduce the drive to establish programs in public policy, but they did alter radically the terms and expectations. The microeconomic theory that dominated the Kennedy School of Government and Carnegie-Mellon's School of Urban and Public Affairs gave way to greater appreciation of the limits of "rational" decisions. The early concern expressed at the University of Chicago for theory yielded in other universities to an emphasis on "implementation" (see e.g. Pressman and Wildavsky 1973 or Bardach 1977).[8] And, most important, the ultimate faith in government of the 1960s dissipated in the next decade. Thus, the fourth approach to the citizen-scholar dilemma was defined as vocational training. Weber's student was to be educated and value-free; Marx's was to be education and value-laden; the American academy's was to be trained and value-less.

Programs became more pragmatic and more utilitarian. History and philosophy fell to quantitative analyses and problem solving. Policy-related programs grew up in schools with technical emphasis, but only to subordinate synthesizing skills to analytical skills and to train specialists who would exercise influence, not generalists who would occupy positions of responsibility. Business schools, especially, developed programs in public management (the Kennedy School and the Harvard Business School established formal links), just as businesses and corporations restored an aggressive approach to the political arena after a decade of political retreat and defensiveness. Business schools offered courses in "Business, Government, and Society," in which ethics lay in the domain of self-regulation once again. Political action committees of American corporate giants told the public that government prevented the private sector from performing its full public service—decontrol oil, they said, to make capital available for more exploration; deregulate civil aviation to improve competitive service, and so forth. The deemphasis once again on foreign affairs, moreover, helped mask the protectionism American industry won from Congress. The private sector, just as Adam Smith had promised, could serve the public better than a large public sector—or so business wanted Americans to believe at the end of the seventies.

This mood has helped divert the curricula of public policy programs to instrumental and vocational directions. Practical experience, in this mood, replaces practical research; problem solving replaces problem definition and inquiry; theories for prescription replace theories for analysis; action replaces thoughtful discussion. Technocrats armed with superior methods supplant generalists with scope, and the ethnocentricity of their preparation is no barrier to making the whole world fall within their expertise.

At the graduate level such developments seem to follow the market realities of scarce jobs and the pressure to provide students with career prospects. Undergraduate education tends to follow the trend of the graduate programs. There is, nevertheless, counterdevelopment (as in the new programs at Oberlin and Dartmouth, Chicago and Lawrence). There do remain pockets of resistance in the liberal arts against the pressures for vocational preparation. They represent the continuing distinction between the undergraduate and the graduate curricula, between the academy and the marketplace.

The swing between confidence in the public sector and confidence in the private sector to solve social problems is not unusual, of course, in the United States. American values oscillate between the individual's right and ability to serve the public good through selfish acts and the community's right and need to dictate individual behavior (or at least to regulate it) in the public interest. The key difference this time around seems to lie in the corporate world's financial and political ability to fill the vacuum of public confidence with its own dominating values, and the inclination of higher education to follow the popular trend instead of trying to generate a trend of its own.

Educational programs, like legislatures, are reactive to the public mood, but the developed programs of the academy bear the burden of social responsibility to lead through informed discussion and the intelligent setting of priorities. The desire for social change which motivated the first public policy programs at Chicago, Berkeley, Michigan, and Carnegie-Mellon seems to be surrendering to the desire of individual students for jobs and prestige and of faculty for access to and the exercise of power.

The line between the citizen and the scholar is fine, and the tendency toward ambiguity is probably greatest in educational programs devoted to the study of public affairs. If the Marxian response to the citizen-scholar dilemma is not to prevail (guided, of course, by a different ideology), vigilance is needed most in these very programs, not only to protect the public's interest in an education which distinguishes facts from values, but also to protect the integrity of the academy in its pursuit of scholarship before influence, truth before power.

Principles and Programs: Education for Citizens

> The task of the teacher is to serve the students with his knowledge and scientific experience and not to imprint upon them his personal political views. (Weber 1958, p. 146)

Neither Weber nor Marx was able to demonstrate convincingly that all personal sympathies can be stricken from the scientific enterprise, but neither wanted, on the other hand, to strike them from the behavior and commitments of responsible citizens. Thus, public policy programs derived from business schools rely heavily on the ideology of capitalism; programs emerging from schools of public administration accept a gospel of bureaucracy; programs rooted in work experience accept the premises of certain institutional arrangements. None acquires the generalist's view because specialists are not asked to distinguish the forest from the trees.

The war in Vietnam proved to Gilbert White that the citizen and the scholar were not always divisible. On other public issues—flood control, civil rights, nuclear safety, education—he could examine legitimate alternatives, organize responses, even marshal resources in support of particular analyses. Initially he sought a professional alternative in dealing with the Vietnam War—major water development on the Mekong River (see vol. I, selection 12). But with the failure of this initiative he was reduced to public protest, simultaneously through his position on the faculty of the University of Chicago and through his activity in the American Friends Service Committee (for the latter, see Jonas 1971). He failed to persuade his Chicago colleagues. The Friends, by contrast, bore witness on the steps of the White House.

White's approach to the citizen-scholar dilemma, through example, practice, and varieties of formal education at different levels, involves two critical assumptions: first, that the individual can sustain several discrete career tracks, which White himself found difficult with respect to Vietnam; and, second, that humanitarianism ultimately motivates every citizen. This second assumption encourages a world view fostered in liberal arts education, and the apparent shift away from the liberal arts to practical training may imply this assumption's decreasing reliability.

Gilbert White may have found a way out of the more unfortunate implications of these assumptions by concentrating on international activities and by adopting a world view shared by citizens from many different countries. He had always repudiated ethnocentrism, and in his concern for environmental protection he found multinational sympathy. Thus, he assumed the presidency of the Scientific Committee on Problems of the Environment (SCOPE), a nongovernmental organization created by the members of the International Council of Scientific Unions—including sixty-six national academies of science and some thirty other international organizations dedicated to international scientific research. He took the lead in organizing an unusually successful United Nations conference in Buenos

Aires on the world's water supplies. His various research efforts in the United States and abroad (reviewed elsewhere in this volume) have taken on a more obvious distinguishing characteristic: he has evolved more prominently into a citizen of the world community.

Because the world's political, social, cultural, and ethnic boundaries are man-made, most social scientists must direct their scholarship to specific countries, regions, or peoples. The geographer is concerned, in contrast, with the relationship between people and the earth. The science governing the earth's patterns and regularities permits the geographer the pursuit of scholarship anywhere. Sociologists, anthropologists, political scientists, psychologists, and economists have all searched for universal theories to explain social organization, economic systems, government, and human behavior: all have found that the range of factors affecting these concerns exceeds man's present capacities to calculate with confidence. There are no constant variables over time except, perhaps, the common biological and physical needs of man. For the geographer, however, there is the constant variable of the earth itself, whose divisions are—in contrast to social, political, and economic organizations—entirely natural.

The potential for a geographer to evolve into a world citizen probably surpasses that of professionals in other social sciences because of the uniqueness of the discipline. Even so, implementing measures of environmental protection and policies for feeding and sheltering the world population depends upon the politics of different countries and peoples. Establishing international credibility depends on more than the acknowledgement of universally applicable skills; it depends, too, on a certain political neutrality in policy choices. Hence, as White observed in one salient example, "meeting the world's food needs is not a simple matter of generously spreading American know-how wherever it is lacking" (1960, p. 70). Technique must be complemented by understanding; policy must be informed by politics; the scholarship of any discipline must profit from the scholarship of every other.

Formal education for public affairs over the last decade has reduced the appreciation for politics to an emphasis on skills and techniques. The curricula generally remain ethnocentric, barring the way to a critical recognition of the limitations in American know-how, both for domestic problems and, even more, for foreign and international ones.

This general direction in the academy raises final questions for the present discussion: Can citizens of any country be educated deliberately to become citizens of the world—responsible in public affairs, caring for the protection of the earth and its people? Can citizens of any country be responsible without an awareness of their membership in a world community which shares the planet? Despite Gilbert White's living example, and despite his efforts to infuse the teaching of public affairs with a comprehensive world view, the challenge to preserve the relationship between the liberal arts and the professions remains; so does the challenge to promote conceptual understandings without sacrificing methodological rigor, and the challenge to teach philosophy as well as politics, ethics as well as technology. Education in public affairs appears to be moving away from these challenges. Confirmation of such a trend will likely make the citizen-scholar dilemma more acute, the White model even rarer, and the supply of world citizens sadly diminished.

Acknowledgements

The author wishes to thank several people for writing or providing lengthy interviews: Alan Altshuler, Brian Berry, Keith Boone, MacAlister Brown, Joel L. Fleishman, Nicholas Helburn, Theodore J. Lowi, Wallace McCaffrey, David McClelland, Duncan MacRae, Sam Natoli, William Pattison, B. Guy Peters, Richard Warch, Martha Weinberg, Anne White, and Laurence Wylie; special thanks to Robert Kates for much-needed guidance in the preparation of this essay.

Notes

1. On concepts and "regularities," see especially White 1967. The concern to undermine ethnocentrism is expressed with particular strength in White 1960, pp. 165–70.

2. Programs for undergraduates receiving early Sloan Foundation support include Duke, University of North Carolina, Swarthmore, Dartmouth, Williams, Oberlin, MIT, and Tulane; Lawrence (Wisconsin) received support sometime later, and Sloan has continued to support new programs at a variety of institutions.

3. Smith reports on an interview with Duke director Joel L. Fleishman: "Undergraduate policy analysis is regarded as a form of

liberal arts curriculum, ideally suited for the future lawyer, physician, etc.—i.e. for those going on to professional training in *other* fields'' (emphasis in the original, p. 37).

4. Brochure of the Institute of Policy Sciences and Public Affairs, Duke University, p. 9; brochure of the MIT Public Policy Program in Political Science.

5. Brochure of the Public Service Studies Program, Oberlin College, 1978–79; catalogue of Dartmouth College, Section on Policy Studies, p. 350.

6. *Announcements* of The College, University of Chicago, 1977–79 (27 September 1977), p. 135.

7. The University of Chicago developed a second undergraduate policy concentration called PERL (Politics, Economics, Rhetoric, and Law), but students tried to convert the program into prelaw training; the faculty suspended admissions in 1980 so that the program could be reassessed, and modifications were introduced to preserve the liberal arts objective.

8. Implementation studies also emphasized distinctions in ''phases'' or ''stages'' of the ''policy process,'' breaking policy down before recognizing its implications as a whole.

References

Bardach, Eugene. 1977. *The Implementation Game*. Cambridge, MA.

Jonas, Gerald. 1971. *On Doing Good*. New York.

Pressman, Jeffrey L., and Aaron Wildavsky. 1973. *Implementation*. Berkeley, CA.

Smith, David G. 1974. ''Policy Analysis for Undergraduates.'' A report to the Committee on Public Policy and Social Organization at the Ford Foundation.

Walker, Jack L. 1976. ''The Curriculum in Public Policy Studies at the University of Michigan: Notes on the Ups and Downs of IPPS.'' *Urban Analysis* 3.

Weber, Max. 1958. ''Science as a Vocation.'' In *From Max Weber: Essays in Sociology,* ed. H H. Gerth and C. Wright Mills. New York.

White, Gilbert F. 1960. ''The Changing Dimensions of the World Community.'' *Journal of Geography* 59, no. 4 (April).

———. 1965. ''Geography in Liberal Education.'' In American Association of Geographers, *A Report of the Geography in Liberal Education Project*.

———. 1967. ''Rediscovering the Earth.'' *Bulletin of the National Association of Secondary School Principals* 316 (February): 1–9.

———. 1970a. "Assessment in Midstream." In Donald J. Patton and others, *From Geographic Discipline to Inquiring Student: Final Report on the High School Geography Project*. American Association of Geographers.

———. 1970b. "Next Steps." In Patton et al. (see White 1970a).

Wildavsky, Aaron. 1979. *Speaking Truth to Power: The Art and Craft of Policy Analysis*. Boston.

8 Awareness of Climate

F. Kenneth Hare and W. R. Derrick Sewell

In the past two decades the scientific study of climate has undergone major changes and has reached higher levels of sophistication in the kinds of question addressed. It has moved from a simple concern for the atmosphere to an analysis of complete systems in which climate is a part—treating its place in natural ecosystems and the human economy. Policymakers have begun to take note of its findings. How have these changes been brought about? Why is there now a greater awareness of climate among professionals and laymen alike?

Such awareness appears to follow a well-defined cycle, rather like that described by Downs (1972) for political problems (figure 8.1).

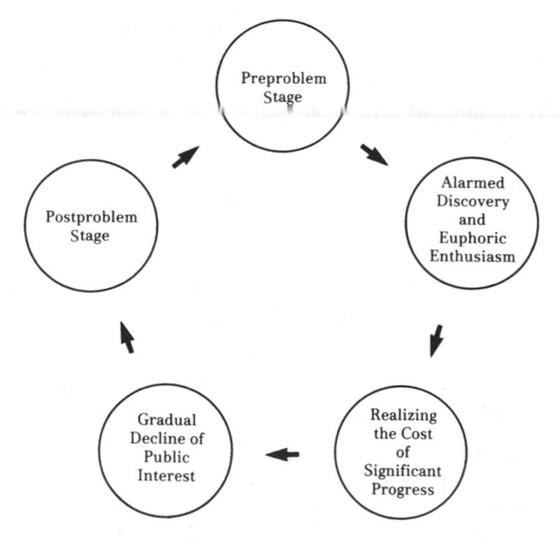

Figure 8.1 The cycle of issue attention in politics (from Downs 1972)

F. Kenneth Hare is Provost of Trinity College and University Professor of Geography, University of Toronto, and Chairman of the Canadian Climate Planning Board. His professional training was as a meteorologist. He is editor of *The Experiment of Life* and coauthor of *Climate Canada*. W. R. Derrick Sewell is Professor of Geography at the University of Victoria, British Columbia. An internationally known authority on the development of natural resources and the management of the environment, he is author, coauthor, or editor of some thirty-four books and monographs, including *National Resources and a Democratic Society: Public Participation in Decision-making* and, most recently, *Water Resources Planning in Australia: From Myths to Reality*.

Such attention cycles begin with the recognition of stress by individuals. This evolves into collective alarm or sensitivity, at which time there are public calls for action. These lead to realization of the costs, whether action is taken or not. There is then a decline of concern, and public attention ultimately passes to some other issue. As we will see later, this cycle has already completed one revolution in water resource policy, thereby influencing the pattern of research quite markedly. The present awareness of climate displays many similarities. It seems now to be well into the second stage of Downs's analysis.

The attention cycle has been at work among professionals as well as the public. The professionals' shift toward a climatic perspective rests on four main bases: the intellectual revolution in the atmospheric sciences brought about by the digital computer, satellite observations, and planetary exploration, all of which have drawn attention to long-term processes; the stimulus provided by a number of leading innovators, including both theoreticians and applied scientists; the impact of a series of climatic shocks on public opinion, leading to political pressure for effective climate programs; and the perceived vulnerability of the world's food system, especially in the Third World, to climatic stress in the presence of heavy population pressure.

Leadership plays a role in all such changes in perception. A handful of individuals in a few countries has played an important part in creating public awareness of climate. Most have been scientists, but some have been journalists. A few have been politicians. The definitive history of the change will probably give pride of place to Reid Bryson, who, a decade ahead of his colleagues, was speaking vehemently about the need to take climate seriously. So was Hermann Flohn—in many ways the founder of modern broad-scale applied climatology—in the Federal Republic of Germany. Among international leaders, the most notable has been Robert M. White, longtime administrator of the National Oceanographic and Atmospheric Administration (NOAA), formerly president of the U.S. Universities Commission for Atmospheric Research (he became president of the National Academy of Engineering in 1983) and the main force behind the World Climate Conference of 1979. And there were several journalists, notably John Gribbin in the United Kingdon and Walter Sullivan of the *New York Times*.

Climatic variation and variability, like other features of the natural environment, present mankind with a range of choices (Kates, Sewell, and Phillips 1968): to do nothing, the fatalistic attitude; to alter activities to accommodate the atmospheric phenomenon; or to alter the phenomenon to suit the activities.

The usual choice has been the second, especially in developed countries. Development of drought-resistant crops, better insulation of houses, and wiser scheduling of field operations are cases in point. In recent decades, however, there has been growing awareness of the third choice. The transformation of nature has been a major objective of public policy in the socialist countries, especially the Soviet Union. To an increasing degree the possibilities of altering the weather are being considered in North America as well. All such choices depend on popular awareness of climate—and on the technical competence of institutions, which in turn depends on professional awareness.

Adaptation to Climate

A large body of technology constitutes a deliberate response to climate, though this fact is taken so much for granted that it is usually forgotten. Agriculture as a whole is an effort to harvest atmospheric and solar resources. It is usually and correctly seen as exploitation of the soil. But it is even more truly the exploitation of atmospheric carbon dioxide, rainfall, and solar and atmospheric radiation. The soil itself is largely conditioned by the climate. These obvious truths have long taken a back seat in political and economic argument about food production. Much the same has been true of energy production and consumption and of transportation. The exploitation of climate as a resource has been left to the farmer, the herdsman, and the geneticist. It has not been clearly articulated in economic or political terms.

This is not the place for a history of technological adaptation to climate. Maunder's *The Value of the Weather* (1970) adopts an economic rather than a technological approach. We do not know of a truly comprehensive history that analyzes the devices whereby mankind has adapted to its climate. Obviously such adaptation has been crucial through the whole history of agriculture, transportation, architecture, and power development of all kinds, from sails to windmills to modern hydroelectric generation. There is a large body of literature that details the history of each of these fields. But rarely do such studies render explicit the adaptation to climate and its variability, even when this is the main objective of the technology—for example, in irrigation or crop hybrid production.

In effect, society's response to climate has been subliminal. We become dimly aware of climate by exposure to it, without articulating our response in a systematic way. From time to time extreme events shatter this complacency, and for a while there is a more conscious attempt to come to terms with climatic adaptation. This was true in the late 1930s, in the wake of severe drought in the crop-producing regions of mid-latitudes. The U.S. Soil Conservation Service and the Canadian Prairie Farm Rehabilitation Administration, which dealt with soil losses in the affected areas, were among the pioneers of climate-related farm practices designed to minimize such damage. The development of new technological methods of avoiding climatic stress was also a major (if elusive) objective of Soviet agricultural policies through the Stalin era and into the 1980s.

Anyone familiar with the western plains of North America knows, however, that the memory of such stresses fades quickly (Saarinen 1966). In the 1970s one could again see in the fields of the Canadian and U.S. grain belts the climatically hazardous systems of tillage and cropping that lead to quick profit in good seasons and to soil losses in bad. The bland weather of recent decades, suitable for good corn and wheat yields, also engendered false optimism in the minds of agricultural economists and plant breeders (Schneider 1976).

The present phase of heightened awareness began with the Sahelian drought and its devastating consequences. Cold winters in eastern and central North America since 1976, coupled with rising fuel costs, also had their impact. So did the 1976 drought in northwest Europe, and that of 1976–77 in California. The awareness, moreover, is now worldwide; the first World Climate Conference was held in Geneva in February 1979. It was well attended.

The political response to the drought-induced food shortages of 1973–74 was a World Food Conference, under full United Nations auspices. Climatic factors, however, received little attention, and the scientific community played only a minor role. In contrast, the World Climate Conference was organized by the scientific and technical communities. The atmosphere, it seemed, had at last been discovered as a resource by those who had been responsible for its study—the scientists themselves. Henceforth it would be regarded by them as a prime factor in human adaptation to the environment.

One aspect of such adaptation that recurs constantly is the means of husbanding rainfall in the interest of human welfare. Water resource policy—the main lifelong concern of Gilbert White—has been much influenced by hydrological and climatological research and by the capabilities of the engineering profession. The history of water resources research has much to teach the climatologist. It has been marked by a broadening of perspective from sole emphasis on physical science and engineering to an acute awareness of human and environmental concern. Many other disciplines have been embraced. The pages of the American Geophysical Union's quarterly, *Water Resources Research*, illustrate this evolution; contributions on ecology, economics, and various environmental aspects are interleaved with articles on engineering.

Crisis had much to do with the change—and still does. The U.N. Water Conference in 1977 was a largely political conference where anxiety about the future was the dominant motive. Impending water shortages, serious declines in water quality, and increasing conflicts in water use have stimulated many governments to broaden and deepen water research, improve the process and content of planning, and make major alterations in legislation, policies, and administrative structures. In several European countries there have been radical changes in approach (Johnson and Brown 1976; Nicolazo-Crach and Lefroy 1977; Sewell and Barr, 1978) involving the adoption of a national perspective and new strategies for dealing with problems of pollution. In North America, too, there have been attempts to broaden the outlook of water management, in part through the enactment of comprehensive legislation, in part through the establishment of water resources research centers, and in part through the introduction of new policies (OECD 1972; U.S. National Water Commission 1973).

There have also been some responses to concerns about climatic resources, but in comparison with action in connection with water resources they are at a much earlier stage in the cycle of issue attention. We believe that, with growing awareness of the former resources, their problems, and their possibilities, high levels of sophistication will appear.

The Growth of a Climatic Perspective among Atmospheric Scientists

"Climate" is a layman's word. Among professional scientists it was accorded scant use until recent years. Meteorologists paid little attention to the longer-term aspects of atmospheric behavior and gave the study of climate low priority. It was taken more seriously by geographers and biologists, who saw it as a central element in the physical environment of mankind and the biota, and who occasionally (as with Griffith Taylor and Ellsworth Huntingdon) expressed strongly deterministic opinions. The major figures of nineteenth- and early twentieth-century climatology tended to come from these fields, which were often regarded with skepticism by the physicists and mathematicians who founded the science of meteorology. The intellectual thread that ran from de Candolle through Köppen to Thornthwaite, Budyko, and Flohn was spun by nonmeteorologists. Geographers have continued to support the cause of climatology and to do excellent work on problems that meteorologists ignored.

Skepticism, faltering but still alive, had its roots in the methods of the mathematical physicist. The working tool of the theoreticians of the nineteenth and twentieth centuries—from Laplace, von Helmholtz, and Kelvin through V. and J. Bjerknes, Jeffreys, and Rossby to modern dynamicists such as Eady, Charney, Leith, Lorenz, and Marchuk—was the differential equation, which specifies instantaneous or local relations between physical variables. The general equations of atmospheric motion include a statement of the balance between forces and changes of momentum that apply at a given instant and a given point (or in a given parcel of air). They are admirably fitted to express the facts about weather, which has little duration in time. But they are much less easily used in the description of long-enduring and spatially extensive systems.

This inevitable concentration upon differential equations, plus the emphasis given by meteorologists to short-term forecasting, assigned intellectual priority to immediate events and distributions. Until recently, however, the usually nonlinear equations defied integration. They were more useful in diagnosis and analysis than in prediction or generalization over time. As recently as the 1930s and 1940s there was no obvious way in which a classically trained meteorologist could approach the climatic problem with any confidence. Those scientists who did, such as geographers and biologists, were written off as amateurs. Those who made climate an integral part of the natural environment were convicted of that cardinal sin among physicists, hand-waving. Meteorologists who saw climate as average weather were viewed as mere statisticians—who were also held in low repute in the early days.

The advent of the computer and numerical analysis since the Second World War has greatly changed these attitudes. Integration of the general equations is now achieved routinely for entire hemispheres, as part of the numerical process of weather prediction. Such methods have also been used to develop general circulation models (GCMs) of the atmosphere and ocean. These GCMs generate synthetic climates that are not unlike those of real time and space. Under the impact of this sweeping change, climatic ideas and concepts have become fashionable among dynamic meteorologists and oceanographers. The numerical modeling of climate is being intensively cultivated where only twenty-five years ago it was an outlandish idea. And statistical methods have been discovered to be integral to the dynamical study of atmospheric motion. The Jeffreys stress tensor, in fact, includes a set of quite ordinary covariances.

Climate is, in essence, an integration with respect to time. To approach it quantitatively one needs a set of integral equations, linked with a set of frequency distributions, a counting of discrete events, and a calculus whereby three-dimensional distributions can be handled as fields. The mathematical tools exist, as do the needed computers. As we have gained experience in the use of such methods, it has become easier to get a grasp on world climate, as distinct from the local perspective typical of the precomputer age.

This grasp has led many meteorologists to move from the view that climate is a purely statistical concept in which it is nothing more than the sum of its parts to the more ambitious stance that makes climate "a basic physical entity, and weather [illegible] the momentary, transient behavior of the atmosphere striving to satisfy the requirements dictated by the climate for horizontal and vertical transfer of mass, momentum, and energy" (Mitchell 1977). In this view, climate—the enduring properties and behavior of the atmosphere—is the central object of study for the atmospheric scientist. Weather events arise from the need to redistribute various quantities; they are, as it were, the molecules of climate.

While this transformation of outlook has been going on, the idea of climate has itself broadened. We now talk of the climatic system, defined as "those properties and processes that are responsible for the climate and its variations" (U.S. National Academy of Sciences 1975b). It is widely recognized that these properties and processes involve the behavior and characteristics of the biota, soil, ice masses, and oceans. To comprehend the causes of climate we must draw on the skills of specialists in these areas. Meteorologists have come to this view cautiously but irreversibly. Biologists, glaciologists, oceanographers, geographers, geologists, geomorphologists, and soil scientists have felt the same pull. Climatology, once the Cinderella of the atmospheric sciences, has become an interdisciplinary field of wide appeal. As this has happened, its intellectual challenge has grown enormously. A major gain has been the creation of interdisciplinary teams such as CLIMAP, in which geologists of the caliber of John Imbrie and geochemists such as Nicholas Shackleton have joined forces to thrust forward the study of paleoclimate, with the aid of meteorologists such as John Kutzbach, Yale Mintz, Jill (Williams) Jäger, and Lawrence Gates (U.S. National Academy of Sciences 1975b).

What brought about this change? One response is that social and political pressure was responsible, following the dramatic events of the 1970s. In our view these events only accelerated a change that would have occurred in any case. Once the digital computer had appeared, entirely new vistas of data management and mathematical analysis opened up. The group of pioneers at the Institute for Advanced Studies at Princeton, headed by John von Neumann and his young colleague Jule Charney, showed as early as 1949 that the new technology would allow numerical integration of the nonlinear differential equations that had so long stood in the way of truly climatological solutions. This path has been followed with immense enthusiasm. The numerical modelers dominate the atmospheric sciences and play an increasing role in oceanography.

In the same way, satellite sensing of the atmosphere, ocean, and continental surfaces has transformed the observational base of each of the relevant sciences, particularly meteorology. The work of Verner Suomi was paramount in bringing this about. An immense flood of new information has invaded the climatological field. Future plans call for an even greater effort. Satellites can see the remote places, including the oceans, the southern hemisphere, and the Antarctic continent. A global view is now a continuous reality, thanks to major international efforts such as the Global Atmospheric Research Program (GARP) of the World Meteorological Organization and the International Council of Scientific Unions. The layman can test this assertion by turning to the daily National Oceanographic and Atmospheric Administration (NOAA) television weather briefing AM-Weather, which transmits movies of the satellite record for the past twenty-four hours over much of the western hemisphere, from pole to pole.

The professional response to these new stimuli has been retarded by organizational conservatism and the lack of able people to use the available resources. One former president of Britain's Royal Meteorological Society said acidly that the combined intellectual achievement of the atmospheric sciences to that time (two decades ago) did not add up to one Nobel prizewinner's output. This was unfair and is certainly untrue. But it underscores a real problem. The expanding horizons of the atmospheric sciences came at a time when the other sciences were also growing rapidly in resources and appeal. The climatic problem has had to compete with space exploration, crustal geophysics, molecular biology, and many other activities that attract the bright young research student. In the circumstances, climate has done quite well. Its high visibility in the 1970s has certainly helped.

There have, moreover, been some long-standing achievements, on which we can now build. The technology of weather modification (discussed below) is one area where the socioeconomic impact of modified weather had to be taken into account. The remarkable and long-sustained work of the Illinois State Water Board, led by Stanley A. Changnon, has had a highly significant impact throughout the Midwest, and further afield. So have the various hail suppression projects. The Soviet Union's scientific community has also taken climate seriously for decades. In all these cases the need to relate climatological knowledge to socioeconomic circumstances has been central—and a difficult hurdle to clear.

This is the new challenge facing the research community. It is one thing to become aware of the impact of climate on society, the economy, and health. It is another thing altogether to develop an intellectually valid approach for research into such questions. Outside hazard analysis, the social sciences have largely failed to develop such an approach. Economists, in particular, have not yet encompassed climatic factors in a useful way. Climatologists themselves are more used to looking at the reverse process—the impact of man on climate. A new field has become necessary—and for the most part remains to be cultivated.

The Shocks of the 1970s

The ultimate conversion of the atmospheric scientists to the view that climate should be a central object of study may have been inevitable. But the conversion was undoubtedly accelerated by the dramatic events of the 1970s, when climatic extremes had a major socioeconomic impact.

The first form of impact to become politically visible was the Sahelian drought, a prolonged episode of rainfall failure that extended across all Africa from Gambia and Senegal to the Ethiopian highlands, Somalia, and parts of Kenya and Tanzania. It has been extensively documented (e.g. Glantz 1976, 1977; Nicholson 1979a, 1979b). Subsequent analysis has usually portrayed it as lasting from 1968 to 1973, with normal rainfall resumed in 1974. In fact, the desiccation of the West African Sahel began as early as 1957 or 1960 (Hare 1977, 1979), and over parts of Africa the dryness continues still (Nicholson, 1982). The drought was the culmination of a creeping failure of monsoon rainfall that evaded attention outside the affected area (figure 8.2). Similarly prolonged desiccation preceded the intense 1972 drought in parts of central Australia, where a few years of good rains beginning in 1973 were succeeded in 1982–83, over eastern districts, by the severest drought on record.

The newly independent (1960–61) republics of the Sahelian and Sudanian belts have thus had to spend the whole of their autonomous history in the grip of a slowly worsening climate. Their agony at the height of the drought has been made visible to the whole world by the media, especially by film and videotape. There has been a severe loss of livestock, destruction of perennial woody vegetation, mass migrations of peoples, with some loss of life, and a drastic disruption of what was left of the traditional forms of pastoral nomadism—a lifestyle unpopular with colonial authorities and new governments alike, though designed to cope with just such situations.

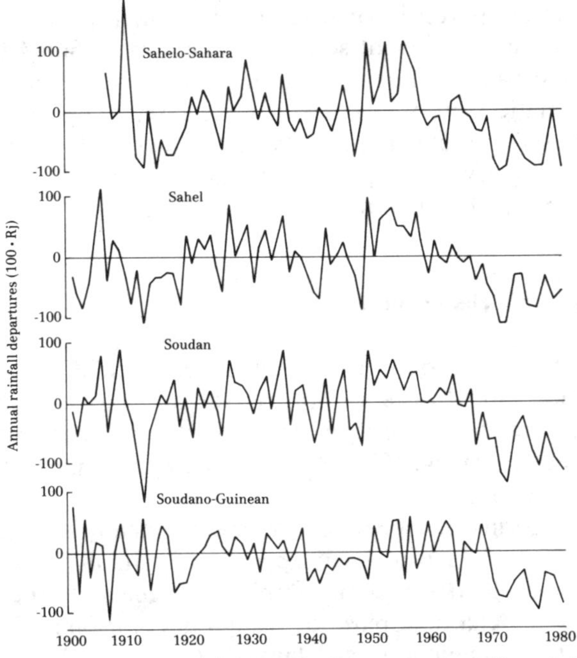

Figure 8.2 Rainfall stations in the Sahel and adjacent zones, with variation of normalized annual rainfall totals since 1900. Note the general decrease in rainfall since the late 1950s in all zones. (From Nicholson 1982).

In the early days of public awareness of the Sahelian tragedy it was generally assumed that drought was solely responsible. The afflicted peoples were seen as the victims of a hostile fluctuation of climate or perhaps even a lasting change for the worse. Evidence was brought forward to suggest that the Sahara had been creeping southward for millennia and that the creep had become a rush. An advance of between 13 and 22½ miles per annum was documented by Rapp (1976) for Sudan.

The view that this acceleration might be due to a change in climate gained support from respected professionals. The most widely publicized early view was that of Winstanley (1973), but this was soon echoed by Lamb (1973). Bryson (1973) suggested that rising levels of carbon dioxide in the atmosphere as well as increased loading of anthropogenic dust might drive the subtropical belt of high pressure southward, thus intensifying the aridity of the Sahel. Further support for the view that the desert might feed itself by positive feedback came through the albedo (reflectivity) hypothesis introduced by Otterman (1974), who noticed that overgrazed land along the desert margins had a higher albedo than protected land (where vegetation cover was more complete). Various numerical modelings of the hypothesis (e.g. Charney 1975) confirmed that heightened albedo intensifies sinking motion in the atmosphere and hence inhibits rainfall. There was thus professional support for two complementary views: that the desiccation of the Sahel might be part of a progressive and irreversible climatic change, and that human misuse of the land might accelerate the desiccation.

So much alarm was created by these and other views that the United Nations convened a Conference on Desertification in 1977, at Nairobi. The agenda was much influenced by Gilbert White, Mohammed Kassas, and Jack Mabbutt. Thanks to their efforts, considerable weight was given to the role of climate, which was the subject of one of the four overview papers before the conference (Hare 1977). By this time, however, it had become apparent that socioeconomic and ecological factors—notably rising population pressure and livestock numbers—were at least as significant as intermittent desiccation in producing desert spread (Kates, Johnson, and Johnson Haring 1977; Warren and Maizels 1977; Hare, Kates, and Warren 1977; see figure 8.3). Moreover, the rains had resumed in some parts of the Sahel. There was some retreat from the position that climate was the dominant factor and that a lasting change was in progress—especially because many meteorologists dissented strongly from the positions taken by Winstanley, Bryson, and Lamb.

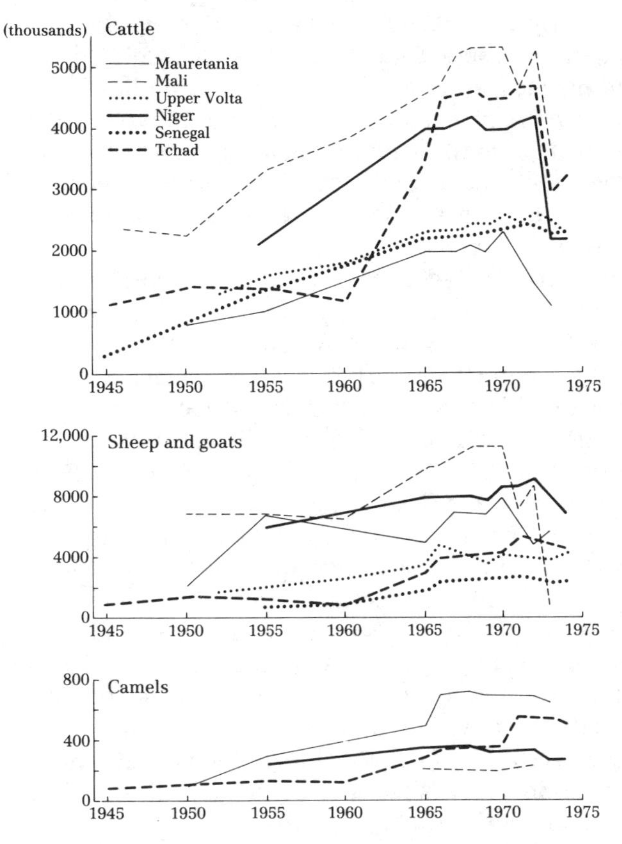

Figure 8.3 Buildup of livestock numbers in the Sahelian countries, 1945–75. Note that numbers increased throughout the desiccation after 1957, implying progressive increase in pressure on pasturage and soils. The catastrophe of the early 1970s speaks for itself. (From: S.E.D.E.S. 1975)

Nevertheless the job of reawakening informed opinion as to the importance of climate had been achieved.

The bad harvests of 1972–73 had a similar effect, and a related sequence of reactions took place. There was at first unwillingness to admit that anything much was amiss. As the scale of crop failures in many countries, especially the Soviet Union, was realized, there was again public concern that an irreversible change toward greater aridity might be in progress. Various professionals also voiced fears of this kind.

A heated controversy ensued between the crop forecasters of the U.S. Department of Agriculture and certain climatologists (e.g. Schneider 1976). A group at the Aspen Institute under W. O. Roberts, with the support of the International Federation of Institutes for Advanced Study, began and has continued an interdisciplinary study of the link between climate and food. A committee of the U.S. National Academy of Sciences (1976a) prepared a widely read review called *Climate and Food.* The Institute of Ecology and the Charles F. Kettering Foundation sponsored two expert conferences on the same theme; it looked at the time as if the assumptions underlying the green revolution—that irrigation water and fertilizers would remain accessible and cheap, and that dryland agriculture had lower priority—might be invalid in a world of harsher climate. The second conference, held at Bellagio in June 1975, was politically influential; its report (Rockefeller Foundation 1976) influenced U.S. foreign policy and was among the stimuli for the 1979 World Climate Conference. Abundant harvests in subsequent years pushed the issue of food and climate off the front pages, but the awareness of climate created by the earlier crop failure did not diminish.

The third major shock of the 1970s operated chiefly in North America. The central and eastern parts of the continent are well adapted to variable climate and severe winter cold. Provision of adequate supplies of fuel for industry and interior heating has been taken for granted and with good reason. Yet the severe winter of 1976–77 brought much of the Ohio Valley and the eastern Midwest to its knees. Ice on the Ohio River halted distribution of coal and oil. The natural-gas pipeline systems could not cope with the greatly increased demand. There was much temporary unemployment, a substantial loss of production, and a great deal of individual frustration and exasperation. Again the media raised the question, Is climatic change in progress? And again there were some professionals who said, "Very likely."

These three shocks, and others of less consequence, revealed an intriguing split within the atmospheric science community. Most meteorologists, and especially the dynamicists who model the atmosphere's behavior on large computers, discounted the phenomena. They were nothing more than extreme fluctuations of an extremely variable system, the conservative argument went (e.g. Landsberg 1975). Yet there were enough radicals within the scientific community who questioned this complacent view to achieve a marked net shift of professional opinion. The media listened to the radicals, and not to the conservatives. Much the same thing has happened throughout the environmental arena. The psychology of natural science permits countervailing biases. Some researchers are biased toward a belief in change; others are biased toward its improbability.

The opinion-making media have a marked bias toward change. It is more newsworthy.

We arrive thus at the interim conclusion that the crises of the 1970s were highly effective in creating public awareness of climate. By the end of the decade, that awareness was probably at its highest-ever level. The uncertainty of the scientific evidence did nothing to impede—indeed, may even have advanced—the process, for the Western media were given the wherewithal of controversy. The more conservative professionals strove to disprove the claims of their radical colleagues. In so doing, some of the conservatives were converted to the view that climate has high social significance as well as being a new-found intellectual challenge.

Institutional Responses

The leading scientific and political institutions have not been slow to respond to the increased awareness of climate. Such responses are, however, by no means new, and past experience of their effectiveness and performance is not encouraging. It remains to be seen whether the recent vogue will endure.

The establishment by the U.S. National Science Foundation of the Task Force on Human Dimensions of the Atmosphere in 1966 illustrated an early attempt to broaden the perspective of climatological research in the United States. The task force was composed of scientists drawn from the physical, natural, and social sciences. It worked for a year, drew up a list of key questions to be answered, and offered suggestions as to priorities for research (U.S. National Science Foundation 1968). The foundation accepted the spirit of the task force's report, but moved slowly in putting the recommendations into action (Sewell 1973). A possible reason was that the organization of the Atmospheric Sciences Division of the foundation, and its review process, remained unaltered; it continued to emphasize the physical sciences. The same was true of one of the major research institutes funded by the foundation, the National Center for Atmospheric Research, though its Advanced Studies Project now includes socioeconomic work. In contrast, several action agencies adopted many of the task force's recommendations and now include human dimensions in their programs of climate-related research, notably the National Oceanographic and Atmospheric Administration, the U.S. Department of Agriculture, and the Office of Emergency Preparedness.

Perhaps the most important result of the task force's work, however, was the long-term commitment of several of its members to this area of inquiry, notably in the legal, economic, social, and environmental aspects of relations between climate and man. In a few cases this work was undertaken on an interdisciplinary basis, involving collaboration between physical scientists and social scientists. For the most part, however, the work has been undertaken within individual disciplines.

The U.S. scientific community sensed a need to review the matter again almost a decade later. In 1977 the National Research Council established a Task Group on the Impact of Weather and Climate on Society. Its report, published in 1978, provides an outline of the key issues and a list of needed research. There is a distinct similarity between that list and the one suggested by the National Science Foundation a decade earlier. The major difference in 1978, however, was that the National Research Council was now officially interested in the human dimensions; ten years before it was not.

The late 1970s and early 1980s have witnessed a number of climate-related problems that demanded institutional response, among which we can list fear of hostile climatic change due to human interference, for example the release of infrared-absorbing gases such as carbon dioxide and the halocarbons; anxiety about the potential health effects of globally dispersed pollutants such as the chlorofluoromethanes (CFMs) or of chemically active oxides of nitrogen that reach the stratosphere; general anxiety about the potential impact of variable climate on the food system and on energy consumption and delivery; and fear of major ecosystem disturbance due to changes in the biogeochemical cycles induced in part by climate-related factors.

There has been a strongly negative quality to this perception. Climate has been seen as a hazard to be endured or a stress to be resisted. The idea that it might be a resource capable of being deliberately exploited, which is the basis of agricultural technology, is almost absent from the literature (but see Taylor 1974). Academic research into the field has been largely concerned with various climatic hazards, typically floods and drought. Many of the contributors to this volume have been involved in such research. It is scarcely surprising that institutional responses have largely addressed climate as a hostile agency—except in certain Third World countries where the optimistic theme of ecodevelopment has taken hold.

In the United States, for example, the National Academy of Sciences has been involved in several analyses of specific climate-related hazards, as have several federal departments and the Office of Science and Technology of the White House. The largest international investigation of a specific man-induced hazard was the Climatic Impact Assessment Program (CIAP) of the U.S. Department of Transportation, which involved the cooperation of more than twenty countries (CIAP 1974). It looked at the potential impact of nitrogen oxides from supersonic transport exhausts on stratosphere ozone, and hence on surface ultraviolet irradiance, and possible effects on human health. A large parallel study was mounted jointly by the United Kingdom and France, which had a strong vested interest in the success of Concorde. Well before CIAP was finished, a parallel analysis of the impact of the release of chlorofluoromethanes was launched in the same countries. These investigations were concerned primarily with impact on health, but they also had considerable implications for world climatology. Several panels of the U.S. National Academy of Sciences sat in judgment on the outcome of these and related studies, and their reports are important milestones along the path toward a full comprehension of the medical and socioeconomic role of climate (e.g. U.S. National Academy of Sciences 1975a, 1975b, 1976a, 1976b, 1977). In a few cases there have been resulting changes in public policy—notably, for example, in the start of measures to phase out the use of the less stable halocarbons in the United States, Canada, and Sweden.

A major consequence of these activities was the involvement of many different disciplines. In the CIAP and halocarbon issues, chemists, economists, epidemiologists, ecologists, agricultural scientists, and engineers worked intimately with meteorologists. Recognition that climate was an aspect of a larger concept, the climatic system, came late to professional meteorologists, who had tended toward a purely physical stance. But it has come in earnest. The major role played by geologists, geochemists, palynologists, and physical geographers in the unraveling of the longer time-scales of climatic variability and variation had a similar effect. The deep sea records of the CLIMAP project, and the use of oxygen and carbon isotope analyses, have made palaeoclimatology more nearly a quantitative science. This makes it more palatable to meteorologists brought up in Kelvin's tradition that "when you can measure what you are speaking about and express it in numbers, you know something about it; but when you cannot measure it, when you cannot express it in numbers, your knowledge is of a meagre, unsatisfactory kind'' (Pettersen 1940).

The U.S. system has contributed notably. In the mid-1970s, suggestions were made (not least by the secretary of state) that climatic instability had become important enough for a world conference analogous to those already held or planned on food, population, water, and desertification. A panel of experts commissioned in 1976 by the secretary-general of the World Meteorological Organization (WMO), a venerable and effective U.N. specialized agency, recommended a two-stage approach: there should be a consultation of experts, followed by a political conference of the usual U.N. sort—if, and only if, the experts could agree on recommendations amenable to political action. This recommendation was accepted, and WMO agreed to organize a World Climate Conference as a consultation of invited experts mainly in the economic and social fields affected by climate. Many other U.N. agencies collaborated, notably UNESCO, FAO, WHO, and UNEP.

The organizing committee of the conference was representative of the intergovernmental organizations, and also of the International Council of Scientific Unions (ICSU). In practice, the planning was largely carried out by a small bureau headed by Robert M. White, later chairman of the conference; James C. Dooge, a distinguished hydrologist who had also served as president of the Irish Senate; Ju. Sedunov, a senior Soviet meteorologist; and F. Kenneth Hare of Canada, a consultant to the secretary-general of WMO. This bureau took very seriously its mandate to search for and involve experts on the socioeconomic impact of climate.

The conference was held in Geneva in February 1979. There was a large attendance of specialists from the economic sector—notably in energy production, water resources, public health, food production, agriculture, land use, forestry, fisheries, marine resources, and economic decision making. Such specialists outnumbered the meteorologists and climatologists by more than two to one. The role of the latter in the formal sessions was to present a balanced view of the nature and scale of climatic variability and change. The nonmeteorologists analyzed specific climatic influences in the various socioeconomic sectors. There was a large representation from developing countries, including the People's Republic of China. The conference adopted a declaration aimed at governments and public institutions and a plan for the forthcoming World Climate Program, to be carried out over several years by WMO (the congress of which adopted the plan in May 1979). A second-stage political conference was discouraged, essentially because it was judged impracticable at this time to formulate a realistic agenda.

WMO has a long experience in such internationally performed research. The Global Atmospheric Research Program (GARP), for example, had been in effective progress for many years. WMO's role has been to persuade national governments to participate (at their own expense) in major world experiments aimed at better understanding of the atmosphere's ways. These have included exercises in the tropical Atlantic, monsoon Africa, and monsoon Asia and a global experiment involving major increases in satellite monitoring, ocean surface observations, and many other developments. These exercises must be ranked among the most effective achievements of international natural science.

The Joint Organizing Committee (JOC) of GARP brought together the two institutional wings of the natural sciences. WMO is an intergovernmental organization, as are the other UN agencies. It is supported by Western and socialist powers alike and by nearly all the nonaligned countries. It relates to the official meteorological services of the member states, which cannot function without worldwide orchestration of observations, communications, and data exchange. But an effort such as GARP required participation also of the nongovernmental scientific communities. The ICSU brings these together, and JOC was a joint body of WMO and ICSU, with a small staff in Geneva headed by the Swedish meteorologist Bo Döös. It was highly effective, though tensions between governmental and nongovernmental inputs were frequent and unavoidable.

The World Climate Program also requires nongovernmental participation, on an even greater scale, since analyzing climatic impact involves the social sciences, most of which are not connected with ICSU. The program consists of four activities. The first two concern the better use of data and the applications of climatic knowledge, both of which are familiar to the governmental agencies and will pose few new problems (though data will have to come from many new fields). The third component activity, dealing with research into climate variations, including man's role, is being planned by a joint WMO-ICSU body, the Joint Scientific Committee (JSC). But the fourth, the analysis of climatic input—poses new and as yet unresolved problems.

Climatic impact on society—climate-society interaction, as it is better called—has been studied academically for decades, chiefly on a sectoral basis. No central body of method has, however, emerged.

Economists, in particular, have given climatic interactions little study. The work of Arrow (U.S. National Academy of Sciences 1975a) and D'Arge (1979) on the CIAP program stands out as exceptions. The study of climatic hazards has been well developed by geographers and others (see for example the authoritative review by Burton, Kates, and White 1978). But a general intellectual framework for a study of climate-society interaction remains to be formulated (Ichimura 1979; Kates 1979).

Hence the conduct of the impact component of the World Climate Program requires new understanding and new institutional arrangements. The UN Environmental Program (UNEP) has agreed to develop these arrangements on behalf of WMO. A major problem is the lack of an organized body of social scientists in any discipline (except geography, a bridge field itself), either national or international, with interest or competence in this field. The intellectual initiative is at present in the hands of a standing ICSU committee, SCOPE (Scientific Committee on Problems of the Environment), the president of which, until 1982, was the geographer honored in these pages, Gilbert White.

An initial attempt to formulate a theory of climate-society interaction was made by a SCOPE Workshop on the Climate/Society Interface in December 1978. Its first report (ICSU-SCOPE 1979) was tabled at the World Climate Conference. Figure 8.4, reproduced from the report, illustrates the themes being pursued. SCOPE has completed a formal project (under Robert Kates) to explore them much more deeply. This work brings atmospheric scientists together with geographers, medical scientists, and others with competence in the study of human response (Kates 1979).

National programs have begun to appear in parallel with the World Climate Program. The U.S. Congress passed the National Climate Program Act in August 1978, calling for a national program bearing a marked resemblance to the world program, but going beyond it in suggesting collaboration with the states. In Canada an interagency Canadian Climate Program is in progress, under the policy supervision of a national Climate Planning Board. Both countries have set up national climate program offices with social scientists on the staff. In the United States a formal five-year plan has been tabled before Congress. In the Soviet Union such activities are long established.

We can say in sum that the heightened public awareness of climate has already prompted significant political and institutional responses. It remains to be seen whether to momentum can be main-

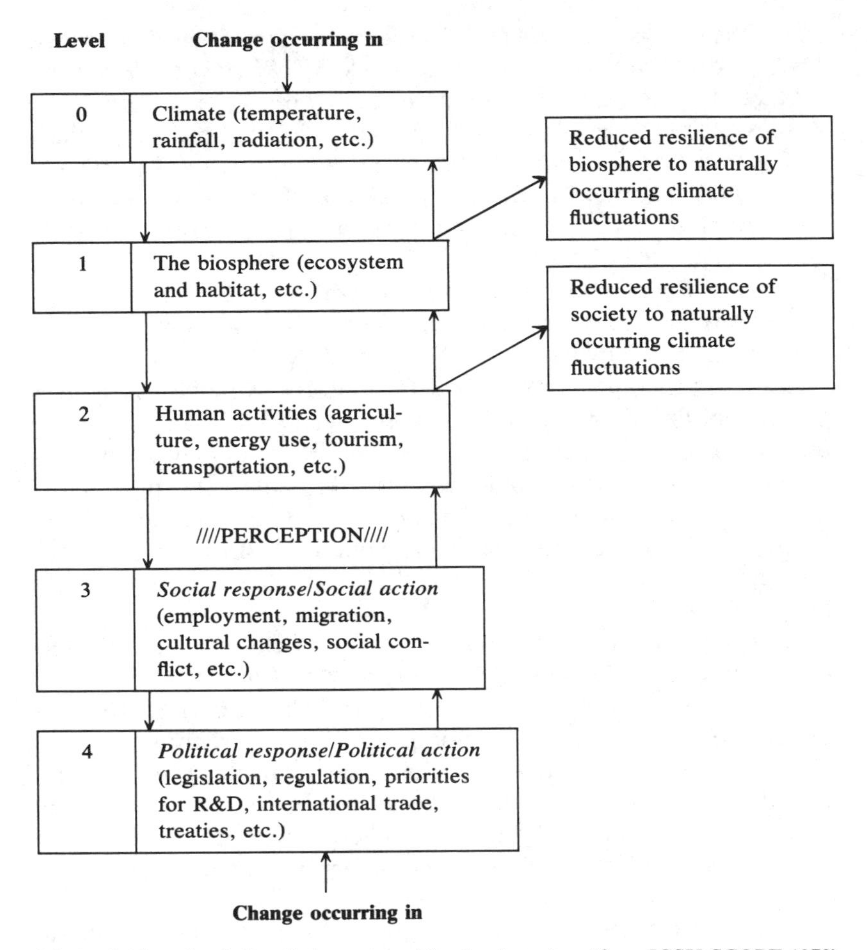

Figure 8.4 Levels of climatic impact, and feedback system (from ICSU-SCOPE 1979)

tained. Political systems adapt most readily to shock and to perceived danger. A decade or two of bland climate might well lessen the present sensitivity and plunge climate into the declining sector of Downs's cycle of issue attention.

The Weather Modification Experience

For the most part, man has accepted climate as given and has tried to alter his activities to accommodate it, either by moving away from less hospitable regions or by developing techniques to cope with particularly extreme climates. There have been attempts throughout history, however, to alter the climate or, more accurately, associated weather events. The purpose has been either to reduce losses occasioned by such events or to provide precipitation, sunshine, or wind in order to improve production. At least three-quarters of the estimated $40 billion a year of global natural hazard costs originate in three major hazards of climatic origin: floods (40 percent), tropical cyclones (20 percent), and drought (15 percent). Accompanying these losses of property and income are losses of life, amounting to an average of 250,000 persons a year in the world (Kates 1979). Occasionally, as in the Bay of Bengal in 1970, a single storm may kill as many as 200,000 people. In the United States alone property losses from climate-related events amount to more than $13 billion a year (U.S. Weather Modification Advisory Board 1978). Not only are such impacts severe, especially in the developing world, but they are growing in magnitude and in the areal extent of their influence.

The other major motivation for altering weather or climate is to improve production by furnishing more of a given climatic element. There are certain activities that critically depend on the timely occurrence of moisture, sunshine, or wind (Maunder 1970). This is especially so in the case of agriculture, but the prospects of activities such as outdoor recreation or movie making might be considerably improved if it were possible to provide more sunshine, snow, or wind at the right times.

Such motivations have stimulated a wide variety of attempts to modify climate and weather since ancient times. These have ranged from magic and symbolism (such as rain dances and prayers to the gods) to the introduction of chemicals into the clouds or attempts to alter the albedo (Halacy 1968; Sewell et al. 1973). The modern era of weather modification, however, dates back only about three decades. It has been described as a scientific era because it has attempted to develop methods grounded in scientific principles. Governments have become involved in its promotion and regulation, notably in the United States, the Soviet Union, Australia, and Israel (Fedorov 1979a, 1979b). Even more important, there has been serious questioning of the merits of trying to tamper with nature. In the United States, at least, weather modification has been proposed as an early candidate for technology assessment.

Scientific research relating to modification of weather and climate has many of the characteristics of research in climatology as a whole. The principal emphasis is on physical processes rather than on human consequences. As a result, the main question posed is "Can it be done?" rather than "Should it be done?" Associated with this emphasis has been a general lack of interest on the part of social scientists in the impact of climate on society. Direction of science policy in this field has been left almost entirely to physical scientists, who until recently did not appear to have more than a passing interest in the human dimensions. Several major changes in outlook, however, began to appear in the 1960s, stimulated by a number of social scientists, notably Gilbert White.

Recognition of the possibility that in certain circumstances it might be possible to increase precipitation by substantial amounts led to the emergence in the United States and elsewhere of commercial operations in the 1940s and early 1950s. There developed, however, some skepticism as to the validity of the claims of success by the weather modifiers. This encouraged the U.S. government to request the National Academy of Sciences to review the situation from a scientific point of view. It established an Advisory Committee on Weather Control in 1953. Its report, presented in 1958 (U.S. Advisory Committee on Weather Control 1958), concluded that in some cases weather modification could indeed be accomplished. Specifically, the committee believed that increases in precipitation of up to 15 percent were possible in some locations, notably the windward side of the West Coast mountains in the United States. It said little about the economic, social, or environmental aspects.

The report was greeted with enthusiasm by the commercial operators who believed that their claims of success were endorsed by its findings. The scientific establishment, however, remained unconvinced and became even more vociferous in its rejection of the conclusions of those involved in the evaluation of weather modification. Meteorologists and others criticized the statistical analyses used in the committee's evaluations, expressed doubts about the data gathered by those outside the scientific establishment but used in the committee's studies, and suggested that the committee had been compromised by the inclusion of nonscientists among its membership. The scientist-controlled Department of Defense was especially critical of the report.

The U.S. government's reaction to the division of opinion among scientists was for Congress to assign an agency, the National Science Foundation, to stimulate research in the field. It began with a budget of $2 million for this purpose, to be allocated to the question "Can it be done?" Rainmaking was the central focus of interest.

In 1961, two events rekindled interest in weather modification in the United States. The first was a suggestion that development of reliable technology could contribute to the solution of several of the major problems facing mankind, namely, losses from natural hazards, inadequate food supplies, and limits on habitation imposed by inhospitable climates. Proposals were made to harness what was regarded as a new resource, "the rivers of the sky." The second was a proposal by the Bureau of Reclamation to begin operation programs designed to increase precipitation in certain arid parts of the country.

Response to the first event was lukewarm, largely perhaps because the public and its political representatives did not visualize the atmosphere as a resource. Response to the latter event, however, was a lively debate in Congress, revealing a major division of scientific opinion as to the extent to which weather or climate could be purposefully altered and whether the Bureau of Reclamation should be the leading agency in promoting the development of techniques to that end.

By 1964 the U.S. National Academy of Sciences was convinced that it was time for a review of the scientific state of the art. Its panel on weather and climate reported in 1966 and reached very similar conclusions to those presented by the Advisory Committee on Weather Control some eight years previously on precipitation augmentation. The panel also made some interesting comments on other forms of weather. At roughly the same time, the U.S. National Science Foundation established a Special Commission on Weather Modification. Its mandate was much broader than that of the panel. While it had to comment on the question "Can it be done?" it was also to reflect on the question "Should it be done?" To that end it gave special attention to the human dimensions of research on the socioeconomic aspects (Sewell 1966; Boyd et al. 1971; Sewell et al. 1973; Haas 1974), legal questions (Taubenfeld 1970; Davis 1974), and ecological considerations (Cooper 1973).

Using data similar to those used by the U.S. National Academy of Sciences, the commission reached more or less the same conclusions about technical feasibility. It broke new ground, however, in suggesting a major expansion and acceleration of effort in research on human and ecological aspects. A major stimulus to a shift in viewpoint was the effort of Gilbert White, a member of the Special Commission. He was instrumental not only in persuading physical scientists to become more aware of possible consequences of an enlarged program of weather modification, but also in bringing several social scientists together to work on these problems.

There were several reactions in Congress to the various reports. The reports resulted in a major expansion of funds for research and development. Support increased from some $7 million in 1966 to nearly $20 million in 1972, since which time it has fluctuated only slowly (table 8.1). Annual funding is now about $21 million. Responsibilities for leadership in research have shifted from the National Science Foundation to an Inter-Departmental Committee on Atmospheric Sciences, and there has been an expansion of research activities and operational programs in several federal line agencies, notably the departments of Commerce, the Interior, Agriculture, and Defense. Cognizant of the need for a concrete program of research on the human aspects, the National Science Foundation established a Task Force on Human Dimensions of the Atmosphere in 1966. Its report, presented more than a year later (U.S. National Science

Table 8.1 The U.S. Federal weather modification research program—allocation of funding (millions of dollars)

Department	FY 72	FY 73	FY 74[a]	FY 75	FY 76	FY 77	FY 78	Congressional budget request[b] FY 79
Agriculture	0.36	0.37	0.27	0.09	0.07	0.05	0.02	0.00
Commerce	3.94	3.77	3.08	2.49	4.30	2.67	3.55	2.7
Defense	1.82	1.21	0.92	1.14	1.41	1.01	0.91	0.7
Interior	6.66	6.37	3.90	4.00	4.65	6.45	7.61	4.5[c]
Transportation	0.40	0.39	—	—	—	—	—	—
NSF	4.94	4.23	4.25	4.70	5.06	4.90	2.40	1.10[d]
Subtotal advertent	18.12	16.34	12.42	12.42	15.49	15.08	14.49	9.0
Thermal fog (DOD)	—	—	—	—	0.39	1.40	2.20	1.2
Total advertent	18.12	16.34	12.42	12.42	15.88	16.48	16.69	10.2
Inadvertent	1.76	3.25	3.84	5.20	4.83	3.69	4.16	?
Total	19.88	19.59	16.26	17.62	20.71	20.17	20.85	?

Source: From the Interdepartmental Committee for Atmospheric Sciences (FY 72–78); FY 79, from telephone survey.

[a] Distribution of funds expended are estimated for DOC, NSF, and inadvertent. Total is from ICAS.

[b] Estimated. (Plans for study of inadvertent modification are still uncertain.)

[c] Congressional add-on is anticipated.

[d] NSF has no specific budget request for weather modification. This is a staff estimate as to how much will be devoted to research directly related to weather modification.

Foundation 1968), contains a list of recommended research topics and actions that were needed to get the work underway.

In the decade since publication of the task force report, operational activities have been widespread (figures 8.5 and 8.6). There has been continued inquiry into the question "Can it be done?" As indicated in the report of the U.S. Weather Modification Advisory Board (1978), the physical sciences still take the lion's share of the expanded research budget. Studies of the human and environmental dimensions have been broadened and deepened, but they have involved only a handful of social and natural scientists and only a few universities. There have been some valuable contributions to the study of the human dimensions and to broadening the general perspective on weather modification. An important illustration of the latter is the work undertaken in connection with the U.S. National Hail Suppression Project (Changnon et al. 1977). In addition, research has continued on the legal, economic, and social aspects (Taubenfeld 1970; Davis 1974; Haas 1974). At present, such research is allocated less than 3 percent of the funds devoted to weather modification investigations in the United States. Proposals for future research suggest a continuation of this proportion (U.S. Weather Modification Advisory Board 1978).

There seem to be two basic difficulties. First, the research-funding agencies have been unable to kindle interest among these branches of science in the weather modification field. For their part, the social and natural scientists have not been convinced that they should turn their attention to these questions. At the same time the funding agencies have continued to believe that the physical dimensions deserve the most attention, and their organizational setup and staffing reinforce this viewpoint. Second, the line agencies have meanwhile pushed ahead with operational programs to increase precipitation for agriculture, reduce hail, redistribute snowfall, and disperse fog at airports, amid opposition from various environmental groups and scientific skeptics. At the same time there are growing concerns about various transboundary effects of particular weather modification activities and about their potential employment as a weapon of war (Taubenfeld 1970; McBoyle 1977).

Just as was the case a decade ago, six major questions remain to be answered about the social desirability of weather modification. Who gains and who loses from a given weather modification activity? How can those who are adversely affected by such operations be compensated by those who gain? Who should decide whether a given weather modification operation should commence (or continue), par-

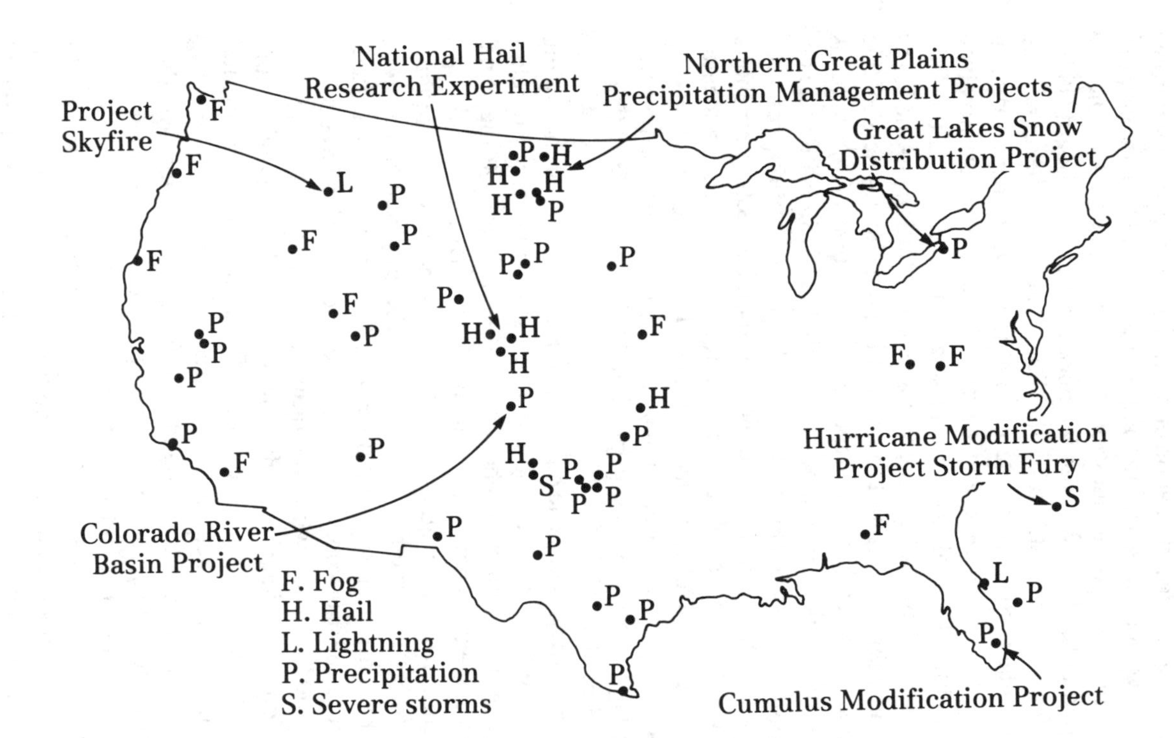

Figure 8.5 Geographical distribution and character of weather modification projects in the United States in 1972

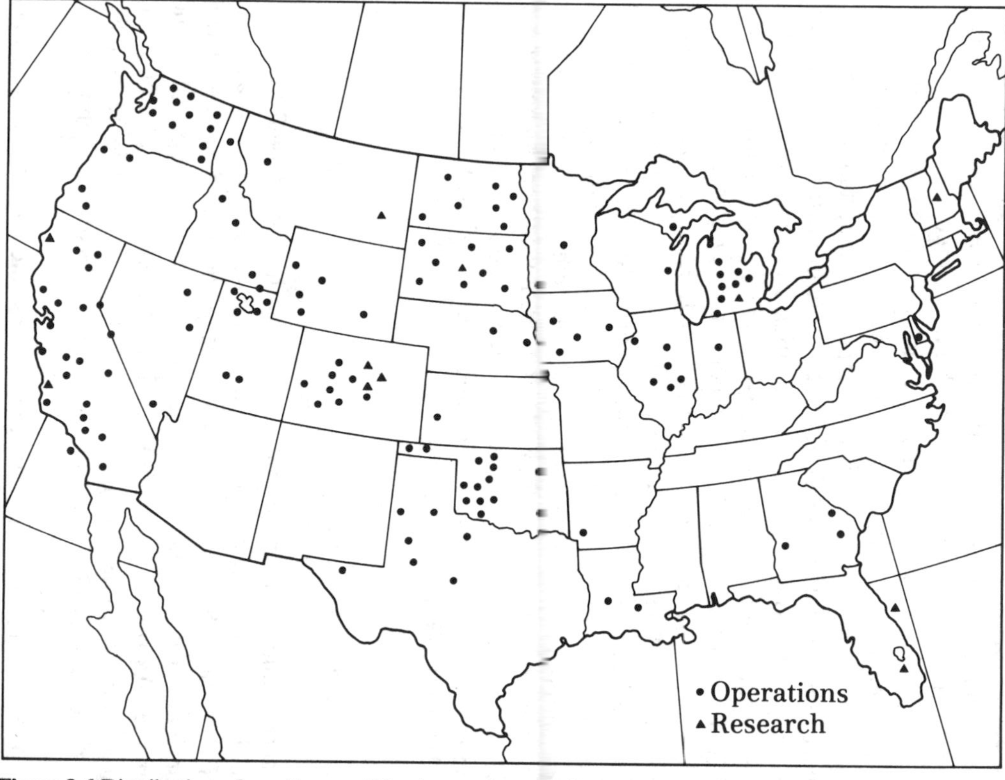

Figure 8.6 Distribution of weather modification projects in the United States, 1973–77. Two projects were in south-central Alaska, and there were no projects in Hawaii. (Source: NOAA, Weather Modification Reporting Program, Rockville, MD)

ticularly when there are fears about its potentially adverse effects? Is weather modification the best way of dealing with the problems it is intended to solve? What is the probable ecological impact of weather modification programs? What kinds of institutional arrangements would ensure that such questions were posed and the public's interest were taken into account in the development of the technology and its application, and that the appropriate level of research were carried out, especially on the human dimensions?

In sum, research on weather modification has made advances in the past decade, particularly in connection with the physical processes involved. There still remain important uncertainties whether, where, and when it can be done. Even more important, the question "Should it be done?" still awaits a satisfactory answer. This is because weather modification is probably only at the preproblem stage of the issue-attention cycle. Although a number of cases have come before the courts in which plaintiffs have tried to sue weather modifiers, and although such activity is banned by law in some states, there has been an insufficient feeling of crisis to stimulate a sustained effort to determine if, when, and where weather modification should be attempted. This experience has undoubtedly had some effect on public awareness of climate, as distinct from weather. As figures 8.5 and 8.6 indicate, most projects have been in areas of climatic hazard. There is not as yet, however, much awareness of the larger question of deliberate climate modification.

Future Concerns

It seems highly probable that crises of various kinds will keep the public very much aware of climatic interactions. One must assume, for example, that the recent sequence of damaging extremes will continue, because there is much evidence that it is normal to the climate. As pressure on land, food, and energy grows, moreover, the economic impact of disruption by climate should increase (Biswas and Biswas 1979). It is the short-term variability of climate that poses this threat.

In the longer term, however, anthropogenic climatic change may become visible. The rise of carbon dioxide concentration in the atmosphere, currently at a rate of 3–4 percent per decade, poses the threat of strongly rising temperatures in the next century. This effect is probably inescapable, since the world is likely to turn from oil and natural gas to coal and synthetic fuels in the next few decades, and only secondarily to nuclear or solar sources. This will tend to accelerate the accumulation of carbon dioxide in the atmospheric reservoir. It seems improbable that we will arrive at either political or technical means of avoiding this increase. Present model calculations suggest that the outcome will be significant changes of world temperatures and rainfall distribution during the twenty-first century.

Such slow change differs fundamentally from the extreme events associated with short-term climatic variability. The latter has to be combated with essentially unchanged technology and economic infrastructure. In the two or three years typical of climatic disruptions of the economy there can be little change in the ability to adapt to the new regime, and no incentive to do so, since the old will presumably reappear. But with slow, consistent change, there will be time to adapt both institutionally and technologically to the new conditions, especially if they are foreseen (Margolis 1979).

All this argues that the new interdisciplinary awareness of climate among professional groups must at all costs be maintained. The world cannot afford to let its climatologists crawl back into the comfortable obscurity in which they worked a few years ago. Both kinds of awareness are needed—the public's concern about what climate may do to its livelihood and personal comfort, and the professionals' sense that the study of the climatic system is a challenging and worthwhile activity. And by professionals one has to mean a versatile, multidisciplinary group, including social scientists who will see climatic interactions as a subject worthy of their time. We have Kates's (1979) assurance that such a group does not exist. So it will be necessary to create it.

There will be formidable obstacles in the way of such changes. Institutions still tend to undervalue the socioeconomic role of climate, in part because their members are often drawn from the physical sciences, and in part because new ideas come hard to collections of individuals. Peer review, for example, tends toward conservative, unventuresome judgment; and funding agencies usually choose peers from among those whose reputations were made years ago and only from disciplines that have seemed relevant in the past. Research funding for the new modes will not be easy to pry loose.

Yet the reality of the challenge is not in dispute. The world faces critical shortages of supply in food and energy, and in both fields climatic factors are crucially important. We conclude that climatology is poised on the threshold of a new era, if its exponents can rise to the opportunity. At present the will is there, but the means are inadequate. There are bold spirits in many countries who intend to see that the means are provided.

References

Arrow, K. 1975. In U.S. National Academy of Sciences, 1975*a*, Appendix K.

Biswas, M. R. and A. K. Biswas, ed. 1979. *Food, Climate and Man*. New York.

Boyd, D. W., R. A. Howard, J. E. Matheson, and D. W. North. 1971. *Decision-Analysis of Hurricane Modification*. Stanford Research Institute, Project 8503. Menlo Park, CA.

Bryson, R. A. 1973. "Drought in Sahelia." *The Ecologist* 3:366–71.

Burton, I., R. W. Kates, and G. F. White. 1978. *The Environment as Hazard*. New York.

Changnon, S., et al. 1977. *Hail Suppression: Impacts and Issues*. Final Report, Technology Assessment of Suppression of Hail, Illinois State Water Survey. Urbana, IL.

Charney, J. 1975. "Dynamics of Deserts and Drought in the Sahel." *Quarterly Journal of the Royal Meteorological Society* 101:193–202.

CIAP (Climatic Impact Assessment Program). 1974. *The Effects of Stratospheric Pollution by Aircraft*. Final Report by A. J. Grobecker, S. C. Coroniti, and R. H. Cannon, Jr. Department of Transportation. Washington, DC.

Cooper, C. F. 1973. "Ecological Opportunities and Problems of Weather and Climate Modification." In *Modifying the Weather: A Social Assessment,* W. R. D. Sewell et al., pp. 99–134. University of Victoria Western Geographical Series no. 9. Victoria, B.C.

D'Arge, R. 1979. "Climate and Economic Activity." In *Proceedings of the World Climate Conference,* pp. 652–81. World Meteorological Organization. Geneva.

Davis, R. J. 1974. "Weather Modification Litigation and Statutes." In *Weather and Climate Modification,* ed. W. N. Hess, pp. 767–87. New York.

Downs, A. 1972. "Up and Down with Ecology—The Issue-Attention Cycle." *The Public Interest* 29:38–50.

Fedorov, E. K. 1979a. "Climatic Change and Human Strategy." In *Proceedings of the World Climate Conference,* pp. 15–26. World Meteorological Organization, Geneva.

———. 1979b. "Climatic Change and Human Society." *Environment* 21:25–31.

Glantz, M. H., ed. 1976. *The Politics of Natural Disaster: The Case of the Sahel Drought*. New York.

———. 1977. *Desertification: Environmental Degradation in and around Arid Lands*. Boulder, CO.

Haas, J. E. 1974. "Sociological Aspects of Weather Modification." In *Weather and Climate Modification,* ed. W. N. Hess, pp. 788–812. New York.

Halacy, D. S., Jr. 1968. *The Weather Changes*. New York.

Hare, F. K. 1977. "Climate and Desertification." In *Desertification: Its Causes and Consequences,* ed. U.N. Conference on Desertification, pp. 63–120. Oxford.

———. 1979. "Climatic Variation and Variability: Empirical Evidence from Meteorological and Other Sources." In *Proceedings of the World Climate Conference,* pp. 51–87. World Meteorological Organization. Geneva.

Hare, F. K., R. W. Kates, and A. Warren. 1977. "The Making of Deserts: Climate, Ecology and Society." *Economic Geography* 53:332–46.

Ichimura, S. 1979. Contribution to ICSU-SCOPE, 1979, q.v.

ICSU-SCOPE. 1979. *Report of the ICSU-SCOPE Workshop on the Climate-Society Interface*. Toronto.

Institute of Ecology and the Charles F. Kettering Foundation. 1976. *Impact of Climatic Fluctuation on Major North American Food Crops*. Dayton, OH.

Johnson, R. W. and G. M. Brown. 1976. *Cleaning up Europe's Waters: Economics, Management and Policies*. New York.

Kates, R. W. 1979. "Climate and Society: Lessons from Recent Events." In *Proceedings of the World Climate Conference,* pp. 682–91. World Meteorological Organization. Geneva.

Kates, R. W., D. L. Johnson and K. Johnson Haring. 1977. "Population, Society and Desertification." In *Desertification: Its Causes and Consequences,* ed. U.N. Conference on Desertification, pp. 261–318. Oxford.

Kates, R. W., W. R. D. Sewell, and L. Phillips. 1968. "Human Responses to Weather and Climate." *Geographical Review* 58: 262–80.

Lamb, H. H. 1973. "Some Comments on Atmospheric Pressure Variations in the Northern Hemisphere." In *Drought in Africa,* ed. D. Dalby and R. J. Harrison Church, pp. 27–28. London.

Landsberg, H. 1975. "Sahel Drought: Change of Climate or Part of Climate?" *Archiv der Meteorologie, Geophysik und Bioklimatologie,* B., 23, pp. 193–200.

McBoyle, G. A. 1977. "Purposeful Weather Modification Activities in Canada: Responses to an Environmental Technique." *Canadian Geographer* 21:81–94.

Margolis. 1979. Contribution to ICSU-SCOPE, 1979, q.v.
Maunder, W. J. 1970. *The Value of the Weather.* London.
Mitchell, J. M., Jr. 1977. "The Changing Climate." In *Energy and Climate: Studies in Geophysics,* pp. 51–58. National Research Council. Washington, DC.
Nicholson, S. 1979a. "Statistical Typing of Rainfall Anomalies in Sub-Saharan Africa." *Erdkunde* 33:95–103.
———. 1979b. "Revised Rainfall Series for the West African Sub-Tropics." *Monthly Weather Review* 107:620–23.
———. 1983. "The Climatology of Sub-Saharan Africa." In National Academy of Sciences, *Environmental Change in the West African Sahel,* pp. 71–72 and esp. fig. 2, p. 5 of main report.
Nicolazo-Crach, J. L. and C. Lefroy. 1977. *Les Agences financières de Bassin.* Paris.
OECD. 1972. *Water Management, Gestion de l'Eau.* Paris.
Otterman, J. 1974. "Baring High-Albedo Soils by Over-grazing: A Hypothesized Desertification Mechanism." *Science* 186: 531–33.
Pettersen, S. 1940. Citation by Kelvin on flyleaf of *Weather Analysis and Forecasting.* New York.
Rapp, A. 1976. "Sudan: Regional Studies and Proposals for Development." In *Can Desert Encroachment be Stopped?,* ed. A. Rapp, H. N. LeHouérou, and B. Lundholm, pp. 155–65. Ecological Bulletin no. 24. Swedish Natural Science Research Council. Stockholm.
Rockefeller Foundation. 1976. *Climate Change, Food Producation and Interstate Conflict.* Report of a Bellagio Conference, 1975.
Saarinen, T. F. 1966. *Perception of the Drought Hazard on the Great Plains.* Research Paper no. 106, University of Chicago Department of Geography.
Schneider, S. H. 1976. *The Genesis Strategy: Climate and Global Survival.* New York.
Sewell, W. R. D., ed. 1966. *Human Dimensions of Weather Modification.* Research Paper no. 105, University of Chicago Department of Geography.
Sewell, W. R. D. 1973. "Public Policy towards Weather and Climate Control." In *Public Policy towards Environment, Annals of the New York Academy of Sciences* 216:30–41.
Sewell, W. R. D., and others. 1973. *Modifying the Weather: A Social Assessment.* Western Geographical Series no. 9, University of Victoria. Victoria, B.C.
Sewell, W. R. D., and L. R. Barr. 1978. "Water Administration in England and Wales: Impacts of Reorganization." *Water Resources Bulletin* 14, no. 2:337–48.

Taubenfeld, H. J. 1976. *Societal Consequences of Weather Modification Activities*. Report to the National Sciences Foundation on Research undertaken under the auspices of Grant No. G1-314 59X1. Institute of Aerospace Law, Southern Methodist University. Dallas, TX.

Taubenfeld, H. J., ed. 1970. *Controlling the Weather*. New York.

Taylor, J. A. 1974. "The Atmosphere as a Resource." In *Climatic Resources and Economic Activity*, ed. J. A. Taylor, pp. 21–45. Newton Abbot.

U.S. Advisory Committee on Weather Control. 1958. *Final Report of the Advisory Committee on Weather Control*. 2 vols. Washington, DC.

U.S. National Academy of Sciences (NAS). Reports as follows:
1975a. *Environmental Impact of Stratospheric Flight*.
1975b. *Understanding Climatic Change*.
1976a. *Climate and Food*.
1976b. *Halocarbons: Effects on Stratospheric Ozone*.
1977. *Energy and Climate*.
1978. *Impact of Weather and Climate on Society*.

U.S. National Science Foundation (NSF). 1966. *Report of the Special Commission on Weather Modification*. NSF 66–3. Washington, DC.

———. 1968. *Human Dimensions of the Atmosphere*. Washington, DC.

U.S. National Water Commission. 1973. *Water Policies for the Future*. Washington, DC.

U.S. Weather Modification Advisory Board. 1978. *The Management of Weather Resources*. Washington, DC.

Warren, A. and J. K. Maizels. 1977. "Ecological Change and Desertification." In *Desertification: Its Causes and Consequences*, ed. U.N. Conference on Desertification, pp. 169–260. Oxford.

Winstanley, D. 1973. "Rainfall Patterns and General Atmospheric Circulation." *Nature* 245:190–94.

9 From Hazard Perception to Human Ecology

Anne V. T. Whyte

"Formation and Role of Public Attitudes": A 1966 Agenda for Research

> At the heart of managing a natural resource is the manager's perception of the resource and of the choices open to him in dealing with it. (White 1966, p. 105)

Gilbert White had been publishing articles on natural resources and public policy for thirty years when he wrote "Formation and Role of Public Attitudes" in 1966. This paper represents a clear landmark in the field of resource perception: it provides a clear statement of White's own position at that time, and sets out an agenda of issues requiring future measurement and conceptual development. It is a touchstone against which subsequent work can be measured.

White's early work on river basin development and human occupancy posed the problems as ones in economics and technical feasibility (White 1936, 1937). During the next twenty years, through his work on river basin development, flood hazard, arid zones, and water supply, White paid increasing attention to the role of individual and collective choice in managing natural resources. It was often the seeming conflicts between the myriad choices of individuals and public policy decisions that he sought to bring to the attention of governments. The image of the individual farmer or the suburbanite watering his lawn, making choices that significantly affect the outcomes of policy decisions, appear frequently in his work (e.g. White 1960).

Gilbert White approached the study of perception from the need to understand the management choices that were having a significant impact on natural resources. For him the ultimate objective of study-

Anne V. Whyte is Associate Professor at the Institute for Environmental Studies, University of Toronto. From 1984 to 1986 she has been working in the Science Sector of UNESCO with the Man and the Biosphere (MAB) Programme. She is the author of *The Use of Land Water Resources in the Past and Present Valley of Oaxaca, Mexico; Guidelines for Field Studies in Environmental Perception;* and (with Ian Burton) *Environmental Risk Assessment*.

ing perceptions is to provide a better basis for individual and collective choice (White 1958). It was his work on changing human use patterns on floodplains in the United States (ibid.) that underscored the differences in decisions, and hence in perceptions, between public and private decision makers. This work was soon elaborated in the work of his graduate students Burton (1962) and Kates (1962), and by other students and collaborators. These people formed the focus of the Natural Hazards Research Group, which became established at the University of Chicago.

In his 1966 paper, White clearly states what has come to be known as the "perception approach"—every perception (opinion, view) has its own validity. No single expert has a prerogative on knowing what kind of development, or response to hazard, is "right." Local people, laymen, individuals may have knowledge unavailable to the "expert": they will certainly have different frames of reference. White warns against the hidden assumptions about public preferences and knowledge that are embedded in administrative decisions, and he sets the tone for much work in environmental perception that has since been performed—the desire to articulate the perceptions of ordinary local people vis-à-vis those of governments and outsiders and thus to draw individuals and communities more into collective decision making.

White also cautions us against normative models of "rational decision-making." The dichotomy between rational and irrational behavior is an artifact of economic models; in the real world, human goals are ambiguous, contradictory, and obscure: "It is enough to struggle for rational, accurate description without seeking or claiming to find rationality in the action itself" (1966, p. 112). In the light of what White and many of us have done since 1966, this statement is particularly interesting because beneath much of the work in natural hazards response are some normative assumptions about "rational" response. Indeed, the notion of the "paradox" of human response to natural hazards in returning to the scene of disaster is predicated on normative ideas of adaptation and maladaptation.

By1966, White and others had had considerable experience of the problems of measuring attitudes and perceptions. He discusses alternative approaches such as choices as revealed by economic measures (land values, prices) or through content analysis of written material; and more direct measurements of (verbally expressed) preferences in experimental situations, field settings, and national opinion polls. These issues are still very real today, although we now have a rich experience of successes and failures on which to draw.

At several points in his 1966 paper, White raises the questions of appropriate models of choice; the relative emphasis to be given to individual characteristics such as sex, age, and personality; and the social context of choice. These questions continue to be asked of perception models, and the amount of criticism that this aspect of natural hazard perception research has generated indicates that our track record has not been very good in studying an appropriate mix of factors. What is at issue is not so much a lack of recognition in hazard perception research of the importance of the social, economic, cultural, and historical contexts of perception, as a general failure to put this recognition into practice in empirical studies.

Growth of Field Research 1966–83

By 1966, some important pioneering field studies in hazard perception and choice had already been completed. These included work on attitudes and knowledge about floods (Roder 1961), farmers' perceptions of flood risk in the United States and the impact on land use decisions (Burton 1962), the perception of hazard and choice in floodplain management by urban residents and businessmen (Kates 1962), Gilbert White's own work (1964) on choice of adjustment to floods, and Sewell's (1965) study of water management in the Fraser River basin in Canada. These studies had their roots in the human ecology (human occupancy) tradition within geography but had expanded the explanatory framework to give prominence to human perception and choice, and thus to behavioral and decision-making models.

Saarinen's (1966) study of perception of the drought hazard on the Great Plains of the United States marked a significant departure from the previous work although it came out of the same school of geography at the University of Chicago. Saarinen studied drought, not floods, and adopted an explicitly psychological explanatory mode—that of the individual farmer's personality in relation to his age and drought experience. The aspects of personality measured were those as revealed by TATs (Thematic Apperception Tests) and included the need for achievement, determination in the face of adversity, and whether man sees himself in control of nature, in harmony with nature, or controlled by nature.

At the same time, studies of resource use were adopting a perception framework in settings as different as peasant agriculture in Mexico (Hill 1964; Kirkby 1973b), industrial water supply in Illinois

(Wong 1968), and wilderness in the Boundary Waters Canoe Area (Lucas 1963). Many of these ideas were brought together by O'-Riordan (1971) and Sewell and Burton (1971). David Lowenthal, who had visited Chicago in the late 1950s, was helping to establish perception geography as a subfield within the discipline (Lowenthal 1961, 1967).

Another significant step in the evolution of hazard perception studies was the extension of the work done on floodplains to coastal storms and hurricane hazards along the shores of the coastal megalopolis of the northeastern United States (Burton, Kates, and Snead 1969). As the authors themselves admitted (p. 5), two of them "have been involved in these floodplain studies and were already engaged in armchair speculations about an extension to other natural hazards." These armchair speculations included two important ideas. First, natural hazards (hazards of geophysical origin) and, especially, extreme events of those hazards had unique properties for human perception and response—namely, they were beyond man's control and thus required a different type of adjustment. Second, natural hazard areas shared common features with respect to human response, one of which was that they were attractive areas for human exploitation (Burton and Kates 1964).

The scene was set for the international comparative case studies of hazard perception and behavior that followed. They sought to bring within a common framework and method hazards as diverse as volcanic eruptions and frost and human use systems as different as rural Bangladesh and urban San Francisco (White 1974b). This international collaboration was achieved through the Commission on Man and the Environment (chaired by Gilbert White) of the International Geographical Union.

The comparative hazard studies involved some forty principal investigators, in twenty countries (table 9.1). The choice of case studies was largely fortuitous, depending on where suitable local investigators could be found. Although it marked a pioneering effort in such work, and produced much useful field data, especially for developing countries, the perception and behavior components of the case studies have been much criticized. The concerns raised, considerably later, by critics such as Waddell (1983), Torrey (1979), and Bunting and Guelke (1979) form part of this discussion.

Table 9.1 reveals another step in the evolution of hazard perception research—the extension of the natural hazard models and approaches to man-made, technological hazards. The first man-made hazards to be included were air pollution in Toronto, Canada (Au-

Table 9.1 Field studies of hazard perception and response within the IGU collaborative programme

Hazard	Country
Air pollution	Hungary
	United Kingdom
	Yugoslavia
Avalanche	Peru
Coastal Erosion	United States
Drought	Australia
	Brazil
	Mexico
	Nigeria
	Tanzania
	Kenya
Earthquake	Canada
	United States
Flood	Ceylon
	India
	Japan
	Malawi
	United Kingdom
	United States
Frost	United States
Hurricane	Bangladesh
	Puerto Rico
	United States
	Virgin Islands
Landslide	Japan
Pesticides	United States
Tornado	United States
Snow	United States
Wind	United States
Volcano	Costa Rica
	Hawaii
	United States

Source: Natural Hazard Research 1973.

liciems and Burton 1970), Hungary, the United Kingdom, and Yugoslavia and pesticide risks in the United States. In its application to air pollution, the original hazard perception model was overextended, and its assumptions and definitions were called into question as a generalized model of risk perception (Kirkby 1973a).

Further work on perception of technological hazards such as nuclear accidents (Hohenemser, Kasperson, and Kates 1977; Whyte 1977), mercury pollution (Lundqvist 1974), highway risks (Bick and Hohenemser 1979), chlorine gas releases from train derailments (Burton, Victor, and Whyte 1981), and pesticides (Tait 1979) has helped to diversify the research models used. This wider range of hazard research has generally moved away from the natural hazard perception model to the adoption of more generalized risk assessment frameworks (e.g. Kates 1978; Whyte and Burton 1980).

This is not to overemphasize the present coherence of the risk perception field but to indicate a historical continuity with natural hazard perception. In practice, as the hazard perception approach was modified to deal with technological hazards, rare events (LOPHIC, or low probability, high consequence accidents), and low levels of contaminants, its methods were profoundly influenced by the new interdisciplinary networks of scientists and social scientists that were generated.

Another development in hazard perception was to study "all-hazards-at-a-place" instead of isolating one major natural hazard and comparing perception of it across different cultural settings. The first study of this kind was done in London, Ontario (Hewitt and Burton 1971), where the probability of natural and manmade hazards, their past distribution, and residents' perception of future risk were all compared. The natural hazards included tornados, hailstorms, floods, icestorms, hurricanes, severe snowstorms, severe windstorms, and drought.

The all-hazards-at-a-place approach has not proved as attractive for field research. Saarinen and Sell (1980) in their review of environmental perception research refer only to one master's thesis on all risks in Atlanta (Trendell 1979) and a methodological doctoral thesis on mapping total risk in urban areas (Wuorinen 1979). The recent UNESCO Man and the Biosphere (MAB) project in the East Caribbean studied the comparative perception of eleven natural hazards in Barbados, St. Lucia, St. Vincent, and St. Kitts-Nevis and compared residents' concern for all natural hazards to other (largely economic) perceived problems (Whyte 1982). This approach is closer to a human ecological approach and responds to some of the major criticisms of hazard perception studies (the lack of local depth and social holism). However, it is more time-consuming in the field and more limited in (policy) implications. It shares some of the strengths and weaknesses of anthropological case studies.

Two other aspects of the changing nature of hazard perception research can be briefly mentioned. First is a trend toward emphasizing the role of information in hazard perception (Committee on Disasters and the Mass Media 1980; Needham and Nelson 1977; Palm 1971; Whyte 1982) and the response of people to hazard warnings immediately prior to an event (Baker 1979; Downing 1977; Waterstone 1978). These studies have clear policy implications for improving warning systems and have provided evaluations of hazard information and education programs.

Second are a group of studies that focus on behavior immediately during and after a disaster has struck. Many of these international comparative field studies have been carried out, using sociological models or group and organized behavior (Barton 1963; Dynes 1974; Quarantelli 1979). A more sociological approach to hazard perception was also shown in the study of the reconstruction following the earthquake that struck Managua, Nicaragua, in 1972 (Haas, Kates, and Bowden 1977).

Changing Research Methods

The history of the research methods used to measure hazard perceptions is characterized by increased attention to research design, especially sampling strategy and a wide-ranging exploration of social science techniques, particularly those of psychology. Four phases can be discerned: early field investigations (1956–66), exploration of methods (1967–70), comparative international studies (1970–74), and social surveys and further case studies (1975 to the present).

Early Field Investigations (1956–66)

The first ten years of hazard perception research were characterized by greater attention to sample design and research methods for the environmental components of the hazard system than for the human populations involved. Burton's (1962) study of flood hazard is explicit in its classification of floodplains and its definitions of environmental factors, but the author summarizes (p. 67) his perception and behavior study methods with statements such as "Ten farm operators were interviewed in the McKinney area, and field examination of the relationship of farm practices to the flood hazard were made."

The study of coastal flood hazard in Megalopolis used a questionnaire format for interviews with residents, but that format is not present in the report (Burton, Kates, and Snead 1969). The tables and figures showing perceptions do not give the questions asked, so that it is difficult for the reader to evaluate the responses tabulated. In the same study, the questionnaire responses for earlier floodplain surveys are agglomerated and compared to those from the dwellers on the coastal plain, without any further assurance of the validity of this process than a remark that "similar interview schedules were used."

It is a measure of the increased methodological sophistication of the field that such scanty attention to questions of population sample frames and questionnaire design would no longer be accepted. Yet these early field studies were innovative in method: they combined traditional geographical study of the physical environment and patterns of human occupancy with nontraditional interviewing of the contrasting perceptions of local people and decision makers.

Exploration of Methods (1967–70)

In September 1967, the first phase of the U.S. National Science Foundation Project on Natural Hazards began. The next two years saw a remarkable exploration of research techniques, including laboratory experiments of psychological concepts and their application to field (real-world) situations where psychologists usually fear to tread. Although many of the techniques examined failed to reach the stage of field testing—usually because they were of the "paper-and-pencil test" format, which is singularly inappropriate for situations in developing countries—this period was one of considerable intellectual energy and creativity. It is important to consider this freshness of vision, for it is all too easy to be wise after the event and to judge merely in terms of results.

The techniques explored included semantic differential tests (Golant and Burton 1969), a modified Rosenzweig Picture-Frustration Test (Barker and Burton 1969), an avoidance-response test (Golant and Burton 1969), photo-choice tests, sentence completions, TATs "Prisoner's Dilemma Experiments," self-anchoring ladder technique to scale perceptions (Moon 1971), and a community disaster game (Natural Hazard Research 1973). Some techniques were not developed further; most were borrowed from social psychology, but the application to hazards and to field settings was new. At the same time, decision models were being adapted for hazard situations.

From today's perspective, and in light of later criticism, it might have been better to open what was essentially a dialogue dominated by psychology to a more interdisciplinary exploration of the concepts and approaches offered within anthropology and sociology. This would have had considerable impact on the explanatory frameworks adopted in the hazard perception studies that followed.

Comparative International Studies (1970–74)

By 1970, *Suggestions for Comparative Field Observations of Natural Hazards* had emerged from all the activity in development of method (Natural Hazard Research 1970). The purpose of this working paper was to provide a standardized format for collecting data on the physical setting, the history of the local human use system (called "sequent occupance" in the working paper), the magnitude and frequency characteristics of the hazard, estimates of hazard damage, and a list of hazard adjustments practiced in the area. The working paper provided also a standardized hazard perception and choice questionnaire (including a sentence-completion test) and suggestions for a quota sampling strategy of heads of households.

The guidelines on methods specified that only one hazard was to be investigated in any one place, and suggested that 120 interviews should be conducted with twenty individuals who had children and whose households fell into one of six classes (grouped according to education and age of children). Although the standardized questionnaire was not regarded less as a blueprint than as a basis for designing questions specifically for the local setting, the working paper did ask international collaborators to return their completed questionnaires to North America for comparative analysis.

The use of this standardized field instrument has been criticized, both by those involved in the international effort and by others. Basically, it is best suited to North America and to hazards that have clear events, such as hurricanes and earthquakes. The researchers who had most trouble using it were those studying perception of air pollution; those working in poor rural areas of developing countries; and Japanese collaborators, who found Western concepts such as "bad," "luck," and "emotion" difficult to translate into Japanese (Saarinen 1974). Saarinen's analysis of the adaptations made in different countries to the standardized questionnaire also shows that the factual questions (age, number of children, etc) were least often altered and the perception questions were most often altered or omitted, thus making comparisons difficult. The important question on adjustments to hazards won the prize for the most frequently omitted item.

The structure of a questionnaire, particularly in its focus on a single hazard and on individual adjustment, also reflects particular emphases in the explanatory framework adopted, which will be discussed later. Although the main purpose of the standardized questionnaire was to help encourage international comparative studies within the framework of the combined NSF and IGU Natural Hazard Project, it is still used by researchers and students around the world.

Social Surveys and Further Case Studies (1975 to the Present)

By 1972, when many hazard perception researchers met at an IGU symposium in Calgary, Canada, concerns were being expressed about paper-and-pencil tests, standardized questionnaires, poor sample design, and lack of in-depth, anthropological approaches. The late 1970s and early 1980s have generally seen a more systematic approach to the definition of populations being studied and appropriate sampling frames and sample sizes (e.g. Downing 1977; Burton, Victor, and Whyte 1981). Some investigations have been able to use experimental and control groups and "before-and-after" research designs, such as in the study of the impact of public information about hazards (Waterstone 1978; Whyte 1982).

Questionnaire interviews are still a primary research tool, but they are usually specifically designed for one case study. In addition, mailed questionnaires and telephone surveys are used in North American studies in an effort to improve the statistical validity of the sample, while keeping down costs. Thus, some hazard perception researchers have moved closer to social survey or opinion poll methods.

At the same time, less structured and more in-depth studies of hazard perception and response have used anthropology-style case studies of families (e.g. Haas, Kates, and Bowden 1977) and tape-recorded unstructured interviews (e.g. Burton, Victor, and Whyte 1981). There is also a continuing interest in developing new techniques for eliciting perceptions, such as scenarios (Erikson 1974, 1975; Downing 1977) and a Repression-Sensitization Scale (Simpson-Housley 1979).

In the extension of hazard perception studies to risk perception, similar trends can be seen. On the one hand, risk perception by the public is measured by sample populations carefully selected to represent the population of whole cities, states, and countries. These surveys are sometimes repeated, much as opinion polls are. On the other hand, studies of the cognitive frameworks in which choices are made tend to use small, experimental designs with volunteer "subjects" (Kahneman, Slovic, and Tversky 1982).

The search for improved techniques and more careful research design continues—a testimony to the continuing significance of hazard perception for public policy, particularly in its metamorphosis into risk perception. What, perhaps, has been too readily left behind is the tradition of human ecology.

The Evolution of Hazard Perception Models

> Were there perfectly accurate predictions of what would occur and when it would occur in the intricate web of atmospheric, hydrologic, and biological systems, there would be no hazard. (White 1974a)

In his 1966 paper "Formation and Role of Public Attitudes," White outlined many of the features of the hazard perception model that was to be more formally described by Kates (1970) and used as the conceptual basis for the international comparative field studies. There have been several statements of the model since, but with little change to its main characteristics (White 1974a; Burton, Kates, and White 1978).

Modeling Choice of Adjustment

As described by Kates (1970), the hazard model of human adjustment has three initial inputs: the physical environment, the human use system, and the characteristics of those people deemed to be managers of the environment at some decision-making level. These state variables interact to produce three processes: the hazard, perception of hazard, and adjustment (figure 9.1). Perception is posed as a filter between the hazard and the adoption of adjustments. Its principal role in the model is to limit the range of adjustment adopted to something less than the total theoretical range. Perception, like the hazard itself, is conceived as a stochastic process, capable of prediction only at a statistical level.

The prime purpose of measuring hazard perception is to help explain the existing choice of adjustments, with an eye to improving adjustments through public policy, including public education. "Essentially, the effort was to: 1. Estimate the extent of human occupance in areas subject to extreme events in nature. 2. Determine the range of possible adjustments by social groups to those extreme events. 3. Examine how people perceive the extreme events and resultant hazard. 4. Examine the process of choosing danger-reducing adjustments. 5. Estimate what would be the effects of varying public policy upon that set of human responses" (White 1974a, p. 4). This approach shows clear continuity with White's earlier work on identifying the range of theoretical and actual adjustments in the human occupancy of floodplains.

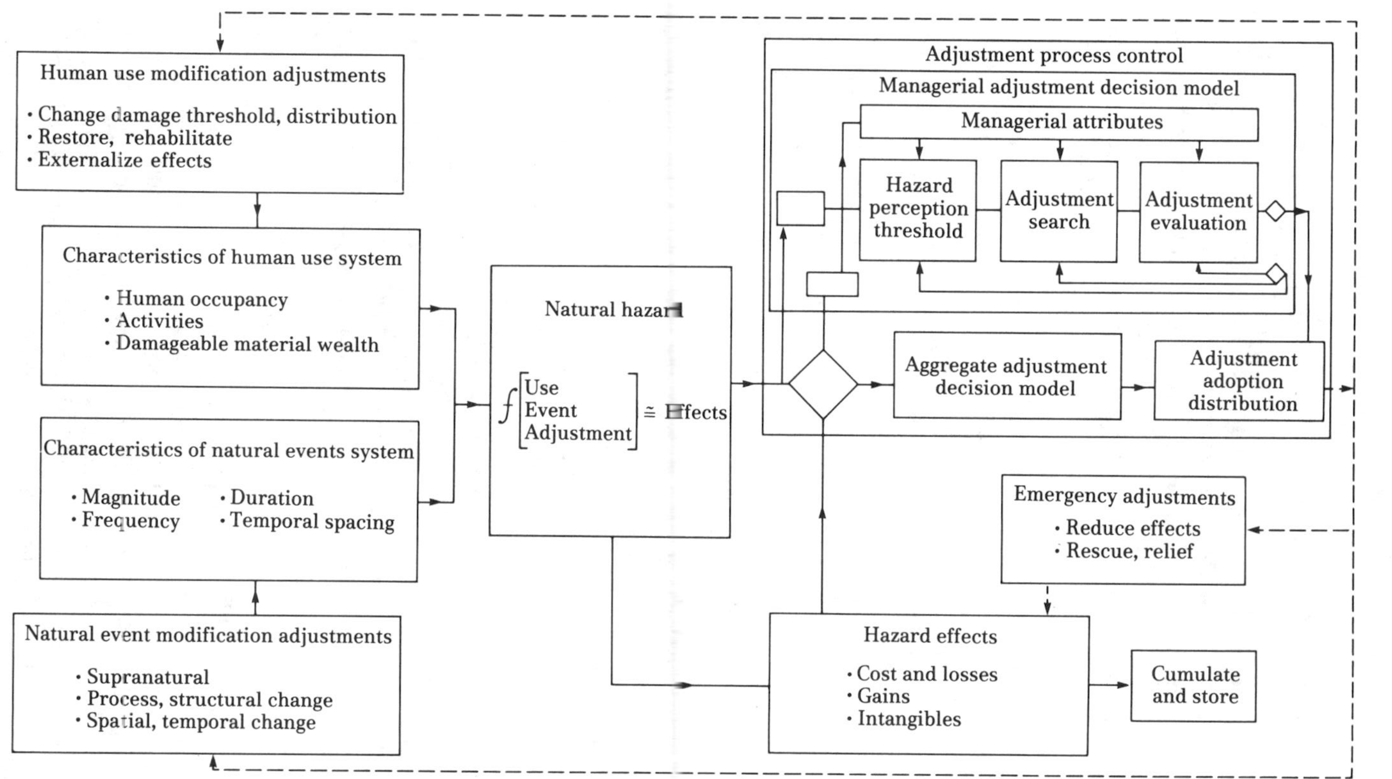

Figure 9.1 General systems model of human adjustments to natural hazards (from Kates 1970, fig. 3)

Bounded Rationality

The underlying model of man is that of Simon's (1956, 1959) "satisficer," or a decision maker who looks for a solution which is "good enough." In Simon's terms, man does not seek to maximize his expected utility (rationality) but is limited by factors such as his knowledge of the options and the time available to search for alternatives (bounded rationality). The model is thus simultaneously normative and descriptive, since it assumes that where constraints to perfect decision making are released, man moves closer to maximizing his expected utility. The path to prediction of man's actual choices is to understand the simplified model of the world that exists in his head—in other words, to describe his perceptions and his cognitive limitations.

For the response to hazards, the corollary is that as bounded rationality (actual adjustment) approaches rational decision making (knowing the full theoretical range of adjustments and maximizing the expected utility within it), hazard losses will disappear (see quotation at head of this section). This is why field studies have focused on the role of perception as an explanatory variable for the differences measured between, on one hand, the theoretical and actual range of adjustments and, on the other hand, past experience and future expectation of hazard; and why experimental studies have focused on limits to cognitive processing (heuristic biases) and on models of attitude adjustment (particularly cognitive dissonance).

Thus perception as modeled in hazard research is, in practice, an acknowledged subset of those factors that might be measured and that are known to influence response to hazard. The concern is not, therefore, whether hazard perception models do not include all relevant factors but whether they take account of the most important ones.

Perception of Probabilities

The early work on perception of floods in floodplains and coastal areas (Kates 1962; Burton and Kates 1964) found that people had difficulty in viewing flood hazard as a probabilistic process but, rather, saw separate events as influencing one another; for example, a recent major flood meant that another flood could not happen soon, or floods come in regular (predictable) patterns. As in other aspects of hazard research, this early work in floods exerted a lasting influence on the perception models adopted, even after they were transferred to hazards with different characteristics—a case of science echoing daily life. As Slobodkin puts it, "People often respond to a new problem as if it were an old problem" (1968, p. 198).

In most areas where they occur (and where, particularly, flood perception has been studied), floods have a recurrence interval that is within the memory, experience, and mobility patterns of residents. Perception of the probability of future flood in relation to past experience is therefore more likely to be a factor in adjustment. Floods also have the advantage, for the researcher, of occurring in recognizable, discrete events.

Over the past twenty years, perception studies have found that perception of hazard probability is less useful than in floods in explaining response to hazards with longer time intervals between events, and particularly rare events such as hurricanes, earthquakes, and nuclear reactor accidents; hazards with less recognizable events, such as air pollution and drought; and hazards perceived as primarily man-made (technological) rather than natural in origin.

Thus, in my view, there has been a growing gap between the psychological-experimental studies of cognitive processing (which are excellently brought together in Kahneman, Slovic, and Tversky 1982, and Nisbett and Ross 1980) and field case studies of hazard perception. The latest work in risk perception has perhaps shown this gap most forcefully.

Risk Perception

A series of experiments conducted in Oregon on risk perception shows that people tend to exaggerate the frequency of some risks and to underestimate others (Slovic, Fischoff, and Lichtenstein, 1982). While the probabilities of death from botulism, tornado, flood, pregnancy, and smallpox vaccination were overestimated, those from heart disease, stroke, stomach cancer, and diabetes were underestimated (figure 9.2). The researchers interpet these results to mean that differences in media coverage and the "memorability" or "imaginability" of the events has enhanced the recall of some risks (cf. the "availability" bias of Tversky and Kahneman 1974).

Heuristic bias is insufficient to explain many people's concerns about nuclear power, toxic waste disposal, the transportation of dangerous

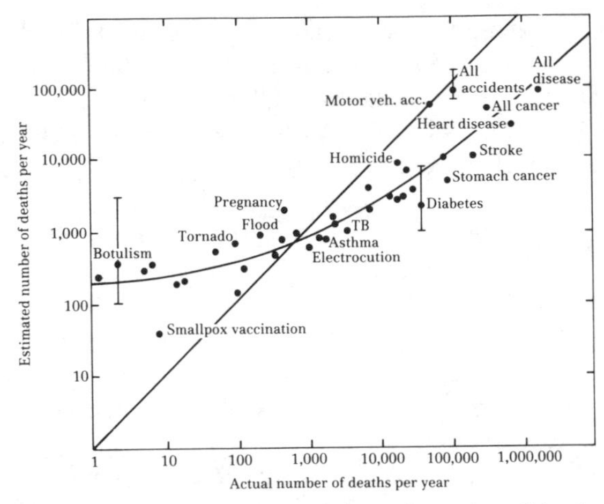

Figure 9.2 Relationship between perceived (estimated) number of deaths per year and actual number for 41 given causes of death. Each point is the average value for the experimental data obtained, and selected bars show 25th and 75th percentiles of the range of perceptions measured. Data are for 111 university students and 77 League of Women Voters in Eugene, OR (from Slovic, Fischhoff, and Lichtenstein 1982, p. 466).

goods, and acid rain. It is altogether too "cognitive" an explanation that plays down the role of the "maximum credible accident" or "worst-case scenario" in people's perceptions and the role of emotional response, or affect, generally (Whyte 1983). One of our present research needs in risk perception is to develop a better understanding of the factors that tend to increase and decrease perceived risk—most of which appear to be related to the characteristics of the consequences of risk rather than to the nature of the probabilities (table 9.2).

The hazard perception model deals with human cognitive processing and reflects Kelly's (1955) model of man as a scientist, rationally trying to make sense out of his world. The world of man's emotions, his attachment to place and to people, his fear of the unknown, his gods and devils do not enter into the perception component of the hazard model. Man processes hazard information, tries to sort out his objectives, brings his attitudes into line by minimizing his cognitive dissonance, and makes choices. Clearly, this is only part of hazard perception, but, equally clearly, no single empirically testable model can do much more. A few field studies have explored other aspects of man's relations to his environment through hazard perception, most notably Saarinen's (1966) study of drought in the Great Plains.

An interesting aspect of the comparative international hazard perception studies was the work on perception of the locus of control and attribution of causality. Locus of control (Rotter 1966) is usually

Table 9.2 Factors tending to increase perceived risk compared to scientific estimates.

Person-related characteristics	Situation-related characteristics	Risk characteristics
Lower educational levels*	Beyond control of individual	Poses immediate threat
Female	Individual at risk involuntarily	Direct consequences to health
Older*	Short time since previous hazard event	Mechanisms not understood
Parent	Children particularly at risk	Probabilities low or uncertain
Anglophone	Inadequate resources available	Unfamiliar "new" hazard
"Anxious" personality	Low credibility in authorities	"Dread" hazard
"External" personality	Scientific controversy	Large number of fatalities per event
	High media attention	Fatalities grouped in space and time
	No risk analysis	

* Varies according to specific cause of risk.

measured along a scale ranging from internal control (belief in one's ability to take decisions into one's own hands) to external control (believing that events in one's life are largely influenced by other people or by fate—god, chance, luck, etc.). Attribution of causality probes the perceived cause of the hazard and probes the important question of the degree to which hazards are perceived as natural or man-made. People with different individual characteristics, roles in society, and cultural backgrounds will assign different mixes of natural versus man-made origin to the same hazard event. This is an area that both field and experimental studies indicate has great potential explanatory power in hazard perception and choice. It can also contribute to our understanding of man-environment relations in the wider context of perceptions of "man over nature," "man under nature," and "man in harmony with nature."

Issues in Explaining the "Hazard Syndrome": A 1983 Agenda for Research

When we look in detail at Mexico City or Manchester, it is the uniqueness of the interaction of natural event and human use in each place that dominates our understanding of the

> response to extreme events. When we step back, however, it is possible to discern a few common elements. (Burton, Kates, and White 1978, p. 203)

The existence of a "hazard syndrome," reflecting a common pattern of human perception and response to hazards as different as hurricane and heat wave, flood and air pollution, was mooted during the 1960s (Burton, Kates, and Snead 1969). It has formed the underlying rationale for much of hazard perception research since, including the international comparative studies. In their summing up of this collective effort, Burton, Kates, and White (1978) conclude: "The theory holds that in using physical resources, people engage in behavior that combines adaptation to extreme events with both purposeful and incidental adjustment . . . Ways of coping are described as adaptation (biological and cultural) and adjustment (incidental and purposeful). Taken together, the many and varied coping actions can be grouped into four modes (loss absorption, loss acceptance, loss reduction, radical change) separated by three recognisable threshold levels—awareness, action, intolerance . . . Although the few modes of coping appear universal, a steady progression through the modes is not universally found" (Burton, Kates, and White 1978, pp. 203–6).

While it might be hard to quarrel with the notion of a hazard syndrome stated with this level of generality, there are some unresolved difficulties in the definitions and assumptions embedded in hazard research to date.

A Question of Salience

> Occupance responds to a variety of conditions among which the flood hazard is not often of primary significance. (Burton 1962)

The problem of the salience of hazards in the lives of ordinary people, even for those living on the sides of a volcano, was recognized in the earliest work on floods. Since then, it has largely been buried as an issue and has not deterred researchers from framing questionnaires as though even the most unlikely natural hazards were part of every family's breakfast table conversation.

Again, to quote Burton (1962, pp. 148–49): "While farmers undertake relatively elaborate and sophisticated estimates of flood hazard, where such hazard is high, the attention given to this problem falls off very rapidly as flood frequency diminishes. Where floods occur as frequently as once a year or every two years, framers tend to engage in careful analysis of risks which typically involve a careful choice of crops and attention to possible seasonal adjustments. As frequency diminishes to 1 in 5 or 6 years, attention to the problem decreases, and beyond this point the flood hazard seems to have little or no significance for them."

Within the framework of the hazard theory, this implies that for most extreme events infrequent losses are absorbed by a largely unaware society. Why then study the relationships between their perception of hazard and their intended (purposeful) adjustments?

At a different level of analysis, Wright and Rossi (1981) have argued, from population, housing, and economic statistics in U.S. disaster areas, that natural hazards are not salient enough at the policy level to make natural hazard policies work.

Salience cannot be deduced from measures of perception and response. Anyone with experience in interviewing will know that respondents will obligingly provide, in response to questions that they may never have considered before, answers that do not match their own views of the world. By putting questions on hazard probabilities and future behavioral adjustments together in the same interview, the researcher can generate what are, in effect, spurious relationships.

Salience has been little discussed in hazard perception since the early search for measurement techniques. It showed up as a problem in the studies of perception of air pollution, partly because of the lack of definable air pollution events to focus on, and partly because air pollution was not considered a real hazard by many respondents (Kirkby, 1972). Salience is a function of hazard frequency and some aspects of hazard magnitude (such as worst imagined future consequences and losses from the most recent event). But even in areas where natural hazards are frequent and can be devastating, such as on the tiny eastern Caribbean islands, studies show that salience is still a problem in measuring perceptions, as indeed, it is for emergency planning policy.

Hazards as Perceived

Our models of human choice are very much framed by the way hazards are defined. Similarly, the evaluation of human response as either an adaptation or an (incidental or purposive) adjustment hinges on the scientist's perception of the hazard being the same as those of the respondents. Yet relatively little work has been done on the validity of the categories we use to define and classify hazards. The methods used in ethnobotany to define folk taxonomies of, for example, weeds and useful plants have rarely been applied to the classification of hazards (Knight 1974).

Why are categories so important? Take the example of drought in southern Mexico. Drought was the hazard as defined by the researcher, but rainfall variability (expressed as wet and dry years and wet and dry places) was the environmental dimension as perceived by the Zapotec peasant farmers (Kirkby 1973b). The difference is more than a question of labels; the farmers' array of individual and community responses to temporal and spatial patterns of rainfall was more complex and strategic than a unidimensional set of adjustments to drought alone could be. By conceptualizing the decision-making task as an annual strategy in matching crops and fields to the pattern of water resources, Zapotec farmers could play simultaneously on the environmental swings and roundabouts on a continuing basis.

A key part of this ongoing hazard response is the substitutions made between space and time. It has often been observed that farmers, as well as stockbrokers, try to hold mixed portfolios of resources in order to spread risk and to capitalize on opportunity. This is clearly a risk perception strategy that is followed by many agriculturalists, nomads, and migrants to the city from poor rural families.

When asked what risk they most feared, Zapotec farmers replied in terms of illness, especially of a working member of the family, which reduced their household labor force; dire poverty; and starvation. Poverty was their real hazard, whether brought about by illness, drought, or economic misfortune.

Somewhat similar discrepancies between researcher and respondent definitions of hazard were observed in the study of air pollution perception in the United Kingdom. For those most at risk in the most polluted parts of inner cities, the perceived problem was not so much air pollution but their own helpless position on the bottom rungs of the British social class system. Again, poverty and social alienation were more real as hazards than was the environment (Kirkby 1972).

The ways in which different cultures classify hazards are an integral part of their cosmologies and belief systems. There are folk taxonomies of hazard that need to be compared with those of hazard research. A short survey of some ethnographic material indicated that in many traditional cultures concern to classify hazards by their different effects is much less pressing than the need to identify the chain of cause and effect. Unless hazard effects can be traced back to their genesis, no response or coping behavior is likely. It may be considered as dangerous to respond to drought as to illness before the cause has been properly identified, because the wrong action can be more harmful than no action (Whyte 1975).

Folk taxonomies also make important distinctions between man-made and natural hazards, although the division between them is not necessarily where we would put it. For example, the Cubeo Indians of the Amazon regard the death of anyone, however old and infirm, as resulting from human enemy action (Goldman 1963). Although the natural hazard model postulates that all hazards result from the ineraction of natural and man-made causes, we have little data on the degree to which people, even in the United States, themselves perceive a hazard as natural or man-made. And we do not know how this judgment affects their response to the hazard.

Daily Life and Extreme Events

> In practice, then, natural hazards have been carefully roped off from the rest of man-environment relations. (Hewitt 1983)

One of the early expectations of hazard perception research was that human behavior in extreme events would throw into high relief man-environment relations that otherwise were hidden from view. In other words, the very extremity of the stress would isolate normally complex processes so that they could be measured.

This hope has now been largely replaced by a recognition that extreme events are, by definition, infrequent; and that despite the considerable forces they exert on society for a short time, they probably have less impact overall on perceptions and behavior than do more frequent happenings. To understand human interaction with the environment, it is more useful, therefore, to look at the whole spectrum of hazard and resource variability in any one place (Hewitt 1983). The research thrust of the international comparative studies of hazard perception was the antithesis of the approach advocated by Hewitt and others. It emphasized the commonalities between extreme events across cultural settings at the expense of the relations between daily life and extreme events within cultural settings. Experience has now shown us that hazard response is so embedded into the structure of society, the economy, the belief system, and everyday life that only a more in-depth approach can elucidate the actual and theoretical ranges of adjustment. This problem has been discussed by several authors, most recently Waddell (1983) and Hewitt (1983).

A related problem in hazard perception research has been the little attention paid to the historical evolution of adjustments and the historical context of perceptions. The early work on floods was more historical in its approach than were the later comparative studies. Regan (1983) convincingly demonstrates the need to understand the historical context of natural hazards in his study of famine in Ireland.

Defining the Hazard Managers

Human perception is influenced not only by the cultural and historical context but also by the roles in society played by the decision maker. We each play a variety of roles in the contexts of our family, neighborhood work, local groups, and the larger society in which we live. Our perception of our role vis-à-vis the hazard profoundly influences the actions that we are likely to take. Hazard perception research has generally erred too much in the direction of defining people's roles from the researcher's point of view rather than from that of the respondent, although there are exceptions, such as Moon's work in London, Ontario (1971). It should be relatively easy to pay more attention to the hazard manager's definition of his or her role, and the results may surprise us.

Hazard perception research has also concentrated mainly on two levels of decision makers: the individual (farmer, resident, floodplain dweller, etc.) and government (usually at the national or state level). Much less attention has been devoted to intermediate levels of decision makers, whether formal, as in municipal governments, or informal, such as community and neighborhood groups.

It is generally recognized that, at each level, decisions are taken within the context of other decision-making levels, whether they are working in the same direction or not. Sometimes the assumptions made by one level about the actions taken at another level prove to be incorrect, but there is usually tacit acknowledgment that other hazard managers exist. Our present perception models do not lend themselves very well to this kind of analysis. They need to be further developed in the direction of "nested" decision (perception) hierarchies, collective behavior, and organizational response (Mitchell 1982).

Although the sampling frames for studies of hazard perception are generally defined as the population at risk (e.g. floodplain dwellers), analysis of responses within samples usually follows a more conventional sociological breakdown—such as age, sex, and education. This approach can fail to capture some important groupings within the study populations. People do not generally respond to hazard as high school graduates or as women (though for the latter, there are exceptions, such as opposition to nuclear power and war). Their self-reference groups are more likely to be defined by ethnic origin, religion, or professional affiliation.

The study of the evacuation of more than 250,000 people in Mississauga, Ontario, following a massive release of chlorine gas in a train derailment in 1979 illustrates the importance of looking beyond standard socioeconomic measures of individual differences. For the authorities, it was important to identify the characteristics of people who used official evacuation shelters, or who refused to evacuate or who returned to their homes illegally. The results showed that evacuation shelter users did tend to come from specific socioeconomic groups, mainly poorer and younger people. However, those who refused to evacuate were more likely to be European immigrants to Canada who had experienced evacuations and other consequences of living in a war zone (on either side of the fighting) during the Second World War. These people tended to dismiss the evacuation order as unnecessary compared to that experience and knew enough about evacuations to wish to avoid them. This finding could not have been predicted from their socioeconomic characteristics as measured in the questionnaire and required in-depth follow-up interviews (Burton, Victor, and Whyte 1981).

Adaptation and Maladaptation

> The lobe-finned fish did not come onto dry land to take advantage of the terrestrial habitat. Rather, relatively minor modification of their fins and other subsystems made them better able to migrate from one drying up stream or pond to another. (Romer's Rule, in Hockett and Ascher 1964)

> Adaptive response sequences have certain important properties . . . The responses most quickly mobilised are likely to be energetically expensive, but they have the advantage of being easily reversible should the stress cease . . . The earlier responses deprive the system of immediate behavioural flexibility while they continue . . . but the structure of the system remains unchanged; thus, they conserve the long run flexibility of the system. In contrast, while the later responses do alleviate the strain of the earlier, they are likely to reduce the long-range flexibility. There is in such a series a continual and graduated trade-off of adaptive flexibility for adapted efficiency. (Rappaport 1977, p. 86)

In biology and cybernetics, the criterion of adaptation is simply a generalized survival, in which the notions of creativity and opportunism play little role. For natural hazards, adaptation has come to mean a reduction in loss of life and overall damage while minimizing human effort (costs) in coping with the hazard (Burton, Kates, and White 1978, pp. 204–23). Thus, to remain in a hazard zone and to bear or share increasing losses when they come is regarded as maladaptive—or the paradox of human response to hazards. All definitions of adaptation and maladaptation are necessarily culture-bound and value-laden, but in this final section I want to examine briefly the specific consequences of the dominant view in natural hazard research for our definition of adaptation.

The idea that bearing losses or returning to live in a hazard zone is somehow maladaptive appears early in natural hazard research. For example, cognitive dissonance was used as one explanatory framework for the "misperceptions" of risk that were found (Burton 1972; Jackson and Mukerjee 1974). Moreover, whether a response is adaptive or not depends very much on the frame of reference. More specifically, it depends on whether hazards fill the entire frame or only one small part of it. A response that may appear maladaptive to hazards (e.g. sitting still and bearing the loss) may be adaptive in the total context of the risks and opportunities available. Thus, adaptive responses may include responses that deal directly with the hazard, responses that are directed elsewhere but maintain flexibility in the face of hazard, and responses that combine both features. Prigogine (1983) has neatly encapsulated the role of flexibility and opportunism in adaptation in his parable of the ants (based on his colleagues' observation of ant behavior). An ant out alone discovers an item of food. It is too large for him to bring back to the nest himself. He returns to the nest and signals to the assembly a message about food. Twenty companions follow him. Five continue to follow directly to the food and help to convey it back. The other fifteen wander around without finding the original food, but are available if, in Mr. Micawber's words, something turns up. Depending on your frame of reference (the transportation task or daily life), this behavior is either adaptive or maladaptive. So also with natural hazards.

Related to the concept of adaptation-response is Waddell's (1977) accusation that natural hazard research is covert environmental determinism. I do not agree with the charge, but I can see reasons for it. First, there is the inappropriateness of the framework selected; that is, the roping off of natural hazards from the rest of man-environment relations. Second, there is insufficient recognition given to the need to maintain flexibility—Bateson's (1972) "uncommitted potentiality for change"—at the expense of "adapted efficiency." Third, despite the inclusion of perception and values in the natural hazards model, not much attention is given to innovation and creativity. Again, this is probably a function of the intellectual separation of natural hazards and resources in perception studies.

Experts and Laymen: Perception in the Service of Participation

One of the enduring themes in environmental perception research has been the contribution of ordinary people's perceptions and knowledge to better decision making and public policy. Gilbert White, and later his wife, Anne, have exemplified this approach in their work (e.g. White, Bradley, and White 1972). In his earlier work on river basin management in the United States, Gilbert White promoted the need to take individual decision makers' behavior and priorities into account, and he railed against the arrogant assumptions about what the public wanted that were hidden in much U.S. public policy.

The past two decades have seen much fieldwork in perception that has articulated, for the "silent majority" as well as for illiterate peasants, their local knowledge and their preferences. In this, perception studies have contributed to the development of community participation both informally, through documenting real-world situations, and formally, through the promotion of community participation and perception studies as part of a development strategy.

Gilbert and Anne White have made significant contributions in their case studies of risk perception in relation to the quality of drinking water in developing countries. They have also played leading roles in persuading development agencies such as the World Bank, the World Health Organisation, and U.S. Agency for International Development (AID) to take a more positive view of social scientists, perception studies, and community participation.

There is a danger that in the effort to dispel "the myth of peasant ignorance" (Waddell 1983) we create a cult of peasant wisdom that denies poor people access to improved health, education, and living conditions as effectively as did former colonial attitudes. There is a fine line to draw between respect for traditional views and wisdom and the need to educate people to new knowledge and technology. Waddell unfairly criticizes Gilbert White for seeing the solution in terms of transfer of information and technology from developed to developing nations (1983). White's more recent work in international development is consistent with his earlier work on river basin management in the United States. For example, in a report to the World Bank, he urged that the keys to improving water supply to the urban poor are to understand their perception of responsibility to help themselves and to promote community education (White 1978). And in their classic work on drinking water in Africa, White, Bradley, and White (1972) show how simply providing Western technology in the form of boreholes can do more harm than good.

Throughout his long career, Gilbert White has reaffirmed the dignity of human choice. His work on human adjustment to hazard has stimulated the development of an entire subfield within geography and has led to an evergrowing body of field studies, literature on method, and even criticisms of the field. All these serve to show that hazard perception research is alive and well, and still with us. The past two decades have provided a wealth of data, some methodological features and successes, and some insights into man-environment relations. What is needed now is a reconsideration of the early assumptions and models—a period of conceptual introspection which can build on two decades of improvisation. It is the sense of this writer that such a reevaluation of the field will lead us back into the fold of human ecology.

References

Auliciems, A., and I. Burton. 1970. *Perception and Awareness of Air Pollution in Toronto. Natural Hazard Research,* Working Paper 13. Toronto.

Baker, E. J. 1979. "Predicting Response to Hurricane Warnings: A Reanalysis of Data from Four Studies." *Mass Emergencies* 4: 9–24.

Barker, M., and I. Burton. 1969. *Differential Response to Stress in Natural and Social Environments: An Application of a Modified Rosenzweig Picture Frustration Test. Natural Hazard Research,* Working Paper 5. Toronto.

Barton, A. M. 1963. *Social Organisation under Stress: A Sociological Review of Disaster Studies.* Washington, DC.

———. 1969. *Communities in Disaster.* New York.

Bateson, G. 1972. *Steps to an Ecology of Mind.* New York.

Berger, E., and T. Luckman. 1966. *The Social Construction of Reality.* New York.

Bick, T., and C. Hohenemser. 1979. "Target: Highway Risks." *Environment* 21: 16–20, 37–40.

Bunting, T. E., and L. Guelke. 1979. "Behavioral and Perception Geography: A Critical Appraisal." *Annals of the Association of American Geographers* 69: 448–62.

Burton, I. 1962, *Types of Agricultural Occupance of Flood Plains in the U.S.* Research Paper No. 75, University of Chicago Department of Geography.

———. 1972. "Cultural and Personality Variables in the Perception of Natural Hazards." In *Environment and the Social Sciences: Perspectives and Applications,* J. F. Wohlwill and M. Carson. ed. Washington, DC.

Burton, I., and R. W. Kates. 1964. "The Perception of Natural Hazards in Resource Management." *Natural Resources Journal* 3, no. 2: 412–41.

Burton, I., R. W. Kates, and R. E. Snead. 1969. *The Human Ecology of Coastal Flood Hazard in Megalopolis.* Research Paper No. 115, University of Chicago Department of Geography.

Burton, I., R. W. Kates, and G. F. White. 1978. *The Environment as Hazard.* New York.

Burton, I., P. Victor, and A. Whyte. 1981. *The Mississauga Evacuation,* Toronto.

Committee on Disasters and the Mass Media, National Research Council. 1980. *Disasters and the Mass Media: Proceedings of the Committee on Disasters and the Mass Media Workshop.* Washington, DC.

Downing. T. E. 1977. *Warning for Flash Floods in Boulder, Colorado. Natural Hazard Research,* Working Paper 31. Boulder, CO.

Dupree, M., and W. Roder. 1974. "Coping with Drought in a Preindustrial, Preliterate Farming Society." In *Natural Hazards: Local, National, Global,* ed. G. F. White, pp. 115–19. New York.

Dynes, R. R. 1974. *Organized Behavior in Disaster.* The Disaster Research Center Series, Ohio State University.

Eriksen, N. J. 1974. "A Scenario Approach to Assessing Natural Hazards: The Case of Flood Hazard in Boulder, Colorado." Ph.D. thesis, University of Toronto.

———. 1975. *Scenario Methodology in Natural Hazards Research.* Boulder, CO.

Golant, S., and I. Burton. 1969. *Avoidance-Response to the Risk Environment. Natural Hazard Research,* Working Paper 6. Toronto.

———. 1969. *The Meaning of a Hazard—Application of the Semantic Differential. Natural Hazard Research,* Working Paper 7. Toronto.

Goldman, I. 1963. *The Cubeo, Indians of Northwest Amazon.* Urbana, IL.

Haas, J. E., R. W. Kates, and M. J. Bowden, eds. 1977. *Reconstruction Following Disaster.* Cambridge, MA.

Hewitt, K., ed. 1983. *Interpretations of Calamity: From the Viewpoint of Human Ecology.* London.

Hewitt, K., and I. Burton. 1971. *The Hazardousness of a Place: A Regional Ecology of Damaging Events.* Research Paper No. 6, University of Toronto Department of Geography.

Hill, D. A. 1964. *The Changing Landscape of a Mexican Municipio: Villa Las Rosas, Chiapas.* Research Paper No. 91, University of Chicago Department of Geography.

Hockett, C. F., and R. Ascher. 1964. "The Human Revolution." *Current Anthropology* 5: 135–68.

Hohenemser, C., R. Kasperson, and R. W. Kates. 1977. "The Distrust of Nuclear Power." *Science* 196: 25–34.

Jackson, E. L., and T. Mukerjee. 1974. "Human Adjustment to the Earthquake Hazard of San Francisco, California." In *Natural Hazards: Local, National, Global,* ed. G. F. White, pp. 160–66. New York.

Kahneman, D., P. Slovic, and A. Tversky, eds. 1982. *Judgement under Uncertainty: Heuristics and Biases.* New York.

Kates, R. W. 1962. *Hazard and Choice Perception in Flood Plain Management.* Research Paper No. 78, University of Chicago Department of Geography.

———. 1970. *Natural Hazard in Human Ecological Perspective: Hypotheses and Models. Natural Hazard Research.* Working Paper 14. Boulder, CO.

———. 1978. *Risk Assessment of Environmental Hazard.* SCOPE 8. New York.

Kelly, G. A. 1955. *The Psychology of Personal Constructs.* New York.

Kirkby, A. V. 1972. "Perception of Air Pollution as a Hazard and Individual Adjustment to It in Three British Cities." Paper presented to Man and Environment Commission, 22d International Geographical Congress, Calgary, Canada, July 1972.

———. 1973a. "Some Perspectives on Environmental Hazard Research." Paper presented at Institute for British Geographers Meeting, Birmingham, January 1973.

———. 1973b. *The Use of Land and Water Resources in the Past and Present Valley of Oaxaca, Mexico.* Memoirs of the Museum of Anthropology, University of Michigan, no. 5.

Knight, C. G. 1974. "Ethnoscience: A Cognitive Approach to African Agriculture." SSRC Conference on Environmental and Spatial Cognition in Africa, May 1974.

Lowenthal, D. 1961. "Geography, Experience and Imagination: Towards a Geographical Epistemology." *Annals of the Association of American Geographers* 51, no. 3: 241–60.

———, ed. 1967. *Environmental Perception and Behavior.* Research Paper no. 109, University of Chicago Department of Geography.

Lucas, R. 1963. "Wilderness Perception and Use: The Example of the Boundary Waters Canoe Area." *Natural Resources Journal* 111, no. 3: 394–411.

Lundqvist, L. 1974. *The Case of Mercury Pollution in Sweden.* Committee on Research Economics (FEK), Report 4. Stockholm.

Mitchell, J. K. 1982. "Hazard Perception Studies: Convergent Concerns and Divergent Approaches during the Past Decade (1972–82)." Paper presented to Annual Meeting of Association of American Geographers, San Antonio, Texas, 25–28 April 1982.

Moon, K. D. 1971. "The Perceptions of the Hazardousness of a Place: A Comparative Study of Five Natural Hazards in London, Ontario." M.A. research paper, University of Toronto.

Natural Hazard Research. 1970. Suggestions for *Comparative Field Observations on Natural Hazards,* revised edition, October 20, 1970. *Natural Hazard Research,* Working Paper 16. Toronto.

———. 1973. *Collaborative Research on Natural Hazards: Progress Report October 1973. Natural Hazard Research,* Boulder, CO.

Needham, R. D., and J. G. Nelson. 1977. "Newspaper Response to Flood and Erosion Hazards on the North Lake Erie Shore." *Environmental Management* 1, no. 6: 521–40.

Nisbett, R., and L. Ross. 1980. *Human Inference: Strategies and Shortcomings of Social Judgment*. Englewood Cliffs, NJ.

O'Riordan, T. 1971. *Perspectives on Resource Management*. London.

Palm, R., 1971, "Public Response to Earthquake Hazard Information." *Annals of the Association of American Geographers* 7, no. 3: 389–99.

Prigogine, I. 1983. "The Convergence of the Sciences: Self-organizing Systems in Physical Chemistry, Biology and the Social Sciences." Address to the University of Toronto, 28 February 1983.

Quarantelli, E. L. 1979. "International Groups for the Study of the Social and Behavioral Aspects of Disasters." Natural Hazard Research Applications Workshop, July–August 1979, Boulder, CO.

Rappaport, R. A. 1977. "Maladaptation in Social Systems." In *The Evolution of Social Systems,* ed. J. Friedman and M. J. Rowlands, pp. 49–70. London.

Regan, C. 1983. "Underdevelopment and Hazards in Historical Perspective: An Irish Case Study." In *Interpretations of Calamity: From the Viewpoint of Human Ecology,* ed. K. Hewitt. London.

Roder, W. 1961. "Attitude and Knowledge on the Topeka Flood Plain." In *Papers on Flood Problems,* by G. F. White et al. Research Paper no. 70, University of Chicago Department of Geography.

Rotter, J. B. 1966. "Generalised Expectancies for Internal Versus External Control of Reinforcement." *Psychological Monographs* 80, no. 1: 1–28.

Saarinen, T. F. 1966. *Perception of the Drought Hazard on the Great Plains*. Research Paper no. 106, University of Chicago Department of Geography.

———. 1974. "Problems in the Use of a Standardized Questionnaire for Cross-Cultural Research on Perception of Natural Hazards." in *Natural Hazards: Local, National, Global,* ed. G. F. White, pp. 180–86. New York.

Saarinen, T. F., and J. L. Sell. 1980. "Environmental Perception." *Progress in Human Geography* 4: 525–48.

Sewell, W. R. D. 1965. *Water Management and Floods in the Fraser River Basin*. Research Paper no. 100, University of Chicago Department of Geography.

Sewell, W. R. D., and I. Burton. 1971. *Perceptions and Attitudes in Resources Management*. Resource Paper 2, Policy Research and Coordination Branch, Department of Energy, Mines and Resources, Ottawa, Canada.

Simon, H. A. 1956. "Rational Choice and the Structure of the Environment." *Psychological Review* 63: 129–38.

———. 1959. "Theories of Decision-Making in Economics and Behavioral Science." *American Economic Review* 49: 253–83.

Simpson-Housley, P. 1979. *Laws of Control, Repression-Sensitization and Perception of Earthquake Hazard. Natural Hazard Research,* Working Paper 35. Boulder, CO.

Sims, J., and D. Baumann. 1972. "The Tornado Threat: Coping Styles of the North and South." *Science* 176: 1386–92.

Slobodkin, L. B. 1968. "Toward a Predictive Theory of Evolution." In *Population Biology and Evolution,* ed. R. C. Lewontin. pp. 187–205. Syracuse, NY.

Slovic, P., B. Fischhoff, and S. Lichtenstein. 1982. "Facts versus Fears: Understanding Perceived Risk." In *Judgement under Uncertainty: Heuristics and Biases,* ed. D. Kahneman, P. Slovic, and A. Tversky. London.

Tait, E. J. 1979. *Measuring Attitudes to Risk: Farmers' Attitudes to the Financial, Personal and Environmental Risks Associated with Pesticide Usage*. University of Toronto, Institute for Environmental Studies, Working Paper EPR-6.

Torrey, W. I. 1979. "Hazards, Hazes and Holes: A Critique of 'The Environment as Hazard' and General Reflections on Disaster Research." *The Canadian Geographer* 23: 368–83.

Trendell, H. R. 1979. "The Perception of Natural Hazards: A Case Study of the Atlanta Metropolitan Area." M.A. thesis, Georgia State University.

Tversky, A., and D. Kahneman. 1979. "Judgement under Uncertainty: Heuristics and Biases." *Science* 185: 1124–83.

Waddell, E. 1977. "The Hazards of Scientism: A Review Article." *Human Ecology* 5: 69–76.

———. 1983. "Coping with Frosts, Governments and Disaster Experts: Some Reflections Based on a New Guinea Experience and a Perusal of the Relevant Literature." In *Interpretations of Calamity: From the Viewpoint of Human Ecology,* ed. K. Hewitt. London.

Waterstone, M. 1978. *Hazard Mitigation Behavior of Urban Flood Plain Residents. Natural Hazard Research,* Working Paper 35, Boulder, CO.

White, G. F. 1936. "Notes on Flood Protection and Land-use Planning." *Planners Journal* 3 (May–June): 57–61.

———. 1937. "Economic Justification for Flood Protection." *Civil Engineering* 7 (May): 345–48.
———. 1958. "Broader Bases for Choice: The Next Key Move." In *Perspective on Conservation,* ed. M. Jarrett. Baltimore.
———. 1960. *The Changing Role of Water in Arid Lands. University of Arizona Bulletin Series* 32, no. 2 (November).
———. 1964. *Choice of Adjustment to Floods*. Research Paper no. 93, University of Chicago Department of Geography.
———. 1966. "Formation and Role of Public Attitudes." In *Environmental Quality in a Growing Economy,* ed. M. Jarrett, pp. 105–27. Baltimore.
———. 1974a. "Natural Hazards Research: Concepts, Methods, and Policy Implications." In *Natural Hazards: Local, National, Global,* ed. G. F. White, pp. 3–16. New York.
———. 1974b. *Natural Hazards: Local, National, Global*. New York.
———. 1978. "Water Supply Service for the Urban Poor: Issues." *Water Supply and Management* 2: 425–54.
White, G. F., D. Bradley, and A. U. White, 1972. *Drawers of Water: Domestic Water Use in East Africa*. Chicago.
White, G. F., W. C. Calef, J. W. Hudson, H. M. Mayer, R. Shaeffer, and D. J. Volk, 1958. *Changes in Urban Occupance of Flood Plains in the United States*. Research Paper no. 57, University of Chicago Department of Geography.
Whyte, A. 1975. "The Cultural Encoding of Risk Assessment." Paper presented to SCOPE/EPRI Conference on Risk Assessment, Woods Hole, MA, 1–4 April 1975.
———. 1977. "Public Perception of Nuclear Risks in Canada, United Kingdom and USA." *Proceedings of Colloque sur les implications psyché-sociologiques du développement de l'industrie nucléaire,* pp. 438–52. Paris.
———. 1982. *Flood Risk Maps and Public Education: An Assessment of the Canadian Flood Damage Reduction Program Pilot Project in Oshawa, Ontario*. Final Report to Inland Waters Directorate, Environment Canada, March 1982.
———. 1982. "Environmental hazards." in *Studies on Population Development and Environment in the Eastern Caribbean: Report to the Governments of Barbados, St. Lucia, St. Vincent and St. Kitts-Nevis*. UNESCO/UNFPA/ISER Man and Biosphere Project, University of West Indies, Barbados.
———. 1983. "Probabilities, Consequences and Values in the Perception of Risk." In *Proceedings of Royal Society of Canada Symposium on Risk*. The Royal Society of Canada, Ottawa.
Whyte, A. V., and I. Burton. 1980. *Environmental Risk Assessment*. SCOPE 15. New York.

Wong, S. T. 1968. *Perception of Choice and Factors Affecting Industrial Water Supply Decisions in Northeastern Illinois*. Research Paper no. 117, University of Chicago Department of Geography.

Wright, J. D., and P. M. Rossi, eds. 1981. *Social Science and Natural Hazards*. Cambridge, MA.

Wuorinen, V. 1979. "A Methodology for Mapping total Risk in Urban Areas." Ph.D. thesis, University of Victoria.

10 Coping with Environmental Hazards

Timothy O'Riordan

Environmental hazards may be natural or man-made. As is often the case, there is a problem of terminology here. In this chapter the term *natural hazard* will be applied to natural events such as floods, droughts, storms, earthquakes, volcanic eruptions, and the like (for a comprehensive catalogue, see Whittow 1980; Perry 1981). For man-made hazards, the terms *technological hazard* or *environmental risk* will be used. This is a sloppy distinction but one that is commonly made. Properly speaking, "hazard" is an adverse event or situation while "risk" is the combination of the nature and consequences of an event and the likelihood of its occurrence. This definition implies that a "hazard" causes or is likely to cause death, injury, or damage to people and to property involving general suffering and distress, and considerable economic cost. It also follows that a natural hazard only becomes a problem if it occurs where people are living and its characteristics are such that they either have not learned adequately how to cope with it or have been unable to maintain coping strategies which once were successful. Environmental risks, however, need not be settlement-specific: they may have regional or global consequences (e.g., CO_2 discharges or enhanced acidity in the atmosphere and in rainfall), and they are characterized by a greater degree of scientific dispute as to cause, consequence, and probability of occurrence than is natural hazard.

Hazard studies, therefore, belong firmly in the behavioral field of human ecology, since by analyzing societal response to environmental stress one should be able to learn how different societies understand and adjust to the forces that help condition their existence. Indeed, one can also learn why people may be encouraged or forced to live in hazardous circumstances. Vulnerability to hazard is not always the result of foolhardiness, ignorance, or misunder-

Timothy O'Riordan is Professor of Environmental Sciences at the University of East Anglia, Norwich, UK, and a European editor of the international journal *Risk Analysis*. He currently heads a research team investigating the workings of the Sizewell B public inquiry into Britain's first pressurized water reactor. He is the author of *Environmentalism* and editor of nine books, all dealing with environmental planning and management.

standing: it may be the result of involuntary pressures forced upon people both with and without their knowledge. The manner in which individuals and various social institutions cope with both predisaster preparations and postdisaster relief should therefore suggest something about the relationship between political power and vulnerability to environmental risks.

This chapter will look at the achievements of both research and performance in reducing the social and economic losses associated with natural hazard and environmental risk. In research terms, considerable progress has been made in developing a theory of hazard adjustment, so that the conceptual understanding of this rather complicated process is now fairly well understood. But in practical terms the record is mixed: societal response is by no means uniform across the globe, even when danger is manifest and lives are being lost. Indeed, despite increasing investments in hazard prediction and prevention, the cost in lives lost, property damaged, and injury to body or mind is still growing (though the relationship between investment and cost varies enormously from country to country).

The link between preventive measures and injury for man-made risks is far less clearly defined. Nevertheless, in most Western democracies people are becoming increasingly alarmed about the dangers they judge themselves to be in as a result of living in a technologically advanced industrial economy. It is unlikely that they are actually in greater danger than their great-grandparents were. But today people know more about environmental risk to the point where safety regulations can be extremely expensive to implement and the safety clearance process enormously contentious and time-consuming. It is not easy directly to compare the coping strategies relevant to natural and man-made hazards, even though researchers continue to strive for a grand theory of hazard response. We shall see that these strategies are not just a function of the characteristics of the hazard; they are also a reflection of the way groups in a society judge government, science, technology, and, indeed, authority in general.

The Achievements of Natural Hazards Research

Prior to the turn of the century, studies of natural disasters were confined to descriptions by travelers and eyewitnesses. Scientific studies of the physical causes of natural disasters were fairly well advanced, but little interest was shown in the social response to such events (for a review of early hazard research, see Westgate 1975; Lewis 1977; Wynne Jones 1976). Curiously, it was a man-caused calamity that stimulated this aspect of research. On 6 December 1917 a French munitions vessel caught fire and exploded in the harbor at Halifax, Nova Scotia. This caused two thousand deaths, six thousand injuries and left some ten thousand people homeless. A sociologist documented the reaction to this disaster, described some of the sociopsychological processes at work and recommended certain basic principles of risk aversion and disaster mitigation (Prince 1968; see also Wynne Jones 1976). This pioneering work led to a fairly substantial literature by sociologists and political scientists into both the social and psychological factors that influence postdisaster emergency operations and the general preparedness of civil defense and like organizations (see e.g. Sorokin 1968).

Although the main emphasis of this early sociological work lay in response to man-caused hazards (fires, explosions, bombing, and the likelihood of a nuclear attack), many of the findings seemed relevant to natural hazards, particularly those of a notoriously calamatous nature (such as tornadoes, hurricanes, and earthquakes). Indeed, the establishment of a Committee on Disaster Research by the National Academy of Sciences in 1953 and the setting up of a Disaster Research Center at the University of Ohio a decade later encouraged the parallel investigation of societal responses to both man-made and natural hazards and gave hazard research a measure of academic respectability (see table 10.1).

In concentrating on how society organized itself in response to possible danger, however, this line of research never fully appreciated the need for fundamental understanding of the physical processes that produce natural disasters. Nor did it adequately recognize the reasons people live and remain in hazard-prone areas. In developed countries, agencies responsible for controlling the likelihood of a natural disaster (e.g. by damming rivers, constructing flood walls, erecting shock-proof buildings) tended to see their role as primarily one of hazard control rather than man-hazard management. This line was very much influenced by the agency terms of reference, the prejudices of the groups benefiting from their investments, uncritical manner in which the budgets and programs of work were accepted, and the lack of any alternative analytical perspective.[1] As a result, there tended to be an undue emphasis on the technologies of disaster prevention and postdisaster relief. This narrow vision made it difficult for policy makers to understand the social forces

Table 10.1 Summary of disaster research

Period	Research activity
Before 1940	Disasters described but not analyzed
1917	First study of a disaster by a sociologist (Prince 1968); ship collision and explosion, Nova Scotia
1920s, 1930s	Widespread damage and distress due to floods and drought in U.S.A.
1940s	
1941	First study of environmental impact (U.S. Bureau of Reclamation 1941)
1942	First publication by White (1942) on floods as hazard
early 1940s	Studies by psychologists and sociologists on the effects of bombing on human behavior and social organization
1950s	
1953	U.S. National Academy of Sciences establishes a Committee on Disaster Studies
mid–1950s	National Opinion Research Council develops basic methodology for disaster research focusing on community and group response in phases following calamity
1960s	
1962	First of a series of publications by students of White looking at individual responses to a natural hazard
1963	Establishment of the Disaster Research Center at the University of Ohio
1966	First of a series of geographical publications involving psychologists and psychological techniques
1966	First of a series of publications on alternatives to the "tech-fix" approach
1968	First of a series of publications on public knowledge and evaluation of alternatives to approaches to hazard mitigation
1970s	
1970	Establishment of the Man and Environment Commission of the IGU with White as Chairman and an emphasis on collaborative research (Publication of Natural Hazards Research Working Paper no. 16)
1972	Calgary meeting of the IGU Man and Environment Commission (White 1974a)
1973	Formation of the Disaster Research Unit at the University of Bradford, and the London Technical Group with a strong interest in disaster mitigation research
mid–1970s	Preparation of summary of natural hazards research (Burton, Kates, White 1978)
1976	Formation of the Clark University Technological Hazards Research Group
1977	Formation of the International Disasters Institute and publication of the Natural Hazards Observer newsletter of the natural hazards research group
1978	Emergence of the comprehensive notion of environmental risk in a management context

that encourage people to face environmental risk and enable them to cope with it when calamity strikes.

In the late 1950s a more behavioral approach to natural hazard research was developed by geographers working under the direction of Gilbert White at the University of Chicago (for a summary, see White 1973). Although initially conducted by geographers working in the United States, this research blossomed into a major global collaboration incorporating psychologists, sociologists, economists, planners, lawyers, and professionals in the fields of hazard prevention and disaster relief. The major stimulus for this work was to overcome the paradox that modern societies, while increasingly able to manipulate nature, are increasingly susceptible to the ravages of natural disasters.

Equally important to these early geographical studies was the interest in developing an ecological approach to this investigation following the pioneering work of Barrows (1923). Barrows visualized geography as human ecology, the interactional adaptation of man to his environment, so it seemed appropriate to look at the way complex social systems responded to natural hazards and, in turn, were altered by the experience of calamity. This created the philosophical setting for the development of theoretical models outlined below.

The five major objectives of this research were (1) to estimate the extent of human occupancy in areas subject to extreme events in nature; (2) to determine the range of possible adjustments by social groups to those extreme events; (3) to examine how people perceive extreme events and resultant hazard; (4) to examine the process of choosing damage-reducing adjustments; and (5) to estimate what would be the effect of varying public policy upon that set of responses.

Occurrence of Natural Hazards

Nowadays much improved records are kept of natural disasters. This in itself is a notable achievement for it requires international collaboration and a sustained commitment from a host of official and voluntary bodies involved with predisaster prediction and preparation and postdisaster relief. Ten years ago such data were not available; today this information provides a vital basis for international aid and the development of natural hazard mitigation efforts. Some of these data are collected by the United Nations Office for Disaster Relief located in Geneva,[2] some by the Natural Hazard Information Center at Boulder, Colorado, established by White in

1968, which publishes a quarterly newsletter, and some by the international and national relief agencies.[3] Many of these records contain valuable details of the nature and extent of damage and the degree to which postdisaster relief was a help or a hindrance.

The general finding is that hazard-prone occupancy appears to be increasing both in developed and in developing countries, despite improvements in hazard-prediction technologies and in the organization of relief agencies. Figure 10.1 illustrates the global picture—a general increase in deaths but no noticeable change in the incidence of hazard. The global data also indicate that loss of life and limb seems to be much greater in the poorer countries, especially those at a mid-stage in economic development, while the richer nations suffer more from economic losses due to property damage and disruption of the economy. Table 10.2, however, suggests that the developing countries suffer more than richer ones from both hazard-inflicted death and hazard-caused economic damage.

The implication here is that the forces encouraging hazard-proneness, which result in increased settlement and/or more property of greater economic value in areas susceptible to natural hazards, are outstripping efforts to predict and warn people of impending disaster and to safeguard their property when natural events strike. This misses the question of whether natural events are becoming more common or whether institutional and political forces are operating in such a way as to make some people more vulnerable to calamity. One obvious point must be stressed: people live in hazard-prone areas because they consider the benefits from so doing to outweigh

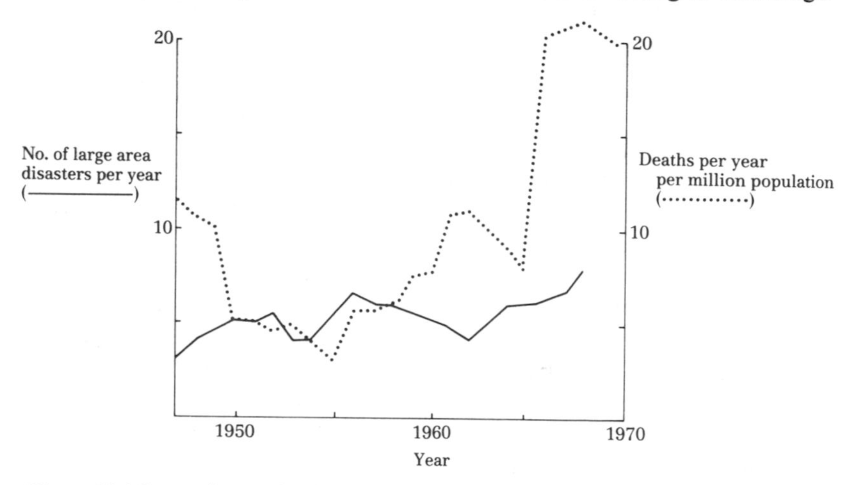

Figure 10.1 Increasing scale of global disasters (five-year moving averages)

Table 10.2 Comparative hazard sources in the United States and developing countries

	Principal Causal Agent[a]			
	Natural[b]		Technological[c]	
	Social cost[d] (% of GNP)	Mortality (% of total)	Social cost (% of GNP)	Mortality (% of total)
United States	2–4	3–5	5–15	15–25
Developing countries	15–40[e]	10–25	?[f]	?[f]

Source: Harriss et al. 1980, p. 105

[a] Nature and technology are both implicated in most hazards. The division that is made here is by the principal causal agent, which, particularly for natural hazards, can usually be identified unambiguously.

[b] Consists of geophysical events (floods, drought, tropical cyclones, earthquakes and soil erosion); organisms that attack crops, forests, livestock; and bacteria and viruses which infect humans. In the U.S. the social costs of each of these three sources are roughly equal.

[c] Based on a broad definition of technological causation, as discussed in the text.

[d] Social costs include property damage, losses of productivity from illness, or death, and the costs of control adjustments for preventing damage, mitigating consequences, or sharing losses.

[e] Excludes estimates of productivity loss by illness, disablement or death.

[f] No systematic studies of technological hazards in developing countries are known to us, but we expect them to approach or exceed U.S. levels in heavily urbanized areas.

the risks. The fact that their judgment of risk differs from some "scientific" analysis is significant, for the nature of the distortion indicates their perception of the benefits to be gained and the help they believe they will be able to obtain should disaster strike.

Theories of Response to Natural Hazards

There have been several attempts to produce a theory of response to natural hazard. No universal theory is yet acceptable to researchers and practitioners, though all attempts recognize the centrality of the relationship between the characteristics of natural event; the geography of settlement and economic activity in hazard-prone regions; and coping strategies—past, present, and likely (management responses and simply bearing the loss).

For the purposes of this review, three theories will be examined: the "transition thesis" developed by Burton, Kates, and White (1978); the "lessening and worsening" analysis tentatively proposed by Kates and his colleagues at Clark University (Bowden et al. 1979); and the "vulnerability thesis" first suggested by Wisner, O'Keefe, and Westlake (1977) but much more substantially argued in a volume of essays edited by Hewitt (1983). There is a lot of common ground within these approaches, so one should be wary of making too much of the differences.

The *transition thesis* proposes an evolutionary or developmental approach to response to natural hazards, evolving from "folk" responses, through technically and managerially dominated "industrial" reactions, to "mixed" adjustments involving a variety of technological, behavioral, and political measures. The criteria for selection relate to economic viability, social and political acceptance, administrative convenience, and environmental suitability. Burton, Kates, and White (1976, p. 120), summarize their perspective as follows: "Common to almost all models of social change is a focus on one-direction development, with an explicit or implicit assumption of growing complexity . . . as societies enlarge their scale of action, differentiate their functions, and integrate their adjustment to hazard changes. As folk societies mix with industrial societies, the number and importance of their cultural adaptations and incidental adjustments diminishes."

This view emphasizes a developmental or maturation approach which may be broadly universal. Ideally, the best characteristics of folk responses (good cultural adaptation, flexibility, low cost) and of industrial reactions (hazard control and protection systems plus zoning and structural improvements) can combine in mixed adjustments coupled with better policy—notably in predisaster planning, insurance, and other loss-sharing schemes, and more coordinated postdisaster relief. Also in policy terms, international disaster relief agencies and national governments should try to understand more about dislocations caused by natural calamities, and should adapt modern coping responses to suit those well established in the community. In many cases, indigenous coping strategies (for example, absorption of the homeless within extended families and community initiatives for reconstruction) are well designed to absorb the worst of posthazard dislocations.

The *lessening or worsening thesis* is hardly a thesis, more a speculation. The theme is that improvements in "mixed" responses as suggested above should reduce the impact of a given natural event so that future events of similar magnitude are less damaging to lives and property. For example, international aid should be so coordinated as to assist and rehabilitate famine victims far more than compared with the past. However, the "worsening" aspect of the analysis proposes that the growing interdependence of world economies could result in more widespread repercussive effects of natural disasters, as hazard-prone economies become more dependent on international aid, and as hazard-created economic damage (e.g. crop losses) create food shortages or price rises within the international economy.

These lines of argument are neither mutually exclusive nor sufficiently detailed to be very helpful. Natural hazards are interesting because they afflict societies and economies which will adjust and respond in a variety of quite individualistic ways. This analysis could be helped by a more critical assessment of the role of postdisaster relief. Unless aid is conspicuously geared to the existing and potentially beneficial coping strategies, of local cultures, it could cause even more economic instability and reliance on external economic assistance in the future. This could worsen the state of hazard vulnerability. Postdisaster relief and long-term rehabilitation cannot be dissociated from the politics and economics of wider developmental processes—which may have to be quite specific to a nation or a region.

The *vulnerability thesis* tackles this line of argument more directly. The view here is that increased vulnerability to hazardous events is due at least as much to the acts of human beings as it is to the acts of God. The man-hazard relationship is seen not in terms of a man-environment interactive relationship but largely in a man-man context, where the environment is an independent phenomenon. It is claimed that those most likely to suffer at the hands of a natural disaster are those with limited choice about where they can live; who are lacking in money, knowledge, and awareness of disaster relief assistance so that they do not know what to do in the event of a calamity and cannot protect themselves; and who often must exist in inadequate shelters without sufficient nourishment.

According to this "vulnerability" school of thought, disaster proneness is a product of a particular form of capitalist economic development which exploits the poor, worsens their conditions so they become involuntarily prone to natural disasters. Furthermore, these forces of "development" may force people to move into areas that were previously uninhabited precisely because of the natural and man-made hazards that existed. Alternatively, people simply are not able to adapt, or to develop appropriate coping strategies. But the analysis need not be so politically motivated. Vulnerability is essentially associated with poverty and powerlessness, which are often linked to ignorance and defenselessness. During the great

American heatwave of 1979 in which over six thousand people died of heat exhaustion, those most in danger were the old who could not afford air conditioning, who had nowhere cool to go to, and who relied on the use of fans, which, under certain extreme conditions of temperature and humidity, created mortal danger.

From a policy viewpoint, the solutions to the problem of increasing hazard proneness lie, in part, in reestablishing indigenous mechanisms of communal self-reliance so as to ensure that vulnerable peoples are provided with the basic knowledge of what hazards they face and how they can cope with them, and encouraging the judicious use of Western technology and organizational experiences as and where appropriate (for an example of this approach, see Davis 1975). But more fundamental questions of economic dependency, power relations and patterns of social change linked to wider educational opportunity and altering the status of women must also be addressed if the "vulnerability" element of hazard-proneness is to be reduced. This is a tall order: current developments do not indicate much scope for optimism (see e.g. Myers 1985).

Models of Coping Strategies

"Perception" is the term employed by most geographers and some psychologists to define the process whereby individuals and groups judge the degree of danger they face in relation to the benefits they enjoy by staying where they are, and hence search for and evaluate various means of reducing that danger should they be motivated to do so. Perception is regarded as the mechanism which links judgment to action. It is also a product of awareness, a function of the character of information provided or sought, and is influenced by the degree of trust in the agencies and mechanisms providing advice and help in disaster probabilities and postdisaster rehabilitation.

Throughout the 1960s and early 1970s a number of studies of hazard-prone dwellers in a variety of countries revealed the following broad findings (summarized in Burton, Kates, and White 1978, pp. 34–52):

1. People did recognize that they were in danger.
2. Their judgment as to the likelihood of damage and/or injury did not noticeably differ from "official" estimates for the more probable events but were more variable for the less probable events.

3. Their response to preventive measures was dependent in part on whether they had experienced a hazard before (or at least lived in a locality that had suffered damage).
4. They tended to make the uncertainties of hazard probabilities more certain by simple devices such as reducing low probabilities to zero, dividing probabilities into absolute factions, passing the blame onto some other agency (a deity or the government), or simply ignoring the notion of uncertainty altogether.
5. They knew of a number of possible preventive measures which could be adopted individually or collectively.
6. Their knowledge and evaluation of the various damage-reducing measures open to them were influenced by such factors as experience, social communication, and their understanding and trust in the accuracy of warnings and the competence of civil defense, police, and postdisaster relief organizations.

The conclusions of these perceptions studies suggest that people living in hazard-prone areas are not likely to respond to preventive measures or loss-sharing strategies (notably insurance) precisely where they should be cooperative—namely in areas where natural hazards are likely but not common, where the benefits of occupancy are great (or where the scope for moving out is restricted), and where modest investments in prevention could be highly cost-effective. To consider how to deal with this kind of problem, we first need to look at three approaches to understanding societal response to natural hazard.

Figure 10.2 simplifies and summarizes the findings with respect to hazard perception. It is believed that there is some kind of threshold of perceived danger (which may or may not relate to perceived gain for staying in situ and doing nothing), which triggers a search for, and possible adoption of, protective actions. Models, of course, are only simplifications of reality: while they may look impressive, field research must support their theses if they are to be usable for policy implementation. Unfortunately, empirical work to date has not yet proved that such a triggering threshold exists.

Looking at figure 10.2 in more detail, we find no clear relationship between a person's propensity for facing or avoiding risks and his/her reaction to danger, yet intuitively we would expect such a relationship to exist. Either the techniques of investigation and measurement require refinement or hazard perception is too complex a process to allow the personal risk propensity to be separately identifiable. One suspects that both factors may apply. The link between

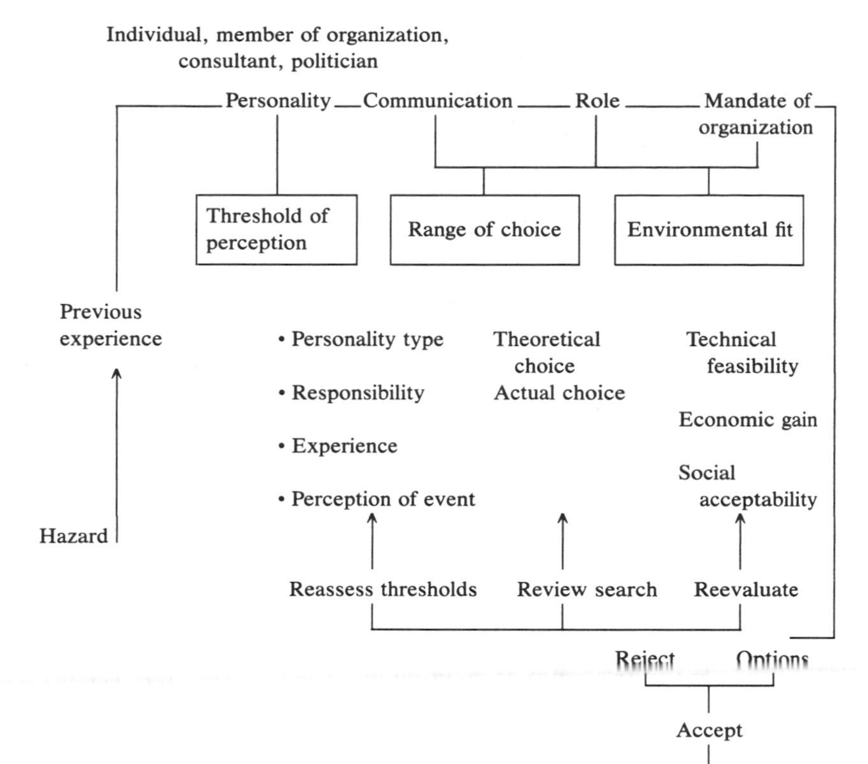

Figure 10.2 Patterns of response to environmental hazard

an individual's sense of responsibility for his or her protection and the adoption of hazard avoidance measures also awaits better examination. Some work has been done to suggest that beliefs about the role of fate and one's own ability to control one's life may be relevant, but the research techniques deployed (largely the use of sentence stem completion) and the size and choice of samples used leave considerable room for doubt. Experience does appear to count a lot, but much will depend on the nature and recency of experience and the manner in which the natural hazard occurs (see figure 10.3). The rapidity of onset, the degree of precise warning available, and the possibilities for undertaking protective measures all seem to be relevant.

Likewise, the range of choice perceived to be available will depend a lot upon communication of information from official and unofficial disaster relief sources and the individual motivation to search long

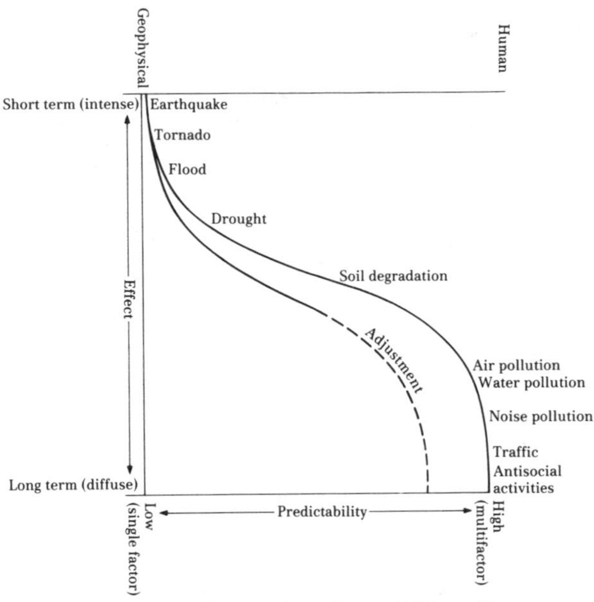

Figure 10.3 The spectrum of hazards (from Perry, 1981, p. 3)

and hard among the options. Kunreuther argues that many factors inhibit this search (see figure 10.4), particularly the belief that a low-probability event will not occur in the forseeable future, and suggests that reinforcement loops may encourage no action (except to await relief) while search loops may stimulate a reappraisal of the threat and an investigation of the most suitable preventive measures (Kunreuther 1974, pp. 204–14; see also Kunreuther 1978). Once again, however, the model behaves better than the practice. The forces acting in an individual to help or hinder him or her in coping with natural disasters are still not sufficiently understood to enable policymakers and hazard relief managers to deal effectively with reducing the danger.

A major problem lies in the quality of the field research undertaken so far. There have been many difficulties in establishing a truly comparable research tool. The questionnaire is notorious for its inadequacies in eliciting information on judgment and motivation, while the choice and size of sample are equally problematical, especially when the researcher is not at all clear as to the scope or intent of the investigation. Obviously the relevant underlying thesis guiding the research influences the methodology. The transition thesis will encourage the researcher to look at people in locations of greater or lesser hazard (as identified by "official" statistics) and also to

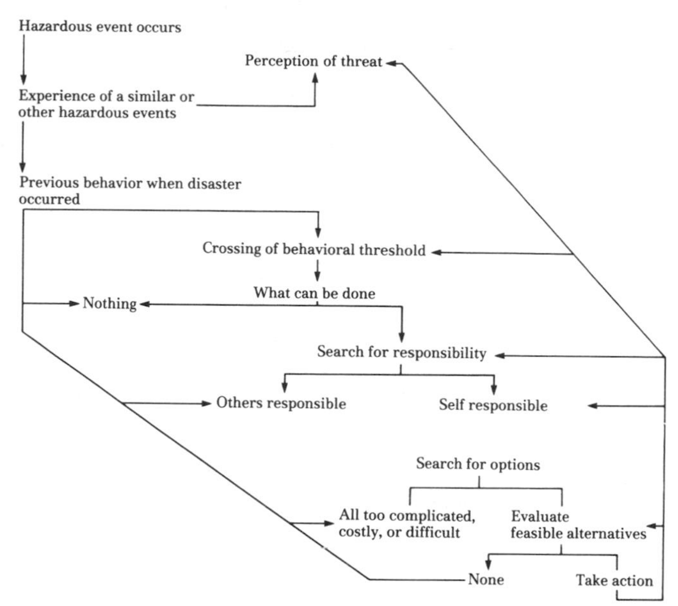

Figure 10.4 Ordered choice under uncertainty. An illustrative sequence of thought and action following the onset of a natural disaster. Note the reinforcement loops that encourage no action (accept/await relief) on the left hand side and the feedback loops that encourage a reappraisal of the threat and a search for the best options available. In the interactional model, the recognition of a threat causes a search for information and a receptivity to the policies being proposed by disaster prevention and damage mitigation organizations.

compare responses in places where pre- and postdisaster preparedness is at different stages of development. But a researcher following up the vulnerability thesis is more likely to concentrate on people which he or she believes to be disadvantaged for various reasons. A more existential (people- and experience-centered) method of investigation will be tried. This relies less on the questionnaire and more on interpretation of how a respondent feels, behaves, and responds to his "lived space." The danger here is that research methodology can become contaminated by the ideology of the researcher. Until now, this particular difficulty has not proved serious, but the fact remains that the three "theses" of hazard coping are neither adequately clarified or properly connected.

A more promising approach is an interactional one, which visualizes perception as a dynamic process of awareness, search, judgment, and action. Figure 10.5 illustrates the bare bones of an embryonic model. This diagram suggests that the perceived occurrence of a natural hazard is distorted by three major influences,

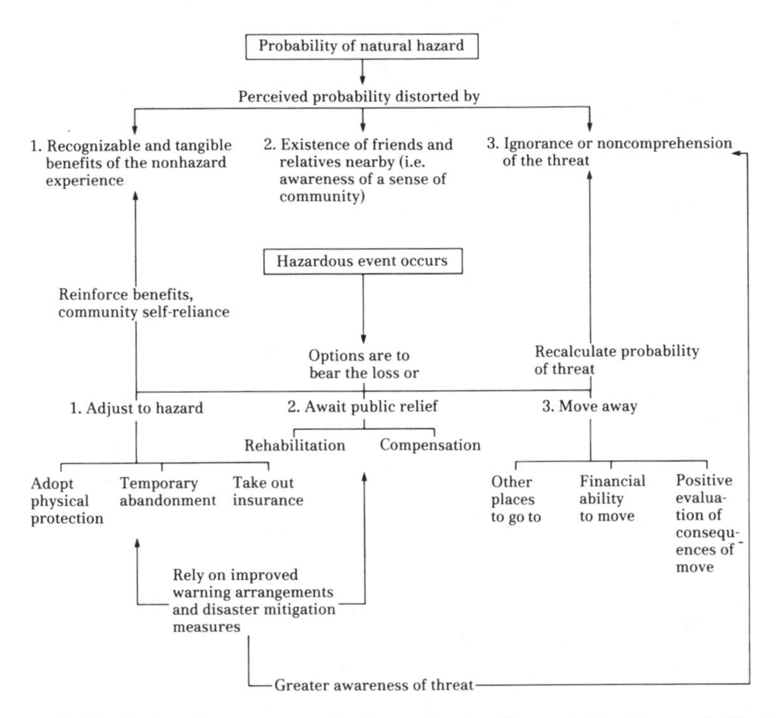

Figure 10.5 Behavioral responses to the threat of natural hazard. The diagram indicates some of the possible options open to a hazard-prone inhabitant and some of the interactive consequences of an actual disaster.

namely, the benefits of continuing to live in a hazard-prone site, the availability of postdisaster community relief, and the level of knowledge about the threat. These often become self-reinforcing influences which lock people into staying put even when life is hard.

The upper part of figure 10.5 also accommodates the vulnerability theses for it implies that people, even when forced to face danger not of their own choosing, can adjust to the threat, mostly through unwilling resignation but sometimes through impressive deployment of collective reliance measures. When a hazard occurs, depending upon its nature as depicted in figure 10.3, people have four options: to adjust to it, await relief, move away, or simply suffer the loss. Each of these responses depends in part on the way in which the "locking in" effect described in the upper half of figure 10.5 operates, in part on the manner and reliability of public relief, and in part on the feasibility of moving away. Those who espouse the vulnerability thesis believe that this last option is not normally available, that the extent of public relief is variable but is improving, and that development of collective self-reliance measures and suffering are the most likely outcomes.

Achievements of Natural Hazards Research

While it may be the case that the range of adjustment to hazard is widening and that anticipatory planning and postdisaster aid are becoming better coordinated, hazard management strategies, no matter how well intentioned, are still poorly integrated. Ways must be sought, then, to give people a greater responsibility in coping before and after natural disasters occur.

Nevertheless, the natural hazards research has achieved some important successes. There is a much closer relationship between research and practice than there was in the past. Many disaster relief agencies fund or conduct their own research and respond attentively to the results of this work. Researchers are also learning a lot about the political and practical difficulties which agency personnel have to face. The relationship between technology, organization, and behavior as suggested in figure 10.5 is now better understood so that more comprehensive strategies are becoming more successfully deployed. For example, agencies which formerly devoted their attention largely to hazard control are now becoming interested in hazard-proofing technologies (such as building design for earthquakes and hurricanes and the siting of valuable stock and equipment above historic flood levels) and the preparation of hazard-prone maps. These technologies guide planners as to what safeguards should be established before new settlement is contemplated and which areas should not be developed. It now seems that a combination of penalties (refusal of grant aid until hazard-zoning plans are completed and building codes enforced) and incentives (special grants for disaster proofing subject to a satisfactory risk appraisal plus variations in insurance premiums), is necessary, with greater emphasis on punitive measures where cost-effective preventive measures are not put into effect. But more research is required, for it is not at all clear how hazard control agencies are responding to a widening of their terms of reference and of their scope for management. Nor is it known how far the affected public, especially in Western societies, is prepared to shoulder responsibility for hazard preparedness and postdisaster reconstruction.

The major disaster-relief organizations have also responded to the need for a package of management measures to cope with natural hazards. The American National Red Cross, for example, has adjusted its former role of being largely a post hoc disaster relief agency and now takes its predisaster responsibilities more seriously. The Red Cross not only lobbies for a national flood insurance program but is currently promoting many of the findings of natural hazards research through its own educational manuals and films, and through its contacts with the national association of voluntary organizations in disaster. In its publication *The Red Cross Disaster Relief Handbook,* the League of Red Cross Societies proposes the formation of national coordinating agencies, each with statutory powers to draw up plans for coping with disasters and to assume leadership if disaster occurs.

In many developing countries a number of researchers (many trained by Gilbert White or his pupils) have encouraged governments to develop better hazard-warning schemes and to improve techniques for postdisaster rehabilitation. In Bangladesh, for example, geographers at the University of Dacca have completed a government-sponsored study of cyclone hazard zones on the coastal strip that will be incorporated in future regional plans. Designation of such zones will not stem the tragedy of flooding and cyclone damage in Bangladesh, but it might at least stop it from getting considerably worse. In addition, the Bangladeshi government is developing, with the aid of a U.S. and World Bank development fund, radio-linked hazard-warning arrangements that will make use of the widespread ownership of cheap one-band radios among the general population. This move is part of current Bangladeshi government policy to develop the nonstructural aspects of hazard damage mitigation. The emphasis on early warning communications is designed primarily to save lives. Community disaster relief organizations with volunteers responsible for assisting specific villages will be linked to the International Red Cross, which has donated two radar stations to vulnerable coastal areas. A basic road network to enable evacuation is being improved, and refugee stations are being established in regions most likely to be flooded.

A considerable amount of research is now being undertaken to combine the current enthusiasm over appropriate technology and the application of "indigenous knowledge" or "folk science" with the growing recognition by disaster relief agencies that conventional postdisaster rehabilitation strategies are not entirely satisfactory. In the past, disaster relief agencies such as Oxfam, Red Cross, and Save the Children Fund have tended to concentrate on immediate relief measures (in the form of medical aid for which there are no long-term replacements, and temporary housing often quite unsuitable for local conditions or cultural habits) and have not always considered how this effort could be translated into longer-term disaster mitigation facilities at little extra cost. These agencies lack the resources either to undertake longer-term rehabilitation measures or to conduct after-the-event evaluations, so they are not always aware of the problems being generated. They may not realize where they can improve their service in the most cost-effective manner. The two most criticized aspects of Western relief efforts are the inappropriate use of emergency shelters, and the uncoordinated application of technology, much of it makeshift, unfamiliar, and frankly unsuitable for prolonged use in the rehabilitation phase. Labor-intensive technology requiring skills that can readily be learned in advance by local people is becoming better deployed (see e.g. Kates et al. 1973; also Davis 1975).

Hazard response is not just a societal phenomenon; it is also a political issue. Governments can stand or fall on the way in which they cope with major calamities. Even when warned of a specific danger, they may not act if they feel no political pressure to take action, as amply demonstrated by the tragic events following the earthquake at Portenza near Naples on 23 November 1980. The Italian government suppressed a report prepared by university geologists who warned of the seismic dangers and of the need for improved disaster-relief measures. Despite the passage of appropriate legislation following the devastating Arno floods of February 1966, the necessary measures were not put into effect in this seismically vulnerable area of southern Italy because it did not seem politically opportune for various Rome governments to do so. In the event, it was seven days before adequate relief was provided by the state. Meanwhile, the international voluntary relief agencies had already been at work for at least three days purely on their own initiative. In addition, many trade union groups, encouraged by the opposition Italian Communist party, moved in from the Communist strongholds of the north to assist in the relief work and to embarrass the ruling Christian Democrats. The Mafia also exploited the plight of the many homeless and destitute families—so much so that three thousand troops were taken off relief work to fight the crime and protection rackets that had almost instantly sprung up. The official relief agencies took longer because the UN Office of Disaster Relief was unable to enter the area without official permission from the Italian government; that permission was long delayed.

There is no doubt that many lives would have been saved and much suffering mitigated if the relief had been better organized, and if building regulations designed to reduce the impact of earthquakes in key buildings such as hospitals and schools had been enforced and shown to have been enforced. In the village of Sant Angelo, eighty-seven people were killed when the wing of a new hospital collapsed; yet this building should have been designed to withstand the tremor. By far the most widespread coping strategy among the local population was to bear the loss and to develop communal assistance largely on their own initiative. As one local resident put it, "What do you mean by 'survivors'? We have all been struggling to survive from the day we were born" (quoted in Page 1980).

While the Italian case is by no means typical, it emphasizes the point that coping with natural hazards reflects both the efficiency of government in its broadest sense, and public faith in authority. Efficiency means preparedness and deployment of prompt and sensible relief measures; authority means the power and determination to do a good job. Natural hazards will always be a feature of human existence. But their destructive potential can be lessened if the research results emerging from White's influence in Chicago and Colorado are heeded, and the wider issues raised particularly by the vulnerability school of thought are fully incorporated into the process of "development."

Societal Response to Environmental Risk

The twin themes of hazard and risk couple well into the current fascination with the potentially dangerous aspects of modern technology. Present-day scares about nuclear power and health dangers of certain chemicals—found as additives in food, used in agriculture, and manufactured in potentially hazardous processes—have done much to make safety and reliability in science and technology a matter of serious public concern. Before analyzing the characteristics of environmental risk, it is worth pointing out that societal response to environmental risk differs markedly from its response to natural hazard.

Figure 10.6 divides society, particularly those living in developed nations, into two major groups: those who have faith in the ingenuity of human beings to manipulate nature on the basis of understanding and technological prowess, and those who have faith in the wisdom of human beings to adapt to the constraints imposed by natural processes and to limit their aspirations of superiority over all animate and inanimate life.[4] It goes without saying that these two world views

Environmentalism

← Ecocentrism | Technocentrism →

Deep environmentalists	Self-reliance soft technologists	Accommodaters	Cornucopians

Ecocentrism (Deep environmentalists and Self-reliance soft technologists):

1 Lack of faith in modern large-scale technology and its associated demands on elitist expertise, central state authority, and inherently antidemocratic institutions

2 Implication that materialism for its own sake is wrong, and that economic growth can be geared to providing for the basic needs for those below subsistence levels

Deep environmentalists:

Intrinsic importance of nature for the humanity of man

Ecological (and other natural) laws dictate human morality

Biorights—the right of endangered species or unique landscapes to remain unmolested

Self-reliance soft technologists:

3 Emphasis on smallness of scale and hence community identity in settlement, work, and leisure

4 Integration of concepts of work and leisure through a process of personal and communal improvement

5 Importance of participation in community affairs, and of guarantees of the rights of minority interests. Participation seen both as continuing education and political function

Accommodaters:

1 Belief that economic growth and resource exploitation can continue, assuming

(a) suitable economic adjustments to taxes, fees, etc.;

(b) improvements in the legal rights to a minimum level of environmental quality;

(c) compensation arrangements satisfactory to those who experience adverse environmental and/or social effects

2 Acceptance of new project-appraisal techniques and decision review arrangements to allow for wider discussion or genuine search for consensus among representative groups of interested parties

3 Provision of effective environmental management agencies at national and local levels

Cornucopians:

1 Belief that man can always find a way out of any difficulties, either politically, scientifically, or technologically

2 Acceptance that progrowth goals define the rationality of a project appraisal and of policy formulation

3 Optimism about the ability of man to improve the lot of the world's people

4 Faith that scientific and technological expertise provides the basic foundation for advice on matters pertaining to economic growth, public health, and safety

5 Suspicion of attempts to widen the basis for participation and lengthy discussion in project appraisal and policy review

6 Belief that any impediments can be overcome given a will, ingenuity, and sufficient resources arising out of wealth

Figure 10.6 The pattern of environmentalist ideologies (from O'Riordan, 1981, p. 376)

represent tendencies, not absolutes, and that no individual necessarily confines his/her beliefs to one or other ideology throughout his or her life or at any one time. The point here is that ecocentrists tend to distrust certain aspects of authority and expertise and hence are suspicious of any technology or "unnatural" substance that is foisted on them by those in authority or by those whose expertise is in dispute. Thus, fear of environmental risks may be as much a failure to accept authority, expertise, technology, and democratic means of adjudicating complex technological questions as it is a feature of the risk associated with a particular technology or substance. Societal response to environmental risk is connected to the manner in which different groups consider how society should run its affairs in the future. It follows that the nature of societal response to environmental risk has important implications for those in power and those who seek to govern.

It is perhaps worth noting that those responsible for designing technologies and manufacturing potentially dangerous substances mostly fall into the technocentric camp, as do those responsible for regulating for safety and ensuring that standards and procedures are observed. Those who accept residual technological risk are therefore generally to be found in the economic mainstream of modern economies—financiers, managers, engineers, accountants, scientists, and production workers. Those wary of environmental risks tend to occupy the economic periphery of a modern economy—academics, clerics, artists, the unemployed, housewives, and adolescents—some of whom are liberals, some radicals, and some simply discontented. While this may appear a simplistic analysis, it does have some empirical support (see e.g. Cotgrove 1982; Milbrath 1984).

The Politicization of Environmental Risk

Why is environmental risk such a contentious issue nowadays?[5] One can advance four general observations and three more specific arguments. First, the general observations.

1. Risk and culture intertwine. All societies acknowledge that life is full of danger, anxieties, and uncertainties. Normally these are incorporated into ways of coping, but if an agency or an event is to be blamed, then risk is a convenient device to vent frustrations. Risk therefore has a socially moral element: exposure to man-caused danger is a feature of societal failure.
2. Fact and value interconnect. Because no science is value-free, risk provides a convenient vocabulary for scientific disputes and a context for intensifying scientific conflicts. Personalities,

egos, agency allegiances, and academic posturing all play a part.

3. Institutions for fusing scientific evaluation with political judgment are proving inadequate. Risk therefore provides an avenue for frustration over secrecy and lack of accountability, and democratic reforms raise doubts about the extent to which interests with the body politic *really* can influence affairs.
4. Internal and external modes of risk analysis do not always connect. How people judge risks is not always distinguishable from their political beliefs or their attitudes to authority, expertise, party loyalty, or national chauvinism. Douglas and Wildavsky (1982) argue that neither external risk estimation nor internal risk perception can adequately reveal how people judge risk acceptability.

The more specific comments are as follows:

1. Some environmental risks are becoming unavoidable: even the wealthy cannot buy themselves out of possible danger. This distributional boomerang means that the more politically influential are becoming more anxious. The "nuclear winter"—a widespread cooling of the globe due to sunlight-excluding dust and smoke following even a modest nuclear exchange—would leave no part of the globe unaffected. Global warming due to CO_2 emissions for fossil-fuel burning and forest depletion could well alter climates to such a degree as to imperil food supplies and increase climate-related natural hazards.
2. The rise of counterestablishment science is a feature of a more pluralistic educated community—a kind of neo-Enlightenment. This is the science practiced by those who challenge accepted scientific views and who expose science to public misgiving and political doubt. It provides attractive terrain for exploration by social scientists and politicians, unencumbered by guilt that they do not understand the highly technical concepts and assumptions so fundamental to informed debate.
3. Environmental martyrdom characterizes the innocent who die or who suffer injury from environmental hazards, about which they know nothing when exposed. The residents of Love Canal or the people living around the Three Mile Island plant, the dying widow of an asbestos worker or children playing with bottles once containing cyanide, are all martyrs whose guiltless

agonies have created much media attention, widespread public alarm, and legislative and/or administrative action.

In short, the political climate is changing to reflect greater public anxieties about the future, doubts over the social benefits of new technologies, concern about how far these new technologies can be brought to heel, disquiet about how far regulators are competent and genuinely independent, and frustration over their seeming inability to change the political ethos of "big" institutions (the military, the nuclear industry, the chemical corporations, the development agencies—even the legislatures). These anxieties express themselves in a myriad of ways, one of the most convenient of which is aversion to environmental risk.

All this means that those responsible for determining socially acceptable standards of safety are faced with a number of difficulties. Environmental risk management consists of four phases of activity—risk identification, risk estimation, risk evaluation, and risk control—which merge science with judgment, morality, and the practicalities of control (see figure 10.7). In the past, the two scientifically analytical tools of identification and estimation have resulted in spectacular advances, but these gains have not been matched by improved rules for defining acceptable standards, enforcing suitable codes of practice or determining the correct balance between environmental risk and social benefit.

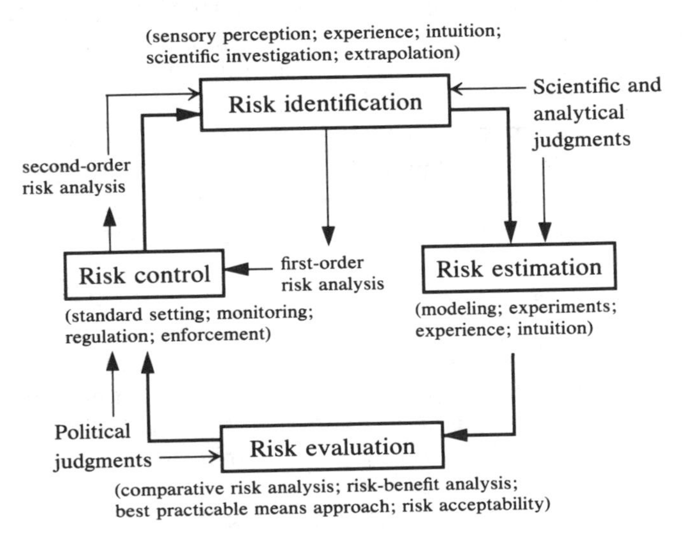

Figure 10.7 Environment risk-management functions (from O'Riordan 1981, p. 392)

In consequence, plant managers tend to rely too much on scientific advice and to accept too uncritically the rationality that underpins that advice. The Kemeny commission (1979), which investigated the causes of the accident to the Three Mile Island nuclear plant in Pennsylvania, in March 1979, commented on the dangers of assuming that because regulations were met, the plant was somehow "safe." It was particularly critical of the complacent notion—which the commission referred to as a "mindset"—that if the equipment worked satisfactorily, human failure could be discounted. Yet human failure appeared to be the main problem: "Our investigation has revealed problems with the 'system' that manufactures, operates and regulates nuclear power plants. There are structural problems in the various organizations, there are deficiencies in various processes, and there is a lack of communication among key individuals and groups" (Kemeny 1979, p. 30).

Managing environmental risk is by no means solely a matter of fail-safe technology. It is crucially a matter of creating public trust in technology, regulation, and the mechanisms by which safety is seen to be assured. In that sense, faith in the authority and efficiency of government and science becomes as much an issue in environmental risk management as in coping with natural hazards.

Yet an adequate theory of environmental risk management still awaits completion. It will not be an easy task, because there are serious academic disputes over the underlying motivations and ideologies, the political context, and the methodologies for tapping risk perceptions. In one of a number of important papers around the general theme of role of social science research in environmental risk management, Otway and Thomas (1982) are anxious not to throw out the conceptually rich baby with the methodologically poor bathwater. They merely want to ensure that researchers are fully cognizant of the political and regulatory contexts of their work.

An interesting feature of current public anxiety about environmental risks (notably toxic chemicals in manufacture, in consumables, and in disposal) is the widely variable "statistical reaction to a life saved." Take the matter of deaths and injury directly attributable to technological risks. From table 10.2 it will be noted that some researchers estimate that as much as 15–25 percent of all deaths can broadly be associated with technology (including automobiles, violence, and cancer that is linked to chemicals and stress). Such risks may "cost" up to $200 billion in the United States alone in lost productivity and medical expenses (Harris, Hohenenser, and Kates 1980). Yet society seems to behave in a fickle way when calculating the value of a life saved. For example, in 1971 the British government decided that drug containers need not be proofed against the inquiring fingers of children, though the cost for each child's life saved thereby would have been only £1000. The following year the same government required very strict building regulations following the partial collapse of a tower block in which no one died but the potential for injury was clearly very great; the cost was about £20 million per life saved. In the nuclear field the industry may invest between £100,000 and £40 million in saving a life, depending on the risk involved (see figure 10.8). We shall see later that this response may not be rational in strictly "economic" terms (rather an artificial distinction) but it reflects public anxiety over "big" and "novel" dangers. Nevertheless, investment in saving a life is distorted wildly in favor of the strange and the terrifying, while small investments in areas of familiar risk may save many more lives—for example, a better ambulance service linked to improved communications between people suffering from heart disease and hospitals could save lives at a cost of around £100 per head.[6]

The problem of opening up scientific and technical analyses to public debate poses a particular dilemma for the authorities in European countries where official secrecy is part of the currency of government. Even in Britain, a nation that has long cherished the advantages of confidentiality, there are signs that senior officials in risk regulatory agencies now recognize the need for fuller and earlier public disclosure.

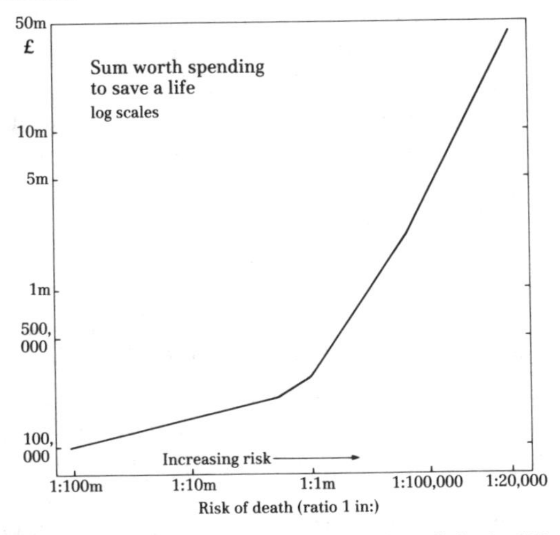

Figure 10.8 An attempt to value life (from National Radiological Protection Board *Cost Benefit Analysis of Nuclear Power,* 1981)

This is a particularly interesting issue for future research. Regulators are under pressure to open up and hence to be more accountable. Politicians cannot bottle up these demands; so a whole arena of risk management is slowly being exposed to informed scrutiny. This may result in a convergence of American and European approaches to risk management. American regulators are attracted to the consultative approaches common in Europe (mediative rather than adversarial mechanisms), whereas European regulatory authorities are becoming interested in more specific standards which, though challengeable, are also implementable. The currently popular European approach to toxic chemical management—to set emissions or exposure to levels "as low as reasonably achievable"—is now being recognized as no longer politically, administratively, or legally feasible.

These then are the major reasons why an adequate theory of environmental risk management is not yet available. Meanwhile, researchers at Clark University are preparing a model of cause, pathway, and consequences that is reminiscent of similar work on natural hazards. Figure 10.9 illustrates the current stage of their thinking. The basic premise is that there are three strategies for reducing the impact of technological hazards: (1) control the event that causes the hazard (by safety measures in process and production); (2) reduce the consequence of the hazard once it has occurred (by safety measures such as seat belts, fireproof materials, detoxicants); and (3) mitigate the actual consequences of the hazard to sufferers (by relief, medical attention, insurance protection). Each of these management functions can be dealt with by attention to the initiating event, the pathway between the event, and the exposed human being and any residual consequences to that exposed person.

It is perhaps worth noting that for many natural hazards, such as earthquakes and cyclones, primary emphasis, notably in developing countries, must be upon the mitigation function. For technological hazards, on the other hand, particularly in developed countries, management strategy is to control the likelihood of the event in the first place. Note that the emphasis of this model generally is on how control interventions develop, where they are applied, and how in many cases they may result in second-generation risks. One of the frustrating aspects of technological risk management is the failure to identify and calculate *all* risks; any attempt to control a risk will almost inevitably set off new risks, probably of a different kind and affecting different groups of people.

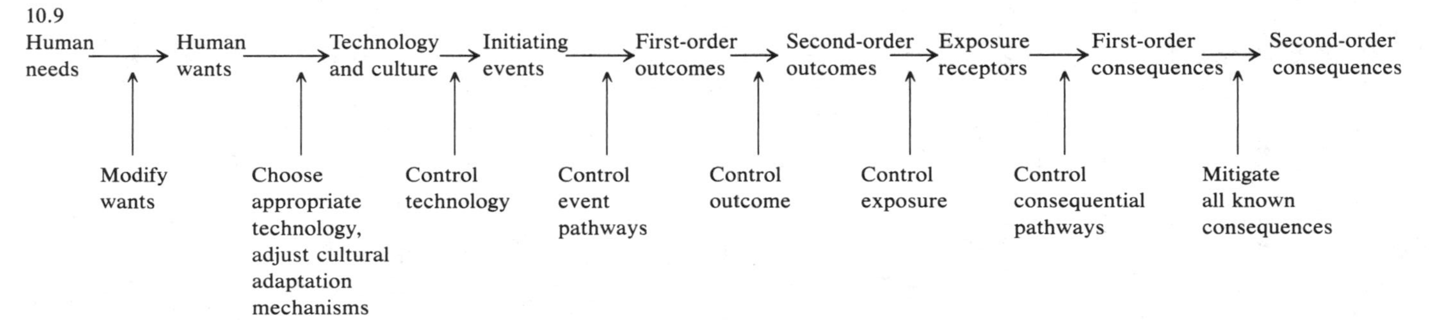

Figure 10.9 Model of hazard evolution with specified control interventions. The model links parts of cause to effects and is based on a series of stepwise control devices, each thought out and applied in the context of previous (incomplete) attempts at control. The process never really stops, for it is one of constant learning and evaluation.

Identifying Acceptable Risk

Because of the political importance attached to new technology, politicians and risk managers want to know how the public perceives and evaluates risk. Surprisingly, there has been relatively little large-scale field research in this crucial area,[7] though a number of interesting laboratory-based studies using selected samples of people and experimental methodologies have been completed.[8] Some of the best work here has been done by the Clark team in cooperation with a group of psychologists (Fischhoff et al. 1979) to discover how ordinary people calculate the likelihood and consequences of quite different kinds of risk; whether these judgments vary for people of different sex, of different educational and economic backgrounds and with different levels of environmental concern; and how these evaluations compare with those of informed experts. This work is an outcome of the research on the perception of natural hazards, though it is rather more sophisticated.

There have been two approaches to the question of determining acceptable risk. The first was based on the concept of *revealed preferences* for risk avoidance, inferring from existing statistical data as to the manner in which people actually expose themselves to risk and seem willing to accept monetary compensation in relation to certain levels of risk. This approach was first developed by Starr (1969), criticized by Green (1980), and subsequently retested by Otway and Cohen (1975) using a slightly different methodology. Starr concluded that people seemed willing to accept a much greater risk if they exposed themselves voluntarily (e.g. driving in cars, smoking, mountain climbing) than if risk was thrust upon them (e.g. living near a nuclear power station or chemical complex, ingesting toxic chemicals in food or from the air). He also discovered that there was a relationship between risk and benefit in the sense that, as benefits were cubed, so risks could be doubled. Finally he assumed that the psychological yardstick for measuring risk was the risk of death from disease that was of the order of one in a million. Risks of less than that magnitude were essentially disregarded.

Slovic and his colleagues were not happy with the revealed performance approach on the grounds that it was not a valid predictor of present preferences and was conservative in that it enshrined current social and institutional arrangements for coping with risk. So they developed the *expressed preferences* technique based on laboratory psychological experiments to test how respondents ranked different kinds of risk and benefit in relation to a base figure and the extent to which they felt the risks could be made greater or lesser (the acceptability criterion).

Their research was designed (1) to discover what people mean when they say that something is or is not "risky" and to determine what factors underline these perceptions; (2) to develop a theory of risk perception that predicts how various groups in the public will respond to new hazards and management strategies (e.g. warning labels, new regulations); and (3) to develop techniques for assessing the complex and subtle opinions that people seem to hold about risk. The approach was based on psychometric studies of judgment and decision making that were developed for utility theory first published by von Neumann and Morgenstern (1947). A major development was the discovery of a set of rules, known as heuristics, which enable people to make sense out of an uncertain world. In the area of risk, it seems that these devices can lead to serious distortions between "reality" as defined by "experts" and the world as recognized by "laypeople." This research has shown that laypeople have difficulty in understanding probabilistic processes; that media coverage biases information about risks in people's minds; that personal experiences and anxieties about the uncertainties of everyday life further distort judgment; that judgments, once formed, tend to be held with unwarranted confidence; and that people tend to perceive too narrow a range of variation between the most and the least risky phenomena. The respondents to the study by Fischoff and his colleagues also appeared to have difficulty in rationally relating risk to benefit (a cherished concept in political decision making), though they could sometimes recognize a link between perceived benefit and acceptable risk.

Additional conclusions from this increasingly varied body of research are:

1. Perceived risk is quantifiable and predictable. However, perceived risk must be related to the socioeconomic characteristics of the population being studied, their occupations, political allegiances, and past experience.
2. Risk means different things to different people, but experts and lay people do not fundamentally differ in their estimates of fatalities for particular categories of risk.
3. Experts and laypeople are in even greater agreement on the characteristics of certain risks, such as controllability, dread, and potential for catastrophe.

4. People tend to overestimate "imageable" risks (such as disease and accidents) and underestimate risks in cases where the cause of fatality is well known (e.g. cigarette smoking, where only lung cancer is commonly perceived as a danger, whereas heart disease poses a much greater threat). They also underestimate the differences in riskiness between the most and the least frequent causes of death.
5. Tolerance of risk is related to judgments about benefits—an issue increasingly found in the nuclear and toxic chemical debates.
6. The "framing" of risk is vital to influencing judgment. A train crash that kills twenty people is more socially acceptable than a nuclear incident that kills no one. Likewise, it is much more common for people to wish to control known dangers than statistical savings of life—even though the outcomes of the two forms of presentation are identical. This means that the pressure will be on to invest more in saving a "statistical" life from being lost than in enabling a "statistical" life to be saved (Rowe 1981).

A splendid summary of all the relevant research leading to these conclusions is available in a recent report by the Royal Society Study Group on Risk Assessment (1983). That report also suggests the following rules for guiding acceptable risk for single individuals:

1. Voluntary risks of less than 1 in 100 should be regarded as unacceptable (the most dangerous activity is professional stunting with a probability of fatality of 3 to 6×10^{-3}).
2. Voluntary and known risks of 1 in 1000 could be acceptable provided they were familiar to the individuals concerned (males between one and twenty years of age face cumulative risks of death of less than 1 in 1000).
3. Fatal industrial accidents of less than 1 in 10,000 are judged to be acceptable.
4. For involuntary death the generally agreed level of acceptability is 1 in 100,000 for "familiar" hazards and 1 in 1,000,000 for "unfamiliar" hazards. However, the latter figure is subject to much criticism as it is not based on empirical evidence, merely on extrapolations from experience and uncritical citation.

This psychometric work has also suggested that "experts" can be prone to many of the same biases as laypeople, notably when forced to rely on their intuitions in the absence of suitable data. Both groups are influenced by what they already know and believe, especially if their existing views are strongly held (a reason why nuclear power decisions will never be universally popular, whichever way they are made). In such cases new evidence appears reliable and informative only if consistent with existing beliefs, while contrary evidence tends to be dismissed as unreliable, erroneous, or unrepresentative. But when people lack strong prior opinions, they become confused and their judgments become fickle. This provides a rich hunting ground for intervenor groups seeking to "capture" neutral opinion over controversial risk issues.

However, one must be cautious about these conclusions. There is an understandable danger of imprinting onto the mind of the subject a rationality that is really that of the researcher. Maybe people just cannot compare risks and benefits on a similar scale. When it comes to such things as toxic chemicals and radioactive substances, people may not be able to conceive of benefits in a formal manner just as they may not readily be able to evaluate the risks. A more existential approach to this kind of investigation might prove more illuminating, that is, a search for meanings and symbols associated with risks and benefits in the context of a person's total experience and expectations. The transactional model of cognition outlined in figure 10.5 might prove of relevance here, because that model is partly based upon a person's sense of political efficacy and judged ability to change one's circumstances or to improve one's well-being. Risk acceptance, therefore, may well be allied to a sense of impotence and resignation over one's ability to alter events: an activated and mobilized citizenry might well be less willing to accept low probability-high consequence risks.

In Western industrial society the concept of risk seems to have taken on meanings which encompass images far wider than safety and technology. It may be necessary to argue that risk assessment and risk acceptability cannot be evaluated in terms of a relationship between probability and consequence. People regard risk as part of a complex of values relating to the very ethos of a society and its institutions.

These conclusions have a number of implications for both research and policy. In research terms, much attention is being given to conflict-resolving mechanisms, such as mediation, value-free exposition, and compensation strategies. These will probably bear some fruit, but not too much should be expected of them. Conflicts over risk issues are becoming rooted in deeper cultural and political soil;

they touch on matters of political representation, accountability, and the legitimacy of government. Future researchers will have to be wary of meddling and manipulation when "real world" examples are used for testing ideas and strategies.

The growing disputes over toxic chemical (including radioactive waste) disposal are in part directed by groups who seek to control "front end" technology by stopping "back end" developments. There may never be social acceptance of certain hazardous processes and waste disposal strategies, and that attempts by governments or the creators of these issues to educate public opinion may well prove counterproductive. Policy options include (1) a recognition that certain risks are going to be very expensive to reduce, highly cost-ineffective in conventional economic terms, but necessary in political terms (e.g. removal of SO_2, NO_2, radioactive safety, carcinogenic chemical limitation, toxic chemical waste disposal); (2) an acceptance therefore that standards of safety may have to be based on what is judged to be politically acceptable rather than scientifically acceptable, with the problem of unnecessary cost burdens on some risk creators; (3) a willingness to establish authoritative, independent scientific review panels to sift the conflicting scientific evidence and to provide an advisory service for the public; (4) a willingness to be completely open about the procedures for determining safety and suitable environmental safeguards so that the public can see how standards are determined; (5) the provision of a pool of "compensation funds" to provide various forms of insurance and compensation for those living adjacent to sites for the disposal of hazardous wastes in cases where such disposal sites must be used in the national interest; and (6) the establishment of appropriate international legal remedies and courts to deal effectively with the control of and remedies for transboundary pollution.

As for the Third World, Western institutions are likely to transfer technological risk into many societies which are ill-equipped to cope with it.[9] Already this is evident in the export of dangerous products such as pesticides, drugs, and milk powder, and it looks as if dangerous manufacturing processes as well as increasingly complex technology will also be transferred, much to the detriment of worker and public safety. At the very least, studies are desperately required to examine the links between Western regulatory institutions, export and trade arrangements, and Third World regulatory activities in relation to importation of products and processes and advice to users and the general public (Bull 1982). All of this relates to the much more fundamental issue of imprinting Western values and dangers upon nations not always receptive to the apparent need for, and unable to handle, such "improvements."

Managing risk in Third World countries might require a totally different approach to the formal regulatory and institutional analysis adopted in the West. As with response to natural hazards and environmental impacts, much needs to be known about the way various societies already perceive and cope with risks and the way they visualize their basic needs before foreign technologies and products are imported and possibly forced upon these populations. Otherwise there is a real danger that traditional coping styles will break down, leaving a gap in the pattern of response that could be subject to all kinds of manipulation and exploitation.

Some Conclusions

Environmental hazard is a circumstance of social existence, the nature and meaning of which depends on experience, stage in economic and social development, and coping strategies. In the field of both natural hazard and man-made risk, despite more organizationally elaborate and economically expensive efforts at predicting, controlling, and mitigating the aftereffects of disasters, the evidence shows that death, injury, and property damage are increasing in the context of natural hazards. At the same time, people's anxieties about man-made dangers are growing, whether their world is actually more perilous or not. Nevertheless, information about the likelihood of danger is now much better developed and coordinated, and international and national relief agencies have proved stalwart. The fact that environmental hazards are now generally regarded as an integral part of the political ecology of civilized evolution has triggered investigations into how and why people from different societies understand and cope with unpredictable calamities. Finally, those in important policy positions are aware of this work, are advising researchers, and are responding to their suggestions.

Nevertheless, both the methodological and institutional problems confronting those genuinely anxious to reduce environmental hazards of all kinds for all peoples are formidable. One remains skeptical that the traditional strategies for economic improvement and compromise-seeking forms of environmental management will result in a net reduction in environmental hazard throughout the globe, though these may succeed in developed countries for a little longer. For Third World countries, more needs to be known about the ways in which different societies get to grips with all kinds of environmental hazard, the quality of their political and social institutions, and the manner in which they evaluate their priorities. Maybe richer does mean safer; and poorer, riskier. Since poverty and hazard cannot be separated, any improvements in coping strategies must involve some transfer of resources from the rich to the poor, and from the powerful to the powerless, combined with opportunities for different groups in various societies freely to develop their own patterns of response. This is wishful thinking indeed. Optimistically, one can say that at least a start has been made. Pessimistically, one fears that there will have to be many more environmental martyrs before the necessary reforms become realizable.

Notes

1. There are numerous studies by political scientists of the way in which disaster control agencies were particularly encouraged to favor engineering options. See especially Maas 1951, Drew 1975; Berkman and Viscusi 1973.

2. The UN Office of Disaster Relief was established in 1972 to coordinate official international effort to alleviate suffering and speed up rehabilitation following major disasters, especially in developing countries. The headquarters in Geneva coordinates activity through a number of field officers. Recently there has been much controversy regarding the management of the office and its effectiveness in coming to the aid of victims in countries where the political and/or military authorities either are not properly in command or do not wish to receive UN aid.

3. The Natural Hazard Information Center is an important source of global hazard information. An equivalent body in London is the International Disaster Institute, 85 Marylebone High Street, London SW6.

4. This is a highly condensed version of what is really a very complex analysis. For the detailed account, see O'Riordan 1981, chap. 1; Sandbach 1980, chap. 1; Cotgrove 1982.

5. This is a well-known phenomenon and is amply covered in the following texts: Lowarance 1976; Council for Science and Society 1977; Kates 1977; Rowe 1977; Kates 1978; Goodman and Rowe 1980; Conrad 1980; Dierkes, Edwards, and Coppock 1980; Burton and Whyte 1980; Schwing and Albers 1981; Royal Society 1981; Society for Risk Analysis 1983; Royal Society 1983.

6. This work on comparative costs of saving life is being conducted by C. Starr, Electric Power Research Institute, Palo Alto, California.

7. The only major study published to date is that of Otway and Thomas (1982), though recently the UK Health and Safety Executive completed a massive survey of public views on risk and safety involving over 2,000 respondents drawn from a carefully chosen sample (Social and Community Planning Research 1982). A Louis Harris poll, sponsored by an insurance firm, found in a sample of 1,500 Americans a preparedness to accept risk even while respondents accepted that knowledge about risks was incomplete.

8. The key groups involved in this work are Slovic, Fischhoff, and Lichtenstein at Decision Research in Eugene, Oregon; Renn at the Nuclear Research Center in Jülich, West Germany; Stallen and Merttens at the Social Psychology Laboratory at the University of Nijmegen, Netherlands; and Vlek at the University of Groningen, Netherlands. See Fischhoff et al. 1982; Vlek 1984.

9. For a revealing account of this problem and the human suffering that it caused, read Medawar 1981.

References

Barrows, H. H. 1923. "Geography as Human Ecology." *Annals of the Association of American Geographers* 13:1–14.

Berkman, R. L., and W. K. Viscusi. 1973. *Damming the West: Ralph Nader's Study Group Report on the Bureau of Reclamation*. New York.

Bowden, M. J., R. W. Kates, P. A. Kay, and W. E. Riebsame. 1979. "The Effect of Climate Fluctuations on Human Populations: Two Hypotheses." Graduate School of Geography, Clark University.

Bull, D. 1982. *A Growing Problem: Pesticides and the Third World*. Oxford.

Burton, I., R. W. Kates, and G. F. White. 1978. *The Environment as Hazard*. New York.

Burton, I., and A. V. Whyte, eds. 1980. *Environmental Risk Assessment*. SCOPE Report no. 15. Chichester, UK.

Cotgrove, S. 1982. *The Environment: Catastrophe or Cornucopia?* Chichester, UK.

Council for Science and Society. 1977. *The Acceptability of Risks*. London.

Davis, I. 1975. *Shelter after Disaster*. New York.

Dierkes, M., S. Edwards, and R. Coppock. 1980. *Technological Risk: Its Perception and Handling in the European Community*. Boston, MA.

Douglas, M., and A. Wildavsky. 1982. *Risk and Culture*. Berkeley, CA.

Drew, E. B. 1975. "Dam Outrage: The Story of the Army Engineers." *Atlantic Monthly,* April, pp. 57–62.

Fischhoff, B., P. Slovic, S. Lichtenstein, S. Read, and B. Coombs. 1979. "How Safe Is Safe Enough? A Psychometric Study of Attitudes towards Technological Risks and Benefits." *Policy Sciences* 9:127–52.

Fischhoff, B., S. Lichtenstein, P. Slovic, S. Derby, and R. Keeney. 1982. *Acceptable Risk*. New York.

Goodman, G. T., and W. D. Rowe, eds. 1980. *Society, Technology, and Risk Assessment*. London.

Green, C. H. 1980. "Revealed Preference Theory: Assumptions and Presumptions." In *Society, Technology, and Risk Assessment,* ed. J. Conrad, pp. 49–56. London.

Harriss, R. C., C. Hohenenser, and R. W. Kates. 1980. "The Burden of Technological Hazards." In *Energy Risk Management,* ed. G. M. Goodman and W. D. Rose, pp. 103–37. London.

Hewitt, K., ed. 1983. *Interpretations of Calamity*. London.

Kates, R. W. 1978. *Risk Assessment of Environmental Hazard*. SCOPE Report no. 8. Chichester, UK.

———. ed. 1977. *Managing Technological Hazard: Research Needs and Opportunities*. Boulder, CO.

Kates, R. W., J. E. Haas, D. J. Amaral, R. A. Olson, R. Ramos, and R. Olson. 1973. "Human Impact of the Managua Earthquake." *Science* 182:981–90.

Kemmeny, J. G., chairman. 1979. *Report of the President's Commission on the Accident at Three Mile Island*. Washington, DC.

Kunreuther, H. 1974. "Economic Analyses of Natural Hazards: An Ordered Choice Approach." In *Natural Hazards: Local, National, and Global,* ed. G. F. White, pp. 206–14. New York.

———. 1978. *Disaster Insurance Protection: Public Policy Lessons*. New York.

Lewis, J. 1977. "Some Aspects of Disaster Research." *Disasters* 1:241–44.

Lowrance, W. W. 1976. *Of Acceptable Risk: Science and the Determination of Safety*. Los Angeles.

Maas, A. 1951. *Muddy Waters: The Army Engineers and the Nation's Rivers*. Cambridge, MA.

Medawar, C. 1981. *Insult or Injury: An Enquiry into the Marketing and Advertising of British Food and Drugs in the Third World*. London.

Milbrath, L. W. 1984. *Environmentalists: Vanguard for a New Society*. Buffalo, NY.

Myers, N., ed. 1985. *The Gaia Atlas of Planetary Management.* London.
O'Riordan, T. 1981. *Environmentalism.* 2d ed. London.
Otway, H. J., and J. J. Cohen. 1975. "Revealed Preferences: Comments on the Starr Benefit-Risk Relationships." Research Memorandum 75–5. International Institute for Applied Systems Analysis, Laxenburg, Austria.
Otway, H., and K. Thomas. 1982. "Reflections on Risk Perception and Policy." *Risk Analysis* 2, no. 2:69–82.
Page, C. 1980. "The Shock That Drew Blood from an Old Wound." *Guardian,* 11 December, p. 17.
Perry, A. H. 1981. *Environmental Hazards in the British Isles.* London.
Prince, S. H. 1968. *Catastrophe and Social Change: Based upon a Sociology Study of the Halifax Disaster.* New York (reprint).
Rowe, W. D. 1977. *An Anatomy of Risk.* New York.
———. 1981. "Methodology and Myth." In *Risk/Benefit Analysis in Water Resources Planning and Management,* ed. Y. Y. Haimes, pp 59–88. New York.
Royal Society. 1981. *The Assessment and Perception of Risk.* London.
———. 1983. *Risk Assessment: A Study Group Report.* London.
Sandbach, F. 1980. *Environment, Ideology, and Politics.* Oxford.
Schnaiberg, A. 1980. *The Environment: From Surplus to Scarcity.* New York.
Schwing, R. C., and W. A. Albers, Jr. 1981. *Societal Risk Assessment: How Safe Is Safe Enough?* New York.
Social and Community Planning Research. 1982. *Public Attitudes to Industrial and Related Risks.* London.
Society for Risk Analysis. 1983. *Real versus Perceived Risk.* New York.
Sorokin, P. A. 1968. *Man and Society in Calamity: The Effects of War, Revolution, Famine, Pestilence upon the Human Mind, Behavior, Social Organization and Cultural Life.* New York.
Starr, C. 1969. "Social Benefit versus Technological Risk." *Science* 165:1232–36.
Vlek, C. A. J. 1984. "Large Scale Risk as a Problem of Technological, Psychological, and Political Judgments." In *Heymans Bulletins Psychologische Instituten R.U. Groningen.* Groningen, Netherlands.
Von Neumann, J., and O. Morgenstern. 1947. *Theory of Games and Economic Behavior.* Princeton, NJ.
Westgate, K. N. 1975. "The Human Response to Disaster." M.Sc. thesis, University of Bradford.
White, G. F. 1973. "Natural Hazards Research." In *Directions in Geography,* ed. R. J. Chorley, pp. 193–216. London.

Whittow, John. 1981. *Disasters: The Anatomy of Environmental Hazards*. London.

Wisner, B., P. O'Keefe, and K. Westlake. 1977. "Global Systems and Local Disasters: The Untapped Power of People's Science." *Disasters* 1:47–58.

Wynne Jones, M. 1976. *Deadline Disaster.* Newton Abbott, UK.

II Energy Policy in Black and White: Belated Reflections on *A Time to Choose*

Kenneth E. Boulding

From 1972 to 1974 the Ford Foundation sponsored an extensive study of energy policy, the major report of which was published in 1974 under the title *A Time to Choose: America's Energy Future*. S. David Freeman was the director of the study, and it had a distinguished advisory board, of which Gilbert White was the chairman. The advisory board consisted of some prominent business executives, especially from the utilities and oil companies, two or three people associated with the environmental movement, and a variety of people from academic life. Gilbert White managed to hold this extremely diverse board together. It produced a brief report on what it had agreed about—which was not very much—and an extremely interesting series of individual comments, all of which are published in *A Time to Choose*.

This was the first in a series of more or less comprehensive reports on the United States or world energy problems. The series now includes the "official" energy plan, The Department of Energy's *National Energy Plan—II* (1979); *Nuclear Power: Issues and Choices*, a subsequent Ford Foundation–supported study prepared through the MITRE Corporation (1977); *Energy, the Next Twenty Years*, a further Ford Foundation — Resources for the Future study (September 1979); individual or group research efforts such as The Harvard Business School's *Energy Future* (July 1979); and perhaps the most extensive and certainly the most expensive of all, *Energy in Transition, 1985–2010*, the report of the National Academy of Sciences Committee on Nuclear and Alternative Energy Systems (CONAES) (December 1979), which struggled with much the same problems for four years.

Kenneth E. Boulding is Distinguished Professor of Economics Emeritus at the University of Colorado in Boulder, where he is also Research Associate and Project Director in the Program of Research on Political and Economic Change at the Institute of Behavioral Science. Among his most recent books are *Human Betterment; A Preface to Grants Economics: The Economy of Love and Fear; Evolutionary Economics;* and *Ecodynamics: A New Theory of Societal Evolution;* and *Stable Peace*.

My own role in the first project was extremely small. I am listed as a consultant, but I contributed very little, if anything. However, I was a member of the Committee on Nuclear and Alternative Energy Systems. Thus, while neither Gilbert White nor I could claim that energy or even energy policy is our major field of competence or interest, we have each gone through a somewhat similar experience of participating in an essentially advisory capacity in an extended study of the problem as a member of a group of very diverse views and interest.

Energy: A Long-term Crisis

There can be little doubt that the Ford Foundation report had an important impact on public thinking and was a major contributor to President Carter's energy policy. It helped to make a policy-making public aware that beyond the Middle East crisis of 1973 there lay a much larger issue. The 1973 crisis was really a minor incident as far as the American economy was concerned; it did little more than ruffle it for a few months. There is, however, a long-term crisis arising out of the fact that our major existing sources of energy—oil and natural gas—will probably not last in their present cheap form for more than fifty years at present rates of extraction. The same is true of nuclear power—still a minor source of energy—in the form of the light water reactor. Even technically, quite apart from its political difficulties, the light water reactor only has a fifty-year horizon because of the scarcity of uranium 235. Hence, either substitutes must be found or energy input into society must be severely curtailed.

It is very hard to sustain political interest, however, in a long-run crisis, especially when it promises to become acute only in the later years of our youngest voters today. The Ford Foundation report, indeed, complains in its very first line that "the energy crisis seems to have vanished as suddenly as it appeared." Five years later it was still difficult to persuade the average American that there was an energy crisis—she turned on the switch and the lights went on, he pulled into the gas station and filled up. Even after OPEC, at least until the present (1985), the retail price of gasoline was relatively little more that it had been fifty years earlier. In dollar terms it was a little more than five times what it had been in 1931, whereas the general price level was at least four or five times higher than the

1931 level. Even in 1985 the Christmas lights were on as if the energy crisis was wholly of the past.

Nevertheless, for those who had eyes to see it, the specter of exhaustion continued to brood over the feast. It reemerged in the consciousness of the American people in the summer of 1979 following the crisis in Iran. Lines formed again at the gas pumps; the price of gasoline went up, edging beyond a dollar per gallon; and for the first time, perhaps, the relative price of gasoline was greater than it had been in 1932. The independent truckers struck over the scarcity and the prices of diesel fuel; fuel was rerouted from agriculture to trucking, portending trouble in agriculture. There were fears of a heating oil shortage in the winter to follow. This crisis, too, died down again.

It certainly looks as if energy crises of the 1973 or the 1979 type will be recurrent and probably of increasing intensity. Even in 1979 the American people on the whole still appeared to believe that the crisis was manufactured by special interests, and they were not willing to adjust to it. At some point, however, reality must break in, and one hopes this point will not be too late.

If anything, the long-run situation in the twenty-first century looks even less promising than it did seven years ago. Little in the way of new technology has emerged. The case that the United States energy policy is in greater disarray today than it was in 1974 is distressingly strong. President Carter's attempt to make energy policy a "moral equivalent of war" certainly did not get very far. President Reagan has not come up with any perceptible policy. One can argue, indeed, that the only positive results of the Ford report have been a slowdown in the breeder reactor program and increasing disarray in the nuclear energy industry—intensified now by the incident at Three Mile Island—and perhaps a little acceleration in the expansion of coal production. We seem to have advanced very little in terms of fusion, solar energy, ocean temperature exchange, wind power, the use of biomass, coal liquification or gasification, oil shale, or any of the solutions that were attracting the attention of concerned people in the early 1970s.

It would be unfair to accuse the Ford report of contributing to the present unsatisfactory situation. The report did a great service in calling attention to the long-run nature of the problem, and it was not intended to provide a particular solution. Nevertheless, there are things it did not say that are important—though the authors are not to be blamed for omitting them—and this paper will deal mainly with these omissions. Gilbert White himself, in his concluding comments on the report (*A Time to Choose,* p. 410) draws attention to two aspects of the problem that are not treated in the report and have not received adequate discussion since. The first is the potential change in human valuations, cultural attitudes, and behavior toward a less energy-using style of life. The second is the report's limited treatment of the global aspects of the energy crisis, a narrowness it shares with most other energy reports.

Change in Values, Culture, and Behavior

The dynamics of value and culture change are mysterious and unpredictable. Such changes may tend to be motivated by symbols and metaphors rather than by strictly rational considerations, but they cannot be neglected in any image of the future. The ultimate efficiency of energy consumption or anything else that human beings consume, produce, exchange, or distribute can only be in terms of the quality of human life. "Quality of life" has become something of a buzzword in recent years. It is extremely hard to identify, measure, or agree about, but it represents an underlying reality of the social system that cannot be set aside. For the individual it is the economist's old friend, cardinal utility—how much "real" satisfaction do we get out of life and out of our activities, environments, productions, and consumptions? In society as a whole we have to coordinate or amalgamate in some way the utilities of various members. That is an even more difficult problem. Nevertheless, we cannot set it aside, for all measures and all considerations that do not include quality of life are imperfect and could easily be misleading.

It is interesting that the Committee on Nuclear and Alternative Energy (CONAES) did sponsor a small study of this problem, titled Energy Choices in a Democratic Society (Synthesis Panel, Consumption, Location, and Occupational Patterns Resource Group, chaired by Laura Nader), a courageous if not very successful attempt to describe the United States in the early twenty-first century as having an energy consumption considerably less even in toto than what we have today, but that might even be a better place to live in. The question of change in lifestyle, therefore, is still very much alive and deserves more extensive study, though it seems like an extreme position and may not have a high probability of being realized in the near future.

Perhaps the closest thing to a large-scale attempt to mold human values and behavior in this direction was Mao's China. While in some respects it had remarkable success, it clearly had such a high social cost and created such serious social and political tensions that China now seems to have turned in quite a different direction, toward getting richer rather than making the most of a decent poverty. In the United States, too, the flavor of the 1970s and 1980s was very different from that of the 1960s—the hippies were much less visible, students went to college because they wanted to get rich, and there seems to have been almost a return to the expansionary philosophy of earlier years, which may even be intensified under President Reagan. This too, however, can change. There is evidence of an underlying movement toward greater simplicity of life, toward making the most of a little rather than striving for much, and there is no doubt that widespread expansion of these values and attitudes in the area of energy would make things much easier. This goes along also with a conclusion stressed in all the studies of energy, in the United States especially: opportunities for energy conservations are very large, even without any fundamental change in known technology or human values and culture. The move toward a "conserving society," as a distinguished Canadian commission has called it (Valskakis et al. 1979), may involve a lucky combination of culture and value change coupled with the capacity to take advantage of certain technological opportunities.

It was no accident that Gilbert White was one of the first to call attention in a public document to this problem. The search for simplicity, difficult and frustrating as it often seems to have been, is an essential part of the Quaker tradition. It is not surprising, therefore, to see White, a Quaker, move in this direction. There is another aspect of White's thought, however, not explicit in his comments on *A Time to Choose* but that undoubtedly underlies his concern for a more holistic, quality-of-life approach to energy problems. As a student of the long-run behavior of rivers, he is strongly aware that river systems occasionally exhibit extreme positions. I learned this from him the first time we became well acquainted, when we were both in 1958 on the Commission of the State of California on the Social and Economic Consequences of the California Water Plan—a commission, incidentally, that the legislature, when it found out what the commission was going to say, abolished before the report could be given.

In connection with this commission, White and I and our spouses drove around California for a week from the Feather River down to Los Angeles, looking at water problems and projects, an experience that changed my whole appreciation of the landscape. Even now, whenever I see a stream or a river, I tend to visualize how it must look in the 300-year flood, however mild and gentle it is today. The social sciences tend to be deficient in what might be called flood theory, that is, the study of extreme positions of a system. The energy field is no exception to this. The social sciences are still obsessed by what is really the bad example of celestial mechanics, which in the present era has no extreme positions, whatever it had in the past. Consequently, we are always ill-prepared for catastrophe, even for the improbably benign, a situation for which there seems no word in the English language. The concern for what may seem at the moment to be improbable solutions to the long-run energy situation is very much in line with the experience of a student of rivers, and it is something that is easily neglected.

Energy and the Really Poor

Gilbert White's second comment on *A Time to Choose* is its neglect, perhaps justified in its rather limited assignment, of the world energy problem, of which the United States is a critical part. First, there is the increasing dependence of the rich countries on imports of oil from the oil-exporting countries, most of which are in the Third World although they are receiving large grants from the rich countries in terms of oil prices far above extraction costs, which make them much richer than their neighbors. Some of them, indeed, with large oil fields and small populations, have per capita incomes above those of the rich countries. As oil gets scarcer, the competition of the rich countries among themselves for oil may result in increasing international tensions. Second, there is the problem of the very poor countries, the Fourth World, poor in natural resources as well as in human and physical capital. In these countries there is little doubt that OPEC, and rising energy prices generally, diminished their chances of development. It is true, as Japan has shown so spectacularly, that know-how over a very large range can be a substitute for natural resources, and that a country with know-how can import both energy and materials, transform them into commodities, pay for them by its exports, and enjoy a very substantial surplus. As the relative price of both energy and materials rises—as it seems almost certain to do over the next decades—this trick becomes harder, and the example of Japan may become more and more difficult to follow. It may be, therefore, that the most adverse effects of the energy crisis will be not on the rich countries, which have great powers of adaptation, but on the very poor countries, the development of which is thereby made much more difficult. A future in which the rich countries of the temperate zone stay rich, though they may not get much richer, adapting by conservation and improved technology to the increasing relative price of energy and materials, while the really poor countries slip further and further down into poverty and destitution, has a probability that is uncomfortably high. This probability is raised even higher by the present "wood crisis" in the poor tropical countries, where the sheer search for firewood, under pressure of increasing population, is denuding the forests and often leading to catastrophic and irreversible erosion.

It is hard to see any solution to the problem of increasing poverty short of a change in values and culture, not wholly unrelated to that which might lead to conserving lifestyles discussed earlier. Rarely in human history have the poor been able to threaten the rich successfully, particularly the rich who live at a distance. While there is some legitimate anxiety about the spread of nuclear weapons and the possibility of terrorist control of them that might lead to some kind of blackmail, I would be surprised if terrorism of this kind turned out to be more than episodic, and to abolish it would make little contribution to the overall solution of the problem of poverty. It is only the development of the moral resource, the slow rise in the area and circle of personal concern, from the family to the tribe to the nation, and finally to the human race as a whole, that seems to offer much hope. Unfortunately, we know very little about the dynamics of this process. But it is certainly implied in Gilbert White's earlier comments and in his own lifestyle. It could well be that looking back on *A Time to Choose* a century from now, White's brief comments will be seen to have had a very important seed of the future in them.

The Economy of Energy

I find it hard to resist the temptation to throw in some of my own conclusions, particularly as an economist. It seems almost impossible for economists to persuade the general public, or even the political and policymaking public, that the structure of relative prices really matters. This difficulty is particularly striking with respect to energy. Attempts to change behavior in desirable directions tend to rely on preaching, prohibition, and subsidy. The structure of relative prices determines the patterns of terms of trade for all individuals and groups. If the price of what I sell goes up relative to the prices of the things I buy, my terms of trade will improve and I will almost certainly perceive myself to be better off. The structure of relative prices, therefore, has a profound impact on the distribution of welfare, particularly among occupational groups. This in turn has a profound effect on behavior in regard to what people buy, consume, produce, or sell. If the price of any item is perceived as low relative to other, similar items, people usually tend to substitute the low-priced item for possible alternatives and to expand their purchases and consumption of it. Those who are producing it will be dissatisfied and will tend to stop producing it and produce something else. This is what supply and demand are all about, and it is a very important aspect of human behavior.

It is true, of course, that the terms of trade are not the whole story. What really determines and changes behavior is our perception of the terms of reciprocity. We give out and take in not only commodities but all sorts of noneconomic inputs and outputs—respect, love, prestige, identity, and so on. It is our perception of the terms of reciprocity, that is, of the total value of what we take in per unit of the total value we give out, that is significant for changes in behavior, but the terms of trade are an important part of this and must not be neglected. It is the difference between trade and reciprocity that opens up the possibility for valuation or cultural change, as we have seen earlier, which will then change people's reactions to terms of trade in the relative price structure. Thus, economics and noneconomics are both part of the total picture, and neither can be understood in the absence of the other.

Perhaps the reason those outside the little band of economists are so unwilling to think about relative price structure is that it has a double function and there may be a conflict between those functions. It plays a central role in the allocation of resources to different industries and occupations. It also plays an important role, however, in the distribution of income and economic welfare, and a change in the relative price structure that moves toward the solution of an allocation problem may move the distribution of income in directions regarded as undesirable. The only answer, I have argued, is a well-constructed "grants economy" of one-way transfers (Boulding 1973). This is an aspect of the total system that even economists have been unwilling to recognize.

The contribution of economics to the energy problem consists of a set of propositions which are almost truisms but which nobody seems to believe. If something gets scarcer, its relative price is almost certain to rise if it is not prevented from doing so by government. This rise in price will have two effects. On the demand side, the scarce commodity will be economized and conserved more, and substitutes will be found for it in consumption. On the supply side, greater effort and more resources will be put into producing it, and substitutes will be found for it in production.

There is some legitimate worry about the extent of these exchanges. The concept of elasticity is relevant here. If demand is inelastic, as it certainly is for many forms of energy (for instance, gasoline), an increase in price up to an amount perhaps two or three times the present price will have little effect on people's behavior, but it will have a very substantial effect on the distribution of economic welfare, redirecting this away from the purchasers and consumers and toward the sellers and producers. Such a redistribution may be unacceptable and steps may be taken to offset it. The market works best when supplies and demands are fairly elastic, that is, when both production and consumption are responsive to changes in the relative price structure. If they are unresponsive, the case for doing something which makes people more responsive becomes much stronger. However, economists are quick to point out that elasticity or responsiveness increases with time. People do not respond quickly in the short run, but they may respond strongly in the long run, simply because it takes time to respond. This time factor is often neglected in political and policymaking behavior.

A principle emerging from these considerations is that if we have something that is plentiful and cheap now but is likely to be scarcer and more expensive later—as has been and probably will be the case with energy—there is everything to be said from the allocational point of view for making it artificially expensive now. This will push people into conserving behavior, for nobody ever conserves anything that is perceived as cheap. It will push them into a search for more efficient forms of energy consumption and into finding substitutes for existing energy resources. And it seems highly probable that energy prices at, say, three or four times the present level would produce enormous changes in the allocation of resources both to production and to consumption, particularly over a period of twenty-five years. However, if energy is allowed to remain cheap, it will not be conserved, alternative resources will not be pursued, technology will not be directed toward desirable ends, and we may find ourselves in fifty years with a very sharp rise in the relative price of energy for which we are totally unprepared and which will be a major catastrophe.

Neither the Ford Foundation report itself nor any of the individual comments included in *A Time to Choose* adequately stresses this vital principle. It may indeed be that our level of political understanding precludes rational policies. Rationality would certainly suggest a heavy tariff on imported oil, the removal of price controls, heavy taxation of oil and gas, coupled perhaps with a system of grants or tax relief to offset distributional impacts on the poor. I have argued, indeed, for shifting the property tax largely onto gasoline and perhaps using politically the moderately legitimate protest against property tax to move the tax structure to a more legitimate base. It seems to me a political principle that we have to find the right excuses to do the things that are reasonable to do. We only do things for excuses, never for reasons, but reason can be used to find the right excuses!

Appropriate distortions of the relative price structure through taxation (and, one should add, through subsidies, which are negative taxes) are not, of course, the only instruments of public policy, though they are perhaps the most neglected and the worst used. We should not exclude rationing in extreme cases, where allowing the relative price system to ration would produce excessive inequities. Rationing itself, however, produces enormous personal inequities and should be regarded only as a last resort. We should not wholly exclude prohibitions for extreme cases, where these can be identified, though the uneasy feeling that regulation is always badly done and that there is usually too much of it is not to be dismissed. But we cannot deny the necessity for some regulation. How to do it well remains one of the more difficult problems of political and social life.

The Ecology of Energy

Another aspect of energy frequently overlooked is its heterogeneity in terms of human uses and human valuations. Energy is not a homogeneous input of ergs or BTUs or quads (the quadrillion BTUs beloved of energy writers). Energy becomes significant only in some particular form, place, time, and application. Electricity, for instance, poses a very different problem from that of fuel. Solid fuel is very different from liquid fuel or gaseous fuel. Even in processes of production, energy has three different functions. First, it is needed to do work, that is, to transport and transform selected materials, guided by a structure of know-how, into improbable structures which constitute products, whether these are horses or human beings, houses, hors d'oeuvres, hats, or history books. Second, it is needed to sustain the appropriate temperatures at which productive transformations can take place, whether the heat of the pottery kiln, the blast furnace, the fusion reactor, the gentle warmth of the human body, or the chill of the freezer. Energy's third function is to transmit information in the form of nerve impulses, sound waves, telephone messages, radio waves, letters carried by pony express, or hormones carried by the bloodstream. Information can be coded both in energy and in materials, but even when coded in materials it takes energy to transmit it.

Different kinds of energy are required for each use and for innumerable sub-uses within them. To think of energy as if it consisted solely of ergs or BTUs is preposterous and misses the whole point of the problem. The different uses of energy, furthermore, are imperfectly substitutable. Electricity is no substitute for gasoline when it comes to driving automobiles. Some forms of energy are much more storable than others. It could be argued, indeed, that the storage of energy is the principal problem we face. In terms of energy flux, the solar energy hitting the earth is ample for all our needs. To translate it into energy, however, when, where, and in the form we want it is a complex problem, which may not even have a solution. The mere presence or quantity of energy is not the question. Difficulties arise because of shifts in the different forms that energy may take. Most energy reports have tended to overlook these variations, concentrating on energy as "a problem." Yet the very heterogeneity of energy demands multiple solutions.

The Moral Economy of Energy

Another conclusion I have drawn from this reexamination is that the area of energy is a particularly good example of the need for a reevaluation and critique of what might be called the "moral economy." Criticism of the values and actions of others, and even of one's own, is widespread in the human race, though it probably has a skewed distribution. A few people are intensely concerned with such criticism, the majority of people only mildly so, and some not at all. The most concerned people tend to be active politically, educationally, and socially and may have a disproportionately large impact on the overall movement of society. The total quantity of moral concern is by no means a constant and varies even within the same society from time to time, as periods of intense moral activity are often succeeded by periods of relative exhaustion and apathy. Moral concern, however, is always scarce. It has an "economy" in the sense that, if directed to one objective, it is likely to be withdrawn in some degree from another. If moral effort is wasted on unworthy or unwise objects, it may not only do more harm than good but may also be diverted from objects where it could be productive.

Energy is an area into which increasing amounts of moral concern have been flowing. The distribution and productivity of this moral effort is therefore of great importance. The movement against nuclear power, particularly against the breeder reactor, which is reflected to some extent in the Ford study, is an important example of the moral economy at work. This movement arouses high moral fervor in its participants, to the point that some of them are prepared to make strong personal sacrifices in terms of going to jail and suffering severe stigma, as in the case of the "clamshell alliance" in New England, which has been attempting to stop construction of the nuclear power plant at Seabrook, New Hampshire, and similar movements elsewhere. When one looks carefully at the alternatives, however, it seems highly probable that the use of coal as a fuel for the generation of electricity is more dangerous to human life and health than is the light water reactor (National Research Council 1979). Even the Three Mile Island accident did not kill anybody directly, though it may eventually shorten and impair some lives. Furthermore, if coal turns out to be the only practical alternative to nuclear electricity for the next hundred or two hundred years, there is an alarmingly high probability—though by no means a certainty—that the increase of CO_2 in the atmosphere will produce large and irreversible changes in the whole ecosystem of the earth, perhaps through a "greenhouse effect," warming the earth up, melting the ice caps, raising the level of the oceans, altering patterns of rainfall, and creating deserts where now there is agriculture—and perhaps agriculture where there is now desert. This result could well diminish, and would almost certainly redistribute, the niche of the human race, even though it might not happen for two or three centuries, though the adverse effects could in part be offset by more rapid plant growth, which in turn might be offset by pest and weed growth. We can say with some reason that the continued use of coal as a fuel is likely to have serious long-term effects.

A perplexing issue here is the noncomparability of the risks of nuclear electricity and of coal-produced electricity. In the case of nuclear power plants there is a very low risk of quite large catastrophes. In the case of coal plants there is certainty of small ones which, in cumulation, assume the seriousness of a large one. The fear of the dramatic somehow becomes greater in people's minds than that of the commonplace. A nuclear meltdown which at some time in the future is virtually certain to happen and which could well kill or injure hundreds or even thousands of people looms large in the imagination over against the mine accidents, black lung, air pollution, and the CO_2 effects that might affect tens or hundreds of thousands or even billions of people, but over long periods and undramatically.

Perhaps the most important impact of the Ford report on actual policy may have been the virtually indefinite postponement of the development of the breeder reactor and of the reprocessing of nuclear waste in the United States, largely for fear of nuclear proliferation and the possibility of sabotage and theft of fissionable materials. Certainly if the breeder reactors are going to be spread around the world, the technical potentiality of nuclear proliferation increases. This is a prospect that no one can contemplate without horror. However, the very fear about the breeder has diverted attention from the danger of large-scale nuclear war that exists right now, quite apart from any proliferation. It seems inconsistent to exhibit extreme moral agitation about a nuclear power plant that is designed not to go off (though there is always a small positive probability that it will release radioactivity) and to shut one's eyes to the overkill in nuclear weapons that are designed to go off. It is hard to justify such a diversion of moral energy. To advise a group of children playing with cigarette lighters in a room already full of dynamite that it would be better if their lighters were replaced by matches strikes one as very poor advice from the moral order.

It is still not wholly clear that the breeder as we know it is economically competitive, at least with existing cheap energy, or even that there are not technical problems involved in fuel reprocessing that will prove intractable. But of all the alternatives to oil and natural gas for electricity production that are now known, the breeder reactor looks like the one with the longest horizon. It could give the whole human race electricity for thousands of years, long after even coal was exhausted, and it would release fossil fuels for uses such as transportation, for which electricity does not seem to be well suited. Furthermore, from a distributional point of view, electricity is a considerable equalizer if it is reasonably cheap, plentiful, and widely available. Fuel, if it becomes scarce and expensive, contributes to widening the gap between the rich who use it and the poor who cannot afford it. Equality is much more a function of the nature of productivity and technology than most people recognize; yet it is seldom considered in this light in the moral economy.

Opponents of the breeder argue that in a generation we will achieve adequate energy with fusion, solar energy, or something unexpected. We will be lucky indeed if we do. A breakthrough in, for instance, the use of direct solar energy to split water into hydrogen and oxygen would give us large, cheap supplies of fuel, which could easily be burned to make electricity, so that we would not have to use nuclear with its risks, its serious waste disposal problems, and its potential use in weaponry. Still, it would seem wise to have some eggs at least in all the baskets available, especially those that look capacious, like the breeder.

Perhaps one of the greatest sources of failure in the moral economy is the illusion, all too prevalent among the morally activated, that to prove something is bad makes a sufficient case against it. To prove something bad achieves nothing unless we show that something else is better and is a feasible alternative. History is full of the wastage of the moral economy in the condemnation of what is perceived as bad without regard to what is better. One thinks of the immense waste of moral effort in promoting both orthodoxy and heresy in religion and politics, of the waste of moral effort that has gone into national defense, into revolution, into alternative economic systems, into prohibition, and into sumptuary laws, and one worries about the cost of similar misdirections of moral efforts.

The impact of *A Time to Choose* on the moral economy is something I would not venture to evaluate. It certainly had large positive elements. It brought to public attention both the importance and the long-run nature of the problem in a way that had not been done before. It emphasized the importance of conservation and of the development of nonexhaustible sources of energy. It created a vivid image of the dangers of nuclear proliferation. On the negative side, it gave inadequate attention to the economics of the problem; and as Gilbert White rightly observed, it neglected culture change and certain global aspects of the energy problem.

Its contribution to the controversy about the breeder reactor seems to be the most debatable aspect both of its content and of its impact. In its defense, one could argue that with an uncertain and perhaps dubious technology there is much to be said for sitting back and letting somebody else be the pioneer—in this case primarily the French and the Russians. From a selfish, nationalist point of view, the United States with its enormous deposits of coal can certainly afford to postpone the breeder for a generation, though the effects of using coal again are, as we have seen, at best uncertain. If, of course, it turns out that by putting resources that might have gone into the breeder into solar energy we could come out with cheap solar electricity by 2020, then indeed the report will have been abundantly vindicated. Even if both solar electricity and fusion fail and the world has to fall back on the breeder, the United States may suffer little for having postponed it and may be able to take advantage of all that the pioneers have learned. Present policy could then turn out to have been a very refined and successful form of national selfishness.

A Time to Disagree

The sad truth seems to be that the uncertainties of the future are so great, especially in technology, politics, and life-style, that we will not be able to evaluate *A Time to Choose* and its impact for at least a century. Even then, what is remembered may be not the content of the report but the method, especially that of the advisory board. The comments of its members might never have seen the light of day but for the efforts of Gilbert White in holding the board together, resisting attempts to reach unanimity, and encouraging the expression of diverse views. I found the comments of the business leaders particularly interesting; they reveal something rarely expressed in the denatured product that usually comes out of committees. The feeling of being unappreciated, misunderstood, and unloved is curiously at variance with the stereotype of the business leader as tough and impervious. This is the angry cry of the "doers," the people who get out the electricity to the homes, get out the gasoline into cars, and give people what they want, who are desperately concerned that there should not be any shortages, and then are confronted with the virtuous and concerned liberal "stoppers," who see the doers' activity as constituting only private interest without adequate concern for the public good.

I confess that the remarks of the environmentalists and even the academics sounded a little thin, even anti-human, as if the human race were an unfortunate nuisance vis-à-vis the divine wisdom of the natural biological order that has existed for so long in its absence. This is not to deny, of course, a great deal of validity in the environmentalists' points of view. The net impact of the report is to present all these different perspectives in creative juxtaposition in conflict rather than as a gray and meaningless consensus.

For the refusal to be content with a bland consensus, and for the clear statement of differences that are both strangely reasonable and yet irreconcilable, we undoubtedly have to thank Gilbert White. The title of this essay is an outrageous pun. It does, however, have important overtones. Under Gilbert White's leadership, black and white did not merge into gray. They became a pattern, perhaps a checkerboard on which the game of the future may be played.

References

Boulding, Kenneth E. 1973. *The Economy of Love and Fear: A Preface to Grants Economics*. Belmont, CA.

National Research Council. 1979. *Energy in Transition; 1985–2010: Final Report of the Committee on Nuclear and Alternative Energy Systems*. National Academy of Sciences, Washington, D.C.

Valskakis, Kimon, and others. 1979 *The Conserver Society: A Workable Alternative*. New York.

12 Global Environmental Prospects

R. E. Munn

Environmental scientists grumble that governments plan seriously only for the immediate future, even though time scales for environmental change are typically measured in decades or longer. Yet the long-term predictions of scientists are not very accurate. How then should society prepare for the future?

There are two states of mind about environmental prospects for the next century. On the one hand, a chorus of pessimists is heard to proclaim that the expectations of society can never be fulfilled, particularly in the Third World. One reason given is that the earth's ecological outer limits are being approached and that the turnaround times required to avoid exponential irreversibility are too long (see e.g. Erhlich and Erhlich 1981).

On the other hand, Gilbert White and like-minded individuals are optimistic. Without discounting the probability that wars, earthquakes, floods, and droughts will cause terrible disasters in many parts of the world, these people remind us that the vigorous application of existing and new knowledge and technology will bring great benefits to society. As a prime example, it may some day be possible to meet the basic needs of the countless millions of the world's poor. White views the future as both a challenge and an opportunity.

This brief essay describes some current thinking on global environmental prospects. It draws heavily upon work undertaken by a small group of people, including White, during 1981–82 in the preparation of a United Nations report; *The Environment in 1982: Retrospect and Prospect* (UNEP 1982). The document was written for the Session of a Special Character of the Governing Council of the United Nations Environment Programme (UNEP), held to mark the tenth anniversary of the UN Stockholm Conference on the Human Environment. The ideas to be elaborated below either are derived directly from the UNEP document or evolved while the document was being written. Although contributions of many persons, includ-

R. E. Munn is Head of the Environment Program, International Institute for Applied Systems Analysis, Lazenburg, Austria. A Specialist in air pollution, climate impact assessment, and optimal design of environmental monitoring systems, he is a frequent consultant to WHO, WMO, and UNEP. Dr. Munn is editor of the John Wiley SCOPE series and author or editor of ten monographs.

ing senior UNEP officers, must be acknowledged, the wise comments of Gilbert White are especially to be remembered.[1]

Methods of Forecasting the Future

If reliable long-range predictions could be made, the problem of environmental management would be simplified. Before considering the general question of environmental prospects, therefore, it will be useful to discuss briefly the methods available for predicting the future. These methods—statistical; modeling; Delphi and other "wise men" techniques; and metaphysical—will be discussed in turn.

Statistical Methods of Forecasting

Two general approaches are available, time-series analysis and extreme-value theory.

Time-series analysis. Time series can be extrapolated forward in time, for example by extending linear or exponential trends, or by harmonic or spectral analysis (Box and Jenkins 1976). In any event, the ability to predict diminishes with increasing intervals of time into the future. Time-series methods of environmental forecasting are often quite useful, provided that externalities do not change. Unfortunately, the dream of steady-state conditions is seldom realized. Continuation of an existing trend is just as unlikely, however, and exponential growth cannot continue indefinitely.

Extreme-value theory. Using historical time series, the probabilities of recurrence of rare events of given magnitudes can be estimated. The results are often expressed in terms of *return periods*. Extreme-value theory developed by Gumbel (1958) and others has been used successfully by engineers in the design of flood control systems, for example. But the method does not include prediction of the actual dates when rare events will occur, and it assumes steady-state conditions in the upper tails of the frequency distributions.

Modeling

A model is a representation, usually simplified, of a system or process. It may be a physical model (e.g. a miniaturized ship to be tested in a tank of water) or a mathematical conceptualization. In the latter case, the various components of the system are represented by equations.

Once the performance of a mathematical model has been tested with an independent set of data, it may be used for making forecasts, including "what if?" predictions for hypothesized sets of conditions that have never been observed. These are called *simulations*.

Mathematical models used for environmental forecasting are of two main types: those greatly dependent on initial conditions and those independent of initial conditions. An example of the former would be the model of weather changes over the next thirty-six hours. An example of the latter would be a model of world climate, which could simulate the climatic effects of a doubled atmospheric carbon dioxide concentration or of deforestation in the tropics.

Beginning with the Club of Rome limits-to-growth projections of the early 1970s (Meadows et al. 1972), global long-term models have been developed in many research institutes. For a review of simulations with environmental components, see Biswas (1979). A special problem with these models is that their predictions must be taken on faith. "What if" forecasts represent combinations of conditions for which there are no historical analogues. However, the projections are useful in alerting scientists to possible long-term consequences of various human interventions. The projections also influence ways in which society thinks about its universe, and may possibly play a role in the future course of history. For a nonspecialist review of environmental simulation modeling, see SCOPE 9 (1978).

Delphi and Other "Wise Men" Techniques

Particularly in cases where environmental issues are strongly dependent on socioeconomic factors, a few wise men may have valuable insights. Several ways of utilizing these human resources are possible, the most widely employed being the Delphi technique. In this approach, specialists are asked individually for their predictions. The results are pooled, and each participant is given the opportunity to revise his or her first estimate. By successive steps, a consensus may be achieved. (For a review, see, for example, Pill 1971.)

Metaphysical Methods of Forecasting

Particularly during periods of socioeconomic stress, predictions made by false prophets gain considerable notoriety. However, those who have forecast that the world would end on a particular day have invariably been wrong (so far at least!). The empirical evidence therefore suggests that such predictions are not to be taken seriously. This, of course, is not to discount the informed hunches of wise men such as Lord Ashby, whose intuition is founded on a sense of history and of current trends not yet perceived by society (Ashby 1978).

Looking Back to the Early 1960s

At this juncture, it may be sobering to review environmental issues of the early 1960s. That was only twenty years ago, but in another twenty years, the next century will have begun, so that the time frame is appropriate for comparison.

A good starting point is Gilbert White's *Social and Economic Aspects of Natural Resources* (1962), which examined U.S. policy on management of natural resources in the global context. White's report predicted that requirements for food, water, energy, and minerals would increase greatly "in response to rising per capita demands of a world population two or three times greater than the current one." The report was optimistic, nonetheless, suggesting that technological development would keep pace, though recognizing that social adjustments may lag.

A similar prospective appeared at about the same time; *Natural Resources for U.S. Growth: A Look Ahead to the Year 2000* (Landsberg 1964). In this study the environment was treated superficially, though continued population growth and resource use (including nuclear energy) were predicted. Thy only specific references to the environment were several mentions of water pollution and two paragraphs devoted to quality of the environment, in which it was predicted that "social scientists are likely to come increasingly to grips with problems of resource quality, as they seek to provide sounder bases for decisions."

These and other publications of the early 1960s reflect a belief that environmental problems would some day be solved by technology. Society would ultimately be able to control pollution and to insulate itself from stresses imposed by the natural environment. For example, crop yields were increasing steadily through the use of irrigation, flood control, fertilizers, and pesticides. In the developed countries, the pollution problems of urban and industrial areas were of course recognized. The London smog of November 1952 had claimed three thousand lives, and in Los Angeles photochemical smog had become an almost daily occurrence. A major issue in the Great Lakes region was lake levels, while flood control and irrigation works were being constructed in many countries.

In developing countries, environmental concerns were the classical ones: droughts, floods, and pests. At that time and even into the early 1970s, representatives from the Third World were suspicious of "environmentalists," fearing that pollution control would be a brake on industrialized development in their countries. There was also a worry that industrialized nations would ban the import of fruit and vegetables that contained trace amounts of pesticides and herbicides.

A major issue beginning in the 1950s and continuing to this day was the threat of nuclear war. Even if such a catastrophe did not materialize, there was still the worry of global fallout from bomb testing. Civil defence organizations were therefore well supported in many developed countries. Nevertheless, there was considerable optimism about peaceful uses of atomic energy. Reporting on the Atoms for Peace Awards in 1960, *Science* stated: "The trustees believe the development of the nuclear reactor is one of the great advances in man's capability for using atomic energy for peaceful purposes. It gives the world a new source of energy with which to meet the growing requirements of modern society for power to run its machines."

To summarize, a major societal goal in the early 1960s was to harness technology in the service of mankind, not only in order to increase the production of food, wood products, and other resources, but also to ameliorate the pollution problems that technology had created. It was believed that with sufficient infrastructure and capital, developing countries could follow the examples set by Europe and North America, for example through the construction of large irrigation schemes.

The Intervening Years 1964–84

The most significant change in people's attitude to the environment is that confidence in technology has faltered. Various environmental accidents (the Torey Canyon oil spill, the 1965 New York power blackout, Minamata mercury poisoning in Japan, the Three Mile Island reactor accident, the Love Canal, etc.) have demonstrated the existence of unexpected side effects of technology.

Some of the particular environmental topics that have gained prominence in the last twenty years include *unexpected effects of toxic substances*—thalidomide, PCBs, DDT, and solid and liquid waste disposal; *resource-related issues*—the Club of Rome predictions, failure of the Peruvian anchovy fisheries, the Sahelian drought, and rising oil prices and the resulting attention given to renewable energy sources; *health-related issues*—eradication of smallpox, baby formula sales in developing countries, and leveling off of populations in some developed countries; *global concerns*—CO_2-induced climate warming, stratospheric ozone depletion, acid rain, mercury in swordfish, and heavy metals in Greenland and Antarctica; *technology development*—satellites and remote sensing imagery, recycling technology, computer development, genetic engineering, and risk assessment; and *general issues*—public participation, the establishment of legislation to require resolution of environmental conflicts (see vol. I, selection 25 for a contribution by Gilbert White to this last topic.)

There is no doubt that in the last twenty years science and engineering have substantially increased our understanding of the environment. But individual branches of science are becoming so technical that scientists sometimes cannot communicate with one another except at the nonspecialist level. The contributions of a few synthesizers such as Gilbert White, Martin Holdgate, and Mohammad Kassas are therefore extremely valuable. The book *Environmental Issues* (SCOPE 10, 1977) is an example of a global synthesis, as is a press release on life-support systems issued jointly by Mostafa Tolba (as executive director of UNEP) and Gilbert White (as president of SCOPE) in June 1979. In that two-page document, biogeochemical cycling was proposed as an integrating framework for a number of seemingly quite different environmental issues: acid rain, CO_2 climate warming, nitrogen fertilizers, and stratospheric ozone depletion (see vol. I, selection 28). The press release is a remarkable quantum jump forward in the setting of environmental priorities.

Planning for an Uncertain Future

That the future is uncertain should come as no surprise, but what is not generally recognized is that there will also be shifts in the goals of society. Changes in values and expectations are even more difficult to forecast than changes that will take place in the environment. Thus, even if we could predict the future with confidence, we could not plan for it in a normative way. For example, a suburban home (away from polluted city centers) was a dream of many North Americans in the 1960s; transportation corridors were planned with this in mind. Today, many people would prefer a townhouse in an older residential neighborhood.

How then can society prepare for the next century? Anyone attempting to answer that question is like the writer of a popular book on how to become a millionaire. Ten chances to one, the author is depending on the royalties from the book to make his first million.

With this in mind, a modest person like Gilbert White would draw upon history. What strategies have been used for centuries to cope with uncertainty? Which have been successful and in what contexts? Many techniques for minimizing risks have been practiced at the family and community levels since prehistoric times and, in fact, are set forth in nursery rhymes, fables, and proverbs (''Waste not, want not''; ''People who live in glass houses don't throw stones''; and so forth).

This historical approach to the management of uncertainty leads to the following check list, with environmental examples in parentheses. Here it should be remarked that Gilbert White's vista was not merely that of cutting our losses but also of improving the qualities of life in an uncertain world.

Group 1. Strategies for eliminating known risks

- 1.1 Removal of risks (banning of hazardous substances)
- 1.2 Risk avoidance (siting towns on high ground to avoid tidal waves)
- 1.3 Risk protection (strengthening the foundations of buildings in earthquake zones)

Group 2. Strategies for coping with known risks

- 2.1 Learning by experience (information retrieval systems)
- 2.2 Diversification (multiple crops, staggered planting)
- 2.3 Small is beautiful (construction of several small dams rather than one large one)
- 2.4 Saving for a rainy day (world food banks; conservation)
- 2.5 Self-sufficiency (recycling; ecodevelopment)
- 2.6 Development of an ability to react rapidly to surprise (adaptive impact assessment; resilience; keeping options open)
- 2.7 Development of early warning indicators (World Weather Watch; tidal wave forecasting systems)
- 2.8 Impact mitigation (the International Red Cross)

Group 3. Strategies for Coping with the Unknown

Numbers 2.2 to 2.8 above, particularly 2.6 and 2.7.

These approaches are all useful in copying with rare events such as earthquakes and droughts. Sometimes, however, the nature of the surprise cannot be foreseen, in which case strategies 2.6 and 2.7 are particularly important. They will, therefore, be briefly elaborated in the following two sections.

Resilience

The word *resilience* was first used in an ecological context by C. S. Holling to denote the ability of an ecosystem to return to an equilibrium condition after being severely stressed. In recent years, the term has been applied to people; in this framework, Holling (1978), Colson (1979), Timmerman (1981), and others have expressed a concern that society is losing its resilience, sometimes in search of reliability.

Burton, Kates, and White (1977) have suggested that civilizations most vulnerable to surprise (least resilient) are neither the least developed nor the most highly industrialized. Instead, the most susceptible are those in transition, in the process of losing their traditional skills for dealing with environmental shocks. Some precolonial civilizations had considerable resilience (Abel and Stocking 1981). These societies were "rich in resourcefulness even though they many have been relatively poor in resources" (Colson 1979). Even today, subsistence farmers in semiarid regions grow quickly ripening crops in preference to species that promise the greatest average return, a hedge against drought that has been practiced for many centuries. Another example worth study is the large city of Kano in northern Nigeria, which has survived for more than two thousand years even though it is in a harsh, semiarid environment.

These cases may not be relevant to industrialized societies, but it seems important to begin structured comparative studies on resilience, particularly of cities, with respect to various environmental stresses (Burton 1983). Anthropologists should be encouraged to participate. Adopting strategies that increase resilience can be recommended, although this simple statement still require considerable elaboration.

Early-Warning Systems

It is clearly desirable to develop a capability to monitor the environment in ways that will provide early warnings of impending change, including the occurrence of rare events. Equally important, of course, one must be able to convince people to act accordingly. Glantz (1981) has said that even if Peruvian anchovy fisherman had been warned of the collapse of their fisheries in 1972–73, they would not have acted any differently than they did because of their large capital investments in boats and gear.

Over the last fifty years there have been significant improvements in the quality of monitoring systems used to provide early warnings of natural disasters. The loss of life in hurricanes and typhoons has diminished substantially, for example. Continued progress in this field is to be expected.

The other type of early-warning monitoring desperately required is a system to detect the imminent onset of irreversible trends. Conventional monitoring systems provide information on performance of various environmental indicators, but the data collected may not be particularly useful in determining whether the outer limits of resilience are being approached. (These outer limits could be exceeded either because the environmental stresses imposed are larger than ever before, or because society has lost some of its resilience.) For example, it has been suggested by Deevey et al. (1979) that the Mayan collapse was instituted or accelerated by a combination of overpopulation and poor soil management, a fatal combination in such a fragile environment: "However visible in historic perspective, Mayan overpopulation and metastability were probably not perceptible to the managerial elite and their advisors." What is not perceptible to us? Are we making the right use of the data available, and are the right data being collected?

Clearly, an early-warning system would be greatly enhanced if the processes likely to contribute to irreversibility had been identified and their interrelations had been modeled. Here the possibility of nonlinear effects[2] should be not be overlooked: These might be predicted by a model well in advance of detection by a monitoring network.

These ideas may be illustrated by the CO_2 greenhouse-warming case. Simulation models predict that if atmospheric CO_2 concentrations doubled, world temperatures would increase—only slightly in tropics but by as much as 8° C in the arctic and subarctic in autumn and winter, with associated shifts in cloudiness and rainfall patterns. There would, of course, be socioeconomic consequences of such a change in climate.

These predictions have sufficient credibility that scientists are beginning to examine global climate monitoring systems in terms of their ability to provide the earliest possible indication of trend. Wigley and Jones (1981), for example, have used the concept of signal-to-noise ratio to explore the efficacy of existing networks, the "noise" being the computed variance of surface temperatures for the years 1941–80, and the "signal" being obtained from CO_2-doubled climate models. Interestingly, the global maps of signal-to-noise ratio indicate that warming is likely to be detected first in mid-latitudes in summer, even though the warming will not be so great as in the arctic. Of course, the climate models contain many simplifying assumptions, so that scientists are presently preoccupied with the strength of nonlinearities in the models. These could cause, on the one hand, a runaway effect or, on the other, a substantially smaller warming.

Environmental Perspectives

A distinction is sometimes drawn between predictions relating to future states in specific environmental sectors (e.g. water, food, energy) and a synthesis of these predictions, however uncertain, that leads to the establishment of priorities. The term *environmental perspectives* is used in this latter regard to denote societal perceptions of impending opportunities and threats. Perspectives, of course, change with time, a fact that can be demonstrated by examining declarations and action plans of sequences of UN meetings such as the 1972 Stockholm Conference on the Human Environment and its 1982 Nairobi counterpart.

In this context, governments should seek not only to improve current prediction capabilities in the environmental sciences but also to give special attention to unraveling the sociological factors that altogether create an environmental perspective. The goal should be to try to predict perspectives at least five years into the future. This capability is urgently needed because institutions tend to be out of step with emerging environmental issues, often seeking solutions to yesterday's problems. For example, university departments and government agencies are frequently structured along traditional disciplinary lines. They are admirably equipped to deal with classical problems at a time when many emerging environmental issues require collaboration by persons in different departments or agencies, usually located in different buildings or even in different cities.

Concluding Remarks

Basic knowledge of environmental processes has increased greatly during the last thirty years. Nevertheless, scientists have considerable difficulty in using the vast store of knowledge now available to resolve current issues, let alone the ones still hidden behind the next hill. The main conclusion to be drawn is that more attention should be paid to the problem of preparing for surprise. In this connection, it should be emphasized that action plans developed to deal with uncertainty require input not only from scientists and engineers but also from social scientists. It is appropriate here to recognize the special contribution made by Gilbert White in drawing people together from the social and physical sciences. This has led to clarification and resolution of a number of difficult environmental issues, particularly those discussed in SCOPE 2 and SCOPE 10.

Acknowledgment

I acknowledge helpful comments from Peter Timmerman.

Notes

1. Gilbert White also coedited a major UNEP report on envionmental trends over the years 1972 to 1982 (Holdgate, Kassas, and White 1982a, 1982b), prepared for the same session.
2. A nonlinear effect is a mathamatical term indicating the possibility of overshoot, feedback, or resonance. Sometimes a nonlinear system returns to a new steady state after stress has been removed.

References

Abel, N. and M. Stocking. 1981. "The experience of underdeveloped countries." In *Project Appraisal and Policy Review,* ed. T. O'Riordan and W. R. D. Sewell, pp. 253–95. Chichester, U.K.

Ashby, E. 1978. *Reconciling Man with the Environment*. Stanford, CA.

Biswas, A. 1979. "World Models, Resources and Environment." *Environmental Conservation* 6: 3–11.
Box, G. E. P. and G. M. Jenkins. 1976. *Time Series Analysis, Forecasting and Control*. 2d edition. San Francisco.
Burton, I. 1983. "The vulnerability and resilience of cities." In *Ecoville: Urbanization in the Context of Ecodevelopment,* ed. R. White and I. Burton. UNESCO. Paris.
Burton, I., R. Kates, and G. White. 1977. *The Environment as Hazard*. London.
Colson, E. 1979. "In Good Years and in Bad: Food Strategies of Self-reliant Societies." *Journal of Anthropological Research* 35:18–29.
Deevey, E. S., D. S. Rice., P. M. Rice, H. H. Vaughan, M. Brenner, and M. S. Flannery. 1979. "Mayan Urbanism: Impact on a Tropical Forest Environment." *Science* 206:298–306.
Ehrlich, P. 1969. "Eco-catastrophe." *Ramparts,* September, 24–28.
Erhlich, A. H. and P. R. Erhlich. 1981. "Dangers of Uninformed Optimism." *Environmental Conservation* 8:173–75.
Glantz, M. 1981. "Considerations of the Societal Value of El Nino Forecasts." In *Resource Management and Environmental Uncertainties,* ed. M. H. Glantz and J. D. Thompson, pp. 449–76. Chichester, U.K.
Gumbel, E. J. 1958. *Statistics of Extremes,* New York.
Hare, F. K. 1980. "The Planetary Environment: Fragile or Sturdy?" *Geographic Journal* 146:379–95.
Holdgate, M. W., M. Kassas, and G. F. White. 1982*a*. "World Environmental Trends between 1972 and 1982." *Environmental Conservation* 9:11–29.
Holdgate, M. W., M. Kassas, and G. F. White, ed. 1982*b*. *The World Environment, 1972–82: A Report by the United Nations Environment Programme*. Dublin, Ireland.
Holling, C. S., ed. 1978. *Adaptive Environmental Assessment and Management*. Chichester, U.K.
Landsberg, H. H. 1964. *Natural Resources for U.S. Growth: A Look Ahead to the Year 2000*. Baltimore, MD.
Meadows, D. H., D. L. Meadows, J. Randers, and W. W. Behrens. 1972. *The Limits to Growth*. New York.
Pill, J. 1971. "The Delphi Method: Substance, Context, A Critique, and an Annotated Bibliography," *Socio-Economic Planning Science* 5:57–71.
SCOPE 2. 1972. *Man-made Lakes as Modified Ecosystems*. Chichester, U.K.
SCOPE 9. 1978. *Simulation Modelling of Environmental Problems,* ed. F. Frenkiel and D. Goodall. Chichester, U.K.
SCOPE 10. 1977. *Environmental Issues,* ed. M. W. Holdgate and G. F. White. Chichester, U.K.

Timmerman, P. 1981. *Vulnerability, Resilience and the Collapse of Society.* Environmental Monograph 1. Institute for Environmental Studies, University of Toronto.
UNEP. 1982. *The Environment in 1982: Retrospect and Prospect.* United Nations Environment Programme. Nairobi, Kenya.
White, G. F. 1962. *Social and Economic Aspects of Natural Resources.* Publication 1000-G. National Academy of Sciences, National Research Council. Washington. D.C.
Wigley, T. M. L. and P. D. Jones. 1981. "Detecting CO_2-induced Climate Change." *Nature* 292:205–8.

13 The Great Climacteric, 1798–2048: The Transition to a Just and Sustainable Human Environment

Ian Burton and Robert W. Kates

Toward The New Global Equilibrium

> Bliss was it in that dawn to be alive,
> But to be young was very heaven.
>
> Wordsworth

In almost all the topical areas covered by previous chapters in this book there has been discussion of the "crisis." Certainly the "energy crisis" was a prominent feature of discussions about man, resources, and environment in the 1970s. Gilbert White himself wrote a paper titled "The Meaning of the Environmental Crisis" (see vol. I, selection 19). The notion of crisis has become fashionable.

These so-called crises and no doubt others that we have yet to learn about are not isolated phenomena. They are part of a broader pattern that is more accurately described as a climacteric (Ashby 1978). A crisis is a situation that will pass and that can be overcome by dint of special efforts or short-lived sacrifice or deprivation. A climacteric, in contrast, is "a critical period in human life" and "a period supposed to be specially liable to change in health or fortune" (*Oxford English Dictionary*). The term is normally applied to the individual; but as applied to population, resources, and environment throughout the world it aptly captures the idea of a period that is critical and where serious change for the worse may occur. It is a time of unusual danger.

Ian Burton is professor of geography and former director of the Institute for Environmental Studies, University of Toronto, and vice-chairman of the International Federation of Institutes for Advanced Study. Robert W. Kates is professor of geography and Research Professor at Clark University and holds a MacArthur Prize fellowship. Burton and Kates, who have published widely, are coauthors, with Gilbert F. White, of *The Environment as Hazard* (1978) and coeditors of *Readings in Resources Management and Conservation* (1965).

Viewed from a global and historical perspective, the period of the present climacteric extends from the early days of the industrial revolution to the middle of the twenty-first century, when the world order will proably have reached a more stable and tranquil period if it has not collapsed in the meantime. The world is now in the climax period of this great climacteric. As this fact becomes widely known and accepted, it generates excitement, especially among the young. But, like the French Revolution, with which the romantic poet Wordsworth identified, it can lead to revulsion at a subsequent reign of terror.

The succession of so-called crises certainly brings danger and the fear of repression or, at the worst, global conflict. It is also a period of challenge and an invitation to both intellectual effort and action. Gilbert White's urgings to attempt an understanding of the global issues of our times and to act accordingly within one's own sphere of life are more pressing in the last decades of the present century than at any time during the great climacteric.

The two major variables that frame the ecology of any species are its numbers and its niche. No less in human ecology, it is population size and dynamics in relation to the determinants of the human niche (resources, environment, science, and technology) that constitute the main concerns of researchers, whether they pretend to ethical neutrality or not (Harvey 1974).

Understanding of these fundamental relationships of human ecology always seems to retreat just beyond our grasp. Even forecasting changes in a relatively short time period is notoriously difficult. Whatever happens in science and technology, however, seems unlikely to alter a growing perception that world population growth will level out to some sort of plateau by the middle of the next century or not long thereafter.

No doubt many arguments will continue to be advanced as explanations for this prospect of population stability. But the question that commanded so much attention at the Bucharest population conference in 1974 appears now to have an encouraging answer. The question was, Can effective fertility control policies be established in advance of general economic improvement? In countries culturally receptive, undergoing development, and with strong population control activity, major reductions in fertility have been experienced. Even without economic development, strong policies could lead to reduction in fertility, albeit at a slower rate, particularly in receptive cultures. Thus, if population stability is achieved, it will come more from collective will than by the blind operation of natural forces of population dynamics. Population will be stabilized because we will it to happen.

The starting point for the recognition of the goal and the growth of the worldwide political will was the publication of Malthus's first *Essay* in 1798. For us this is the symbolic starting point of the period of the climacteric. An even 250 years takes us to 2048, close to the midpoint of the next century.

Optimism over the adequacy of the natural resources base comfortably to sustain human life and society has shifted sharply in the years following the Second World War. As students of Gilbert White, we were both witnesses to, and participants in, the profound changes in thinking that were taking place. In the late 1950s when we began graduate work, the first turning point could be observed away from immediate postwar concerns about the adequacy of the resource base toward a series of professional appraisals and accompanying theory that was redirecting attention to environmental quality rather than resource quantity. Initially, amenity and recreational resources subject to degradation by an expanding urbanized society were a focus of concern; later, attention centered on the quality of an environment subject to pollution and destruction by an industrialized economy.

Writing in 1964, still in the midst of the transition, we found that our mentors and peers ranged between two poles: "In its extreme form, one pole is determinist in its view of nature, Malthusian in its concern with the adequacy of resources, and conservationist in its prescription for policy. The opposite pole is possibilist in its attitude toward nature, optimistic in its view of technological advance and the sufficiency of resources and generally concerned with technical and managerial problems of development" (Burton and Kates 1965, p. 2).

To document these poles, we listed some of the "best-known 'neo-Malthusian' works on the scarcity of resources and growing population written in polemical style": Brown, *The Challenge of Man's Future* (1954); Osborn, *Our Plundered Planet* (1948) and *The Limits of the Earth* (1953); and Vogt, *The Road to Survival* (1948) and *People: Challenge to Survival* (1960). Little known today, they seemed to us in Rosenwald Hall at the University of Chicago as Carson, Commoner, Ehrlich, Hardin, and Meadows would appear to another generation, two decades later.

In contrast, we cited the recent findings of three major studies prepared by Resources for the Future which we had reviewed under the title of "Slaying the Malthusian Dragon" (Burton and Kates 1964). These influential studies (Potter and Christy 1962; Barnett and Morse 1963; Landsberg, Fischman, and Fisher 1963) concluded that "technology can overcome increasing shortages of natural resources ad infinitum" (Potter and Christy 1962).

More recently we have seen still another shift in the balance of influence between the poles. The question of the adequacy of material resources for developed economies has given way to deep-seated fears for the environment in the industrialized world, for food adequacy in the developing world, and, finally, for the global biosphere and its basic systems of life support. Limits to growth (Meadows 1974) and the steady-state economy (Daly 1973) again became fashionable intellectual concepts.

These shifting tides of optimism and despair may be inherent in human nature or perhaps merely in social scientists (Luten 1980). Tiger (1979) argues that optimism is a distinctly human quality and the key to our survival and evolutionary success. The idea of progress is deeply rooted (Lovejoy and Boas 1935; Van Doren 1967; Nisbet 1980); yet a pessimism that focuses on golden ages past and sees a better time in a simpler, more "natural" world is as old as Hesiod. Thus, our professional oscillations between the neo-Malthusian poles may reflcct these deeper currents of human behavior and cognition. But their modern expression must surely begin with Robert Malthus.

The Principle of Population: Dragon or Dilemma?

Robert Malthus begins with biological verities: hunger and sex and the ability to satisfy the latter with greater ease than the former. The discrepancy would inevitably lead to a more rapid growth in numbers of persons than in the means of subsistence, sex still being linked to reproduction. Numbers increasing geometrically and subsistence arithmetically cannot continue unchecked. "Positive checks" in the form of vice or misery recurrently reduce population, and (in later revisions of the essay) "preventive checks," primarily delayed marriage, slow the growth of population.

It is not obvious what the sources of inspiration were for Malthus. The literature on his life and times has been enriched by two new major studies (Petersen 1979; James, 1979) but these contain no clear-cut answer. Malthus published his *Essay on the Principle of Population* (1798) anonymously, at age thirty-two while serving as curate at a small country chapel in Oakwood. James called the decade preceding 1798 the fallow period in his life. His only previous written work was never published.

He had been born in a remarkable century, in which England and Wales almost doubled in population from 4–5 million to 9–10 million. But, until the first census of 1801, knowledge of the event was lacking, and a major dispute on the "political arithmetic" of Britain raged in intellectual circles, with strong voices expressing belief in the depopulation of Britain at precisely the time its population was doubling.

At Oakwood, a steady stream of baptisms may have heralded this change in numbers, and the realities of hunger were always known, even to the sheltered parsons of the "higher ranks" of society. Adam Smith, the most powerful intellectual influence on Malthus, wrote that

> in some places one-half the children born die before they are four years of age; in many places, before they are seven; and in almost all places before they are nine or ten. . . . Every species of animals naturally multiplies in proportion to the means of their subsistence, and no species can ever multiply beyond it. But in a civilized society it is only among the inferior ranks of people that the scantiness of subsistence can set limits to the further multiplication of the human species and it can do so in no other way than by destroying a great part of the children which their fruitful marriages produce. (Quoted in Petersen 1979, p. 40)

The "provocation" for the *Essay* itself, of course, was the utopian visions of perfection of Condorcet and Godwin, which Malthus felt called upon to refute by the hard arithmetic of the principle of population.

Malthusian Forecasts

As Petersen (1979) labors to demonstrate, Malthus was an inspired scientific observer. He constantly collected new data, revised his statement of theory in the course of examining them, and allowed for feedback mechanisms that would serve to dampen the tendencies he forecast; a life of vice and misery for the lower classes pressing against the levels of available subsistence. But in the final analysis, in England and Wales, where he eagerly joined the debate on the poor laws, and in Europe, where he both studied and traveled, Malthus was wrong.

Only seventeen years after the *Essay* was first published, a great natural experiment occurred that stressed the subsistence system of both North America and Europe, providing an effective test of Malthusian theory. The test took place in the form of a major diminution of food supplies accompanied by price rises in the wake of widespread crop failure. The great volcanic eruption of Mount Tomboro in Indonesia in 1815 had resulted in a dust cloud and temperature drop. Carried by the prevailing westerlies, the dust led to two successive years of cold, wet growing seasons, beginning with the "year without summer" in 1816. Like all major natural calamities it was further aggravated by a major and painful economic readjustment following the conclusion of the Napoleonic wars. Widespread suffering ensued (as Post 1977 has documented), but unlike earlier famines of the eighteenth century it did not entail significant mortality, except on the outer marches of the Western world—Ireland and Transylvania—whose people were viewed by the powers that were as beyond the pale. Instead, for the first time, massive efforts at relief took place, moving food from the grain bins of Russia and the Baltic states to Britain, France, the Netherlands, and Germany. There was widespread suffering but little death (the great positive check) even under great stress, meriting Post's accurate label (1977), the "last great subsistence crisis of the Western world."

Thus, the ink had scarcely dried when the forecast faltered. It would take another half-century to extend food security to Ireland and Transylvania; and, in the meantime, conditions gave birth to Marxism. And it would take another twenty years in Britain and parts of Europe to reach a steady-state population. Twenty-seven countries currently have a rate of reproduction equal to or below that required to reproduce their population. But the process was underway before 1798 in France and well underway elsewhere even as Malthus wrote and rewrote, argued with Ricardo, and inspired Darwin.

Malthus overestimated the rate of population increase: in 1798 it was thought to be 3 percent; it dropped steadily from that time. He underestimated the resource base; he thought it consisted basically of undifferentiated agricultural land located where the population lived. Later, agricultural land was to become a small part of natural resource requirements, and its productivity was to have major additives in the form of machinery, cultivation techniques, fertilizer, improved seeds, herbicides, and pesticides. The agricultural base would expand to all corners of the world through trade and transport.

Technological progress was grossly underestimated, as well as the ability to undertake corrective or ameliorative action.

The Dragon's Teeth

Nonetheless, Malthus's theory still is influential, despite all our attempts to slay the Malthusian dragon. Despite repeated refutation, both of its theory and observation, the *Essay on Population* arises anew in fresh guise in each successive generation. A sequence is sketched in figure 13.1. Changes in the numerator (resources) are matched by changes in the denominator (population).

In 1798, the numerator of the Malthusian equation was food and agricultural land, not because Malthus was ignorant of other natural resource needs but because food requirements so dominated other needs (Barnett and Morse 1963). But by the 1850s. food requirements were expanded to include other energy and material resources, marked by the classic volume of Jevons (1906) on coal. Food, how-

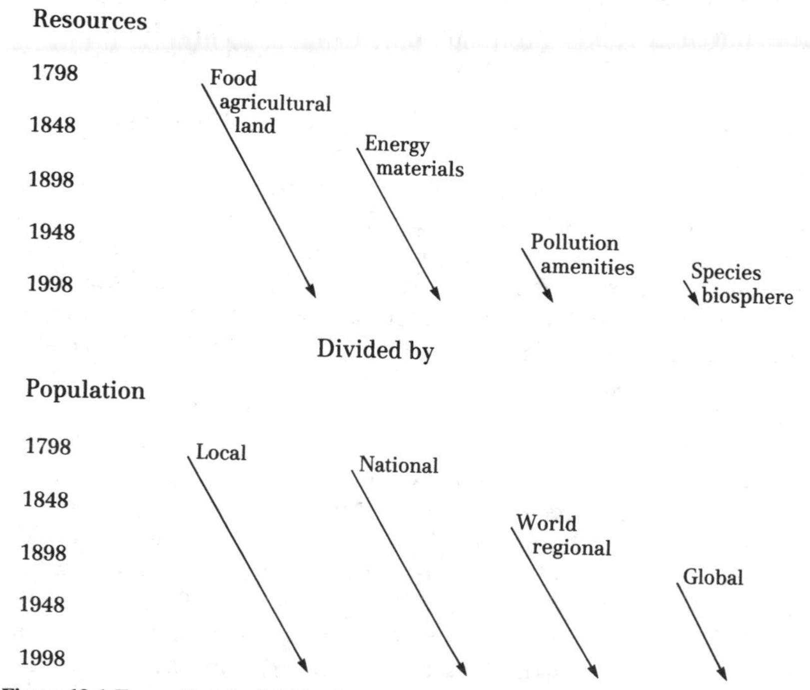

Figure 13.1 Expanding the Malthusian equation: resources divided by population.

ever, did not disappear from Malthusian concerns; it persisted on a wider scale (along the diagonal of increasing scale shown in figure 13.1). In the United States, the plea of the American Association for the Advancement of Science (1890) for forest conservation marks the nineteenth-century stage in the evolution of neo-Malthusian thought in the United States, which develops through both the first and second conservation movements.

The postwar United States reached a new turning point with the report of the President's Materials Policy Commission (1952), discounting fears as to resource scarcity and laying the groundwork for a new definition of the Malthusian numerator involving amenity resources and the pollution-absorbing capacity of the environment. The Stockholm Conference on the Environment in 1972 enlarged these concerns to a global scale, with a resulting concern with the biosphere and the basic life support system of biogeochemical cycles. The latest extension is the current concern with extinction (Myers 1979; Ehrlich and Ehrlich 1981) and predictions of massive species destruction over the next several decades.

Each of the earlier definitions persists. Concern with food adequacy gets new life in the context of the high population growth rates of the developing world, and concern with the adequacy of energy and material resources is revived by increases in the 1970s in some commodity prices and the recognition of the limitations of oil reserves.

Like the earlier Malthusian predictions, the neo-Malthusian predictions also fail. A classical case is the predictions about coal made by Jevons (1906). An extrapolation from the time Jevon's book was written is shown in figure 13.2 together with the actual course of British coal production. This pattern is analagous to the figure published in the *Global 2000 Report* (Council on Environmental Quality 1980), which comprises an extrapolation of 1950–75 oil production with King Hubbert's projection of U.S. oil production.

Reflecting on the "slain dragon" of the 1950s and 1960s brings to mind the myths of the dragon's teeth, which, when sown, give forth men armed to engage each other in combat at the drop of a stone. The persistence of Malthusian food and population issues along with the continuing redefinition of the resource numerator in the equation arises primarily from the extraordinary character of our times. For within the short span of memory of our grandparents and great-grandparents, humankind has become engaged in one of the three great transitions.

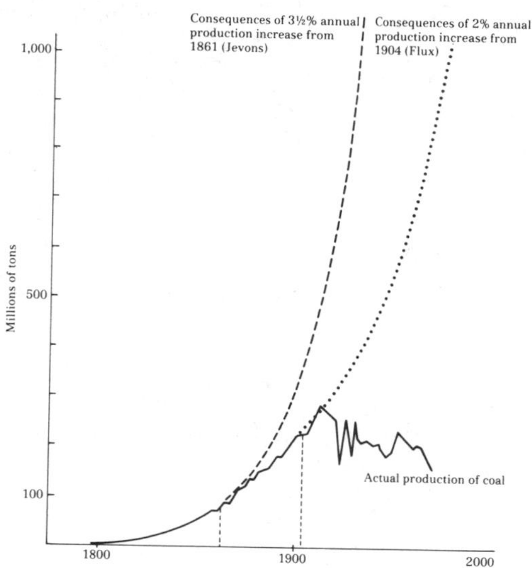

Figure 13.2 Graphs plotted by Jevons and Flux to illustrate the implications of sustained growth in coal consumption, superimposed on a graph showing actual consumption of coal in Britain (compiled from the frontispiece to Jevons 1906 and incorporating more recent statistics).

The New Global Equilibrium

In an insightful analysis, Edwin Deevey changed our perception of population problems, Writing in 1960 he noted: "The commonly accepted picture of the growth of the population out of the long past takes the form of the graph [in figure 13.3]. Two things are wrong with this picture. In the first place the basis of the estimates, back of about A.D. 1650, is rarely stated. One suspects that writers have been copying each other's guesses. The second defect is that the scales of the graph have been chosen so as to make the first defect seem unimportant. The missile has left the pad and is heading out of sight" (1960, p. 11).

Deevey collected available estimates of population over a longer period and then compiled figure 13.4 on a logarithmic scale, explaining that:

> The stepwise evolution of population size, entirely concealed with arithmetic scales, is the most noticeable feature of this diagram. For most of the million-year period, the number of hominids, including man, was about what would

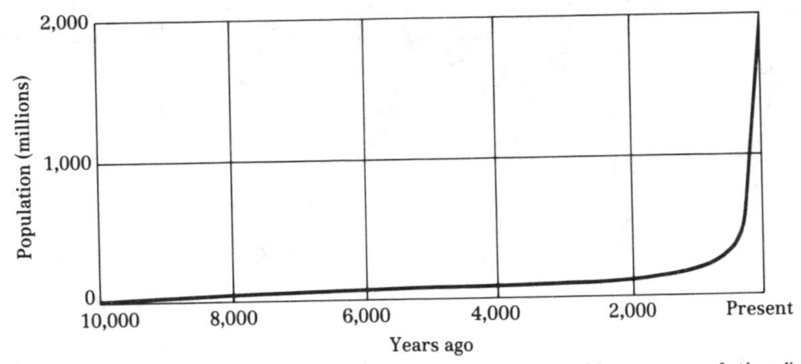

Figure 13.3 Arithmetic population curve plots the growth of human population from 10,000 years ago to the present. Such a curve suggests that the population figure remained close to the base line for an indefinite period from the remote past to about 500 years ago, and that it has surged abruptly during the last 500 years as a result of the scientific-industrial revolution. (From Deevey 1960.)

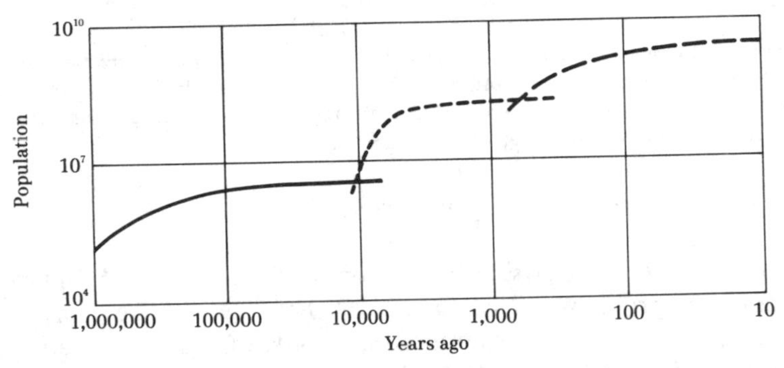

Figure 13.4 Logarithmic population curve makes it possible to plot, in a small space, the growth of population over a longer period of time and over a wider range (from 10^4, or 10,000 to 10^{10}, or 10 billion, persons). Curve, based on assumptions concerning relationship of technology and population as shown in figure 13.2, reveals three population surges reflecting tool making or cultural revolution (solid line), agricultural revolution (short dash) and scientific-industrial revolution (long dash). (From Deevey 1960.)

be expected of any large Pleistocene mammal—scarcer than horses, say, but commoner than elephants. Intellectual superiority was simply a successful adaptation, like longer legs; essential to stay in the running, of course, but making man at best the first among equals. Then the food-gatherers and hunters became plowmen and herdsmen, and the population was boosted by about sixteen times, between 10,000 and 6,000 years ago. The scientific-industrial revolution, beginning some 300 years ago, has spread its effects much

> faster, but it has not yet taken the number as far above the earlier baseline. The long-term population equilibrium implied by such baselines suggests something else. Some kind of restraint kept the number fairly stable (1960, p. 13).

Deevey's insight seems further substantiated twenty years later. Significant downturns in population rates have been recorded in many developing countries, giving rise to an optimistic prediction of a steady-state world by the middle of the next century.

The World in 2048: The Consequences of Stability

If industrial civilization survives to 2048 without major collapse—and we expect that it will—what will be the character of the world? This question is another way of asking what we must do to make our way successfully through the great climacteric.

Five characteristics of that world seem inescapable. First, a major successful effort to reduce the nuclear arms of superpowers and other powers and to limit their proliferation will have been made. We anticipate at least partial success in limiting the share of gross world product spent on armaments. Whether such a shift will come only as a result of a nuclear exchange short of full-scale war is an open question. We are not optimistic that all use of nuclear weapons can be prevented before the world comes to its senses.

Second, the global population will have begun to level off. The rate of increase will have slowed down over large areas of the world to that prevailing in the most stable (in population terms) countries of today.

Third, the wealth of nations must be much more equal than today. A pattern analogous to the distribution of wealth among people in today's more egalitarian states such as Sweden or Britain seems possible and indeed essential. This would mean a very few wealthy states (perhaps fortuitously well endowed with a natural resource in high demand), a small number of very poor states (those to achieve population stability last), and a very large "middle class" of nations.

Fourth, a much greater degree of social discipline and control will exist, especially in those states where it is now lacking. These are of two types—developing countries where the apparatus of state bureaucracy is not yet fully established, and the wealthier industrial nations of today that are attempting to combine a high degree of social discipline in some areas while preserving a high degree of freedom of choice in others.

To much conventional Western thinking, the more disciplined and bureaucratically controlled society of the future sounds like a society without freedom where the choices of individuals and organizations will be everywhere constrained. This is instead a challenge of the new society of 2048: to find ways of management that ensure the necessary control while preserving the individual sense of freedom and avoiding a fatal overweight of bureaucracy.

Fifth, there will be a new relationship with the environment, in which decisions are guided by a concern for the long-term viability of the biosphere and including a recognized set of "rights" for the natural environment, especially other living organisms.

In our view, these five essential characteristics of the world of 2048 are attainable and will be attained. It is beyond our courage to attempt any description of the pattern of events that might lead to their achievement. (What will probably be crucial is the interaction between nuclear weapons, population, distribution of wealth, the discipline of society, and the biosphere.) Instead, we limit our speculations to the nexus of population change and land resources as expressed in population densities. The actual level of population that the world will have to cope with ranges according to current estimates from 8 billion to 11 billion.

Three independent forecasts of population are in general agreement that the world population will level off, or be seen to be leveling off, by the middle of the next century (figure 13.5). This is perhaps an agreement on assumptions rather than the product of convergence in "scientific forecasting." The forecasts are as follows: 8.1 billion in 2048 (Bogue and Tsui), 9.8 billion in 2090 (World Bank), and 11.0 billion in 2130 (UN). Important consequences of this leveling off are to be seen in the changed distribution of the size of nations and the people-land ratios.

The Balance of Demographic Power

The Population Reference Bureau estimates for mid-1980 list sixteen countries with a population in excess of 50 million. These are listed in rank order in table 13.1, together with estimated density per square mile. The population densities of these countries exhibit a wide range, from lows of 37 and 40 in Brazil and the Soviet Union to a high of 1,630 in Bangladesh. This ordering of demographic power is

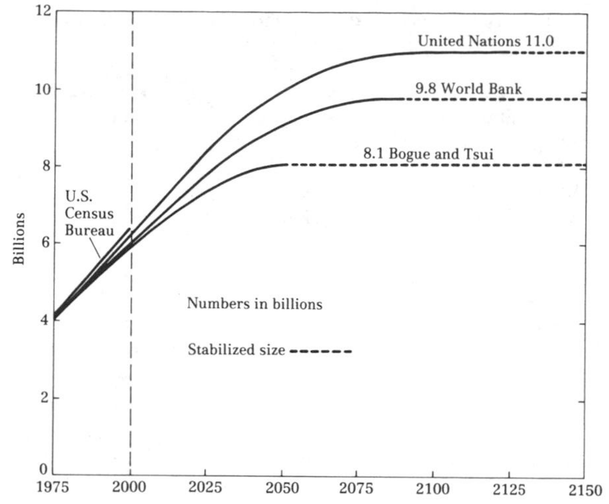

Figure 13.5 *Population projections*. Sources: U.S. Bureau of the Census, *Illustrative Projections of World Populations to the 21st Century,* Current Population Reports, Special Studies Series, P. 23 no. 79, January 1979. United Nations, *Prospects of Population Methodology and Assumptions,* Population Studies no. 67 (New York 1979). World Bank—K.C. Zachariah and My Thi Vu, *Population Projections, 1975, and Long-Term (Stationary Population)* (World Bank, July 1979), unpublished tables. Donald J. Bogue and Amy Ong Tsui, Community and Family Study Center, University of Chicago, "Zero World Population Growth," *The Public Interest,* Spring 1979, pp. 99–113.

Note: The U.S. Census Bureau has not published projections beyond 2000. The World Bank publishes only one series. The other projections shown are medium series (between high and low variants).

in the course of major change. In table 13.2 we compare population on a continental basis for 1800 and 1980 with the "ultimate" population. This latter figure is based on calculations made by the Population Reference Bureau. The "ultimate size" is that level of population at which growth will have virtually ceased, regardless of the date at which this will be. On the basis of the data in table 13.2 we have drawn figure 13.6.

In Malthus's time there were fewer than a billion people alive, 90 percent of them in Europe and Asia. Today there are well over 4 billion, of whom only 75 percent are found in Europe and Asia; and in the ultimate world of perhaps 10 billion, only 65 percent will be in Europe and Asia. In 1800, 1 in 5 humans were Europeans, 1 in 14 Africans; in the ultimate population the proportions will be reversed.

This ultimate population size leads to some remarkable changes in ranking. Without a doubt the most dramatic is that of Nigeria,

Table 13.1 Countries with populations of more than 50 million in 1980

Rank	Country	Total population (millions)	Density per square mile	Rank order for ultimate size
1	China	975.0	263	2
2	India	646.2	509	1
3	Soviet Union	226.0	40	4
4	United States	222.5	62	9
5	Indonesia	144.3	196	5
6	Brazil	122.0	37	6
7	Japan	116.8	813	13
8	Bangladesh	90.6	1630	8
9	Pakistan	86.5	279	7
10	Nigeria	77.1	216	3
11	Mexico	68.2	89	10
12	West Germany	61.1	636	30
13	Italy	57.2	492	29
14	United Kingdom	55.8	592	32
15	France	53.6	254	31
16	Vietnam	53.3	419	11

Source: Population Reference Bureau (1980).

Table 13.2 Changing world population (in millions)

	1800		1980		Ultimate	
	Number	Percentage	Number	Percentage	Number	Percentage
Africa	70	7.8	472	10.7	2,051	20.9
America						
North	6	0.7	247	5.6	296	3.0
Middle and South	18	2.0	360	8.1	955	9.7
Asia	625	69.3	2,663	60.3	5,743	58.4
Europe	180	20.0	650	14.7	750	7.6
Oceania	2	00.1	23	00.5	37	0.3
Total	901	100.0	4,415	94.9	9,832	99.0

Sources: 1800, C. McEvedy and Richard Jones, *Atlas of World Population History* (Harmondsworth, U.K., 1978); 1980 and Ultimate, Population Reference Bureau, 1980 World Population Data (Soviet Union population distributed between Asia and Europe, 1980: 100/166, ultimate: 170/190).

from a 1980 population of 77 million and a rank of tenth largest nation to an ultimate count of 434 million and a rank of third, well over the population of the Soviet Union, though only a quarter the size of the population of India and China (see table 13.1).

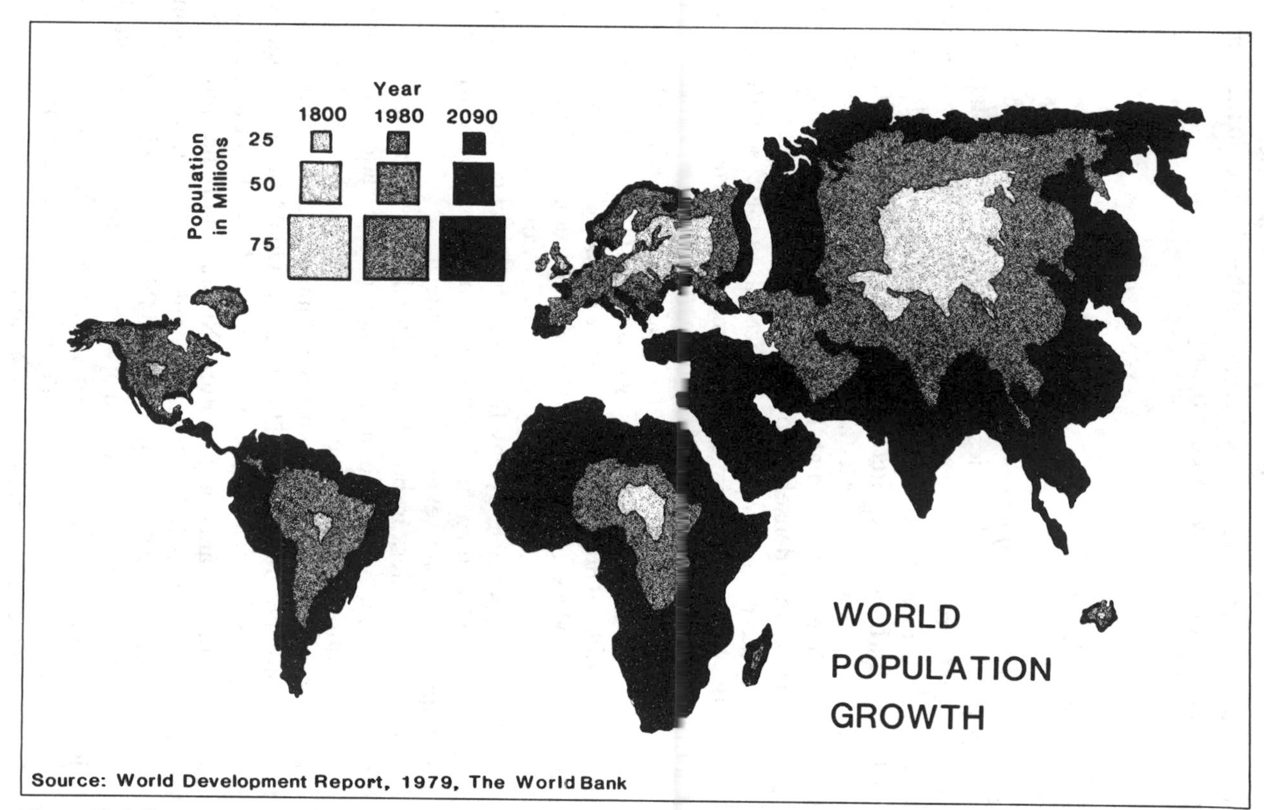

Figure 13.6 Changing world population.

The rise of Nigeria to the status of world (population) superpower is part of a trend perceptible throughout much of Africa. As table 13.2 makes clear, the climacteric involves major changes in the world population balance between countries.

Man-Land Ratios

The comparison of population density is more striking than that of total population. Peak densities occur in a few small city-states and islands. The city-states are typically supported by heavy trading activities and local manufacturing. The cases of two small islands, Malta and Barbados, are interesting because on a very limited land area (122 square miles for Malta, 166 for Barbados) both have managed to obtain a relatively high level of development although both are considered very densely populated and approach their limits of carrying capacity. Clearly this is not necessarily the case if they can achieve the kind of economic development experienced in Singapore, for example.

The most desperate case of people-land ratio is Bangladesh. Even now it has almost double the density of Belgium (1,630 to 840 persons per square mile) while being five times the size of Belgium in area (55,000 to 11,000 square miles). The ultimate density forecast for Bangladesh of over 6,000 p.p.s.m. seems by any of today's standards absolutely unsupportable by any form of agricultural economy. There is no other large country that in terms of people-land ratio anywhere nearly approaches the severity of Bangladesh.

Other world leaders in population size all have much more land per capita. Even though this neglects any consideration of quality of land, an ultimate population density for Nigeria of 1,189 p.p.s.m. is less than that of Taiwan for today. Similarly, even India with a forecast ultimate density of 1,294 p.p.s.m. would be only slightly more densely populated than Taiwan today and less than Bangladesh today. Both these giants of the future, Nigeria and India (judged by population size), will have ultimate population densities less than the density that exists today in Bahrein, Barbados, and Malta.

While a national population total of 1.6 billion (India) or 434 million (Nigeria) appears intolerable by some criteria, there are examples elsewhere, albeit on a smaller scale, of the implied densities managing to survive and to prosper reasonably well.

Middle-Class Status

What are the constraints likely to be on such countries as India and Nigeria in the next seventy years as they seek to negotiate the great climacteric? For if India and Nigeria can do it, and emerge as middle-class countries by 2048, then there can be little doubt that many other countries will have been able to do likewise.

There appear to be two major routes to middle-class status for such countries. They can accumulate capital wealth by the sale of some fortuitously possessed natural resources that command a high price on international markets. This was the case with petroleum in the 1970s, and Nigeria had an opportunity to accumulate capital to support its needed development program. Much the same possibility opened up for Mexico (ultimate population 203 million; ultimate density 267 p.p.s.m.) and Venezuela (40 million; 114).

It may well be that other countries will be helped by the discovery of a rich resource endowment that can be sold profitably and thus power the development process. Of course, such countries have to face the problem that their resources will not last indefinitely, that even badly needed commodities such as oil cannot indefinitely sustain high prices, and that they can provide only a breathing space in which the necessary industrial infrastructure will be constructed and trained manpower can be built up to earn its livelihood in the world without a rich resource endowment.

But what of the countries that lack such an endowment in the first place? How will they achieve "operation bootstrap"? The economic miracle of Japan shows that it can be done. To go from a virtually feudal society to a leading industrial power in a period of about a hundred years is a spectacular achievement. As recently as the 1930s few in Japan or elsewhere believed that such a thing could be achieved without territorial expansion and the control of additional resources (e.g. Manchuria). Nevertheless, Japan today contemplates a future with an ultimate population of 133 million and an ultimate density of 928 p.p.s.m.—no great increase on today's density of 813 p.p.s.m. and less than South Korea has now.

Japan is not an isolated case. Similar transformations are taking place in South Korea and Taiwan, and the countries of the Association of South-East Asian Nations (ASEAN) promise to follow the same path.

There is a major difference between these two paths, however. The path of resource windfall is likely to become more available as world demand for natural resources increases. Where these windfalls will occur is not easy to forecast, however, and there is no guarantee that they will fall according to need. Some countries with very small populations happen to have enormous oil reserves. Britain, a country that for many decades thought of itself as resource-poor and had to rely on manufacturing and trade, has now dramatically become resource-rich again with the discovery of North Sea oil.

Although resource windfalls are likely to become more available, they do not add greatly to the total of global wealth. The ability to fuel development through high-export revenues from the sale of natural resource commodities will be more a means of redistributing wealth than increasing the total. Competition for resources therefore promises to become more of a zero-sum game in which there will be winners and losers.

The route of indigenous industrialization (Japan, South Korea, Taiwan, etc.) is likely to become more difficult. However, this development may increase total global wealth if it does not occur simply at the expense of exporting industrial nations.

Migration

Along with national development, greater individual movement is likely to occur. Major movements of people will take place, reproducing the migrations already taking place within Europe and North America. It is safe to predict massive African outmigration, with Africans occupying the migrant-worker status of Hispanics in North America and southern Europeans in Europe.

Such massive movement, which will persist despite constant efforts to limit immigration and to control so-called guest workers, will bring fresh energy and new culture to the older industrialized societies as well as the old and painful issues of power, exploitation, and racism to most of the wealthy world. If the currently estimated flows of legal and illegal immigrants increase by only a third or a half, one recent projection foresees an absolute majority of Asians, blacks, and Hispanics in the United States by 2048 (Davis 1982).

A Global Plan?

Any global plan for development should therefore recognize that help to the underdeveloped to industrialize is likely to be beneficial to all. What are the prospects of the adoption of a global development plan by 2048? Is some sort of planning necessary for the great climacteric?

We are convinced that it is. To some extent this is already happening and will continue to do so. We neither propose nor expect to see mandatory planning at the global level. What we do foresee is a gradual emergence of a global consensus that will lead nations voluntarily to adopt policies conducive to a pattern of development that allows all a share in the process. The power and vested self-interest of nation-states is strong and appears likely to remain so. However, the ship of national sovereignty is leaking, and recognition of the interdependence of the global system will produce actions to preserve the stability of the world.

Stability may come about in two ways. First, the growth of shared perceptions of the global predicament will lead nations to policies that coincide with the interests of the whole. Just as world population is being stabilized by a shared consciousness and not by a world population plan, so too the issue of equity will be addressed via enlightened self-interest rather than global planning. We will not be surprised if the system hovers perilously close to the brink of collapse. The requisite steps are not likely to be taken until perceived as an urgent necessity. No doubt the negotiations will be long and difficult, with the outcome seeming constantly in doubt, as demonstrated in the Law of the Sea conference.

One great opportunity probably lies in an international and global agreement on the future of Antarctica. Despite some long-established imperial claims to sovereignty, there is a strong moral case to be made for recognizing Antarctica as part of "the common heritage of mankind" and using the benefits of its potential accordingly.

Beyond enlightened self-interest and the concerted action that occurs as a result of increase in shared perceptions is the second way—response to crisis. In developing ideas on how to make the transition to a more just and sustainable human environment, we see no reason to expect change in the traditional crisis-response pattern of international decision making. A question to be asked about the recent series of "crises" therefore is: To what extent have they helped or hindered in making the transition through the great climacteric? The energy crisis of the 1970s undoubtedly helped to stimulate research and development on alternative energy sources, including wind and solar energy. It also spurred more vigorous steps at energy conservation. It is difficult to view these developments as anything but highly desirable, and we hope that the effort will be maintained as the immediate crisis passes by.

As the energy crisis fades (for the time being) due to the combination of increased supply and lowered demand (both from conservation and the economic crisis), it is the sluggish economies of East and West, North and South that command attention. Their plight will be acute if the current universal downturn ushers in the decades-long decline of a Kondratiev long-wave (Freeman, Clarke, and Soete 1982). Ironically, enlightened self-interest may well call forth new North-South monetary agreements and financial interdependence. With the powerful banking interests in danger of collapse, the ensuing sense of interdependence may lead to a new economic order that shifts the balance of resource exchanges somewhat more in favor of the developing nations of the world.

In other areas a sense of crisis is more difficult to establish. One response of the international community has been to establish specially recognized decades or to develop action plans, as seen, for example, in the United Nations Drinking Water Decade and UNEP Action Plan on Desertification. Where queues at the gas station or other dramatic events are not present to drive the message home, then crises have to be engineered and made dramatically visible in other ways.

This type of thinking has led to a sequence of forecasting or predictive studies including the well-known *Limits to Growth* (Meadows 1974) and, more recently, *Global 2000* (Council on Environmental Quality 1980). Such documents have been useful in alerting governments and people to the possible consequences of inaction. Their essential message is: If you don't act, this will happen. The time seems ripe to change the form of the message. The international community can and should begin to formulate goals of where we wish and expect the world to be at the end of a series of future plan periods, and to develop ways of getting there. The essential message would become: Here are some goals or targets, and this is the action required to accomplish them.

Epilogue

A concern for future food supplies, resources, and environment in relation to human needs inevitably has a strong materialist flavor. Without in any way diminishing the importance of meeting the basic needs of the human family, and successfully negotiating the great climacteric in material terms, we believe that those goals cannot be achieved without a corresponding growth in the human spirit.

Throughout his life and work, Gilbert White has devoted much energy to enlarging the human spirit. Although his works deal mainly with the material needs of man, his creed has never been a materialistic one. Making the transition through the great climacteric will change human beings and human society. The more stable and tranquil period that we dimly foresee will provide new opportunities for the human species to grow and fulfill its nonmaterial potential. It may well again be bliss in that dawn to be alive, but it will not be paradise on earth.

References

American Association for the Advancement of Science. 1890. *Proceedings American Association for the Advancement of Science* 39:28.

Ashby, E. 1978. *Reconciling Man with the Environment*. Stanford, CA.

Barnett, H. J. and C. Morse. 1963. *Scarcity and Growth: The Economics of Natural Resource Availability*. Baltimore, MD.

Brown, H. 1954. *The Challenge of Man's Future*. New York.

Burton, I. and R. W. Kates. 1964. "Slaying the Malthusian Dragon." *Economic Geography* 40, no. 1:82–89.

Burton, I. and R. W. Kates, eds. 1965. *Readings in Resource Management and Conservation*. Chicago.

Carson, R. 1962. *Silent Spring*. Boston.

Commoner, B. 1971. *Closing the Circle*. New York.

Council on Environmental Quality. 1980. *The Global 2000 Report to the President*. Washington, D.C.

Daly, H. E. 1973. *Steady State Economics: The Economics of Biophysical Equilibrium and Moral Growth*. San Francisco.

Davis, C. 1982. "The Future Racial Composition of the United States." *Intercom* 10, no. 9/10:8–10.

Deevey, E. S., Jr. 1960. "The Human Population." *Scientific American* 203 (September).

Ehrlich, P. R., and A. H. Ehrlich. 1981. *"Extinction: The Causes and Consequences of the Disappearance of Species*. New York.

Freeman, C., J. Clarke, and L. Soete. 1982. *Unemployment and Technical Innovation*. London.

Harvey, D. 1974. "Population, Resources and the Ideology of Science." *Economic Geography* 50, no. 3 (July):256–77.

James, P. 1979. *Population Malthus: His Life and Times*. London.

Jevons, W. S. 1906. *The Coal Question*, 3d ed., ed. A. W. Flux. London.

Landsberg, H. H., L. L. Fischman, and J. L. Fisher. 1963. *Resources in America's Future: Patterns of Requirements and Availabilities 1960–2000*. Baltimore, MD.
Lovejoy, A.D. and G. Boas. 1935. *Primitivism and Related Ideas in Antiquity*. Baltimore, MD.
Luten, D.B. 1980. "Ecological Optimism in the Social Sciences." *American Behavioral Scientist* 24, no. 1 (September/October): 125–51.
Meadows, D. H. 1974. *Limits to Growth*. New York.
Myers, N. 1979. *The Sinking Ark: A New Look at the Problem of Disappearing Species*. Elmsford, NY.
Nisbet, R. 1980. *History of the Idea of Progress*. New York.
Osborn, F. 1948. *Our Plundered Planet*. Boston.
———. 1953. *The Limits of the Earth*. Boston.
Petersen, W. 1979. *Malthus*. Cambridge, MA.
Population Reference Bureau. 1980. *World Population Data Sheet*. Prepared by C. Haub and D. W. Heisler. Washington, DC.
Post, J. D. 1977. *The Last Great Subsistence Crisis in the Western World*. Baltimore, MD.
Potter, N. and F. T. Christy. 1962. *Trends in Natural Resource Commodities*. Baltimore, MD.
President's Materials Policy Commission (Paley Commission). 1952. *Resources for Freedom*. Washington, DC.
Tiger, L. 1979. *Optimism*. New York.
Van Doren, C. 1967. *The Idea of Progress*. New York.
Vogt, W. 1948. *The Road to Survival*. New York.
———. 1960. *People: Challenge to Survival*. New York.

Index